Developments in Mathematics

Volume 51

More information about this series at http://www.springer.com/series/5834

Yong Zhou · Rong-Nian Wang
Li Peng

Topological Structure of the Solution Set for Evolution Inclusions

Yong Zhou
School of Mathematics and Computational
  Science
Xiangtan University
Xiangtan, Hunan
China

Li Peng
School of Mathematics and Computational
  Science
Xiangtan University
Xiangtan, Hunan
China

Rong-Nian Wang
Mathematics and Science College
Shanghai Normal University
Shanghai
China

ISSN 1389-2177   ISSN 2197-795X   (electronic)
Developments in Mathematics
ISBN 978-981-10-6655-9   ISBN 978-981-10-6656-6   (eBook)
https://doi.org/10.1007/978-981-10-6656-6

Library of Congress Control Number: 2017953788

Mathematics Subject Classification (2010): 34G25, 37C70, 34K09, 35R70, 60H15

Printed on acid-free paper

This Springer imprint is published by Springer Nature
The registered company is Springer Nature Singapore Pte Ltd.
The registered company address is: 152 Beach Road, #21-01/04 Gateway East, Singapore 189721, Singapore

Preface

A lot of phenomena investigated in hybrid systems with dry friction, processes of controlled heat transfer, obstacle problems, and others can be described with the help of various differential inclusions, both linear and nonlinear. The theory of differential inclusions is highly developed and constitutes an important branch of nonlinear analysis.

To the best of our knowledge, there were very few monographs concerning the topological theory and dynamics for evolution inclusions. This monograph gives a systematic presentation of the topological structure of solution sets and attractability for nonlinear evolution inclusions and its relevant applications in control theory and partial differential equations. The materials in this monograph are based on the research work carried out by the author and other excellent experts during the past four years. The contents of this book are very new and rich. It provides the necessary background material required to go further into the subject and explore the rich research literature. All abstract results are illustrated by examples.

This monograph deals with the focused topic with high current interest and complements the existing literature in differential equations and inclusions. It is useful for researchers, graduate or Ph.D., students dealing with differential equations, applied analysis, and related areas of research.

We acknowledge with gratitude the support of National Natural Science Foundation of China (11671339, 11471083).

Xiangtan, China   Yong Zhou
Shanghai, China   Rong-Nian Wang
Xiangtan, China   Li Peng

Contents

Introduction

Since the dynamics of nonlinear and hybrid systems is multivalued, differential inclusions serve as natural models in many dynamical processes. In addition, differential inclusions also provide a powerful tool for various branches of mathematical analysis. In the past twenty years, theory of differential inclusions has been developed very rapidly. The several excellent monographs by Aubin and Cellina [20], Benchohra and Abbas [34], Borisovich et al. [42], Bothe [46], Deimling [80], Djebali et al. [89, 90], Dragoni et al. [96], Górniewicz [113], Graef [116], Hu and Papageorgiou [125], Kamenskii et al. [130], Kisielewicz [135, 136], Mahmudov [141], Smirnov [176], Tolstonogov [185], Vrabie [189], and Zgurovsky et al. [207] summarize a lot of important works in this area.

Since a differential inclusion usually has many solutions starting at a given point, new issues appear, such as investigation of topological properties of solution sets. In the study of the topological structure of solution sets for integral/differential equations and inclusions, an important aspect is the R_δ-property. Recall that a subset of a metric space is called an R_δ-set if it can be represented as the intersection of a decreasing sequence of compact and contractible sets. It is known that an R_δ-set is acyclic and, in particular, nonempty, compact, and connected. From the point of view of algebraic topology, an R_δ-set is equivalent to a point, in the sense that it has the same homology group as one-point space.

For the Cauchy problems of ordinary differential equations having no uniqueness, Kneser [137] proved in 1923 that the sets of their solutions are at every fixed time continua, and then, Hukuhara [127] showed that the solution set (on a compact interval) itself is a continuum (i.e., closed and connected). Later, Yorke [203] improved this result in the sense that solution sets are R_δ-sets. Let us also mention that by using topological degree arguments, Górniewicz and Pruszko [115] proved that the solution set (on a compact interval) of a Darboux problem for hyperbolic equation is an R_δ-set; an analogous result was also established by De Blasi and Myjak [79] by using a different approach and recently, by means of the theory of condensing mappings and multivalued analysis tools, Ke et al. [133] investigated the R_δ-structure of the solution set for an abstract Volterra integral equation without uniqueness on a compact interval.

Another interesting aspect connected to a deeper understanding of topological structure is differential inclusion (a relation associated with set-valued mappings), which appears as the natural framework for describing hybrid systems with dry friction, processes of controlled heat transfer, obstacle problems, and others (see, e.g., [4, 70, 74, 130, 157, 163] and references therein). Notice in particular that Aronszajn [19] carried out a systematic study for the topological properties of solution set of a differential inclusion on a compact interval, where he showed that solution sets are compact and acyclic, and he in fact specified these continua to be R_δ-sets. Since the work of Aronszajn, there have been published, up to now, numerous research papers concerning topological structure of solution sets for differential inclusions of various types; see, e.g., Bothe [45], De Blasi and Myjak [78], Deimling [80], Hu and Papageorgiou[126], Staicu [177], and Zhu [210] and references therein for related results of differential inclusions defined on compact intervals.

Topological structure of solution sets for differential equations or differential inclusions on noncompact intervals (including infinite intervals) has been studied by various techniques; see Andres et al. [11] for boundary value problems of differential equations and inclusions, Bakowska and Gabor [25] for differential equations and inclusions in Fréchet spaces, and Sĕda and Belohorec [173] for initial value problem of second-order ODE with time delay. Particularly, Gabor and Grudzka [106] recently treated an impulsive abstract Cauchy problem governed by a semilinear differential inclusion involving a family of time-dependent linear operators in the linear part. What has been considered by them is the R_δ-structure of solution set on a noncompact interval. For more details on this topic, we refer reader to, e.g., Gabor [105] and O'Regan[159] and references therein.

The characterizations of solution sets (including compactness, acyclicity, and R_δ) are useful in the study of the qualitative theory for deterministic problems. Here, we sketch some references, but not a list of all references is included. Bader and Kryszewski [24] proved that the set consisting of all mild solutions to a constrained semilinear differential inclusion on a compact interval is a nonempty, compact, and R_δ-set and gave its applications to the periodic problem and to the existence of equilibria. Making use of R_δ-property of the solution set to a semilinear Volterra integral equation as well as the fixed-point theory for ANR-spaces, Ke et al. [133] established the invariance of a reachability set for the corresponding control system. More recently, Andres and Pavlačková [14] studied the R_δ-structure of the solution set for a fully linearized system of second-order ODEs and then obtained an existence result for the corresponding semilinear system by using information about the structure and a fixed-point index technique in Fréchet spaces.

Let us note that when dealing with topological structure of solution sets to the problems on noncompact intervals, one of the key tools is the inverse limit method, sometimes is also called the projective limit, was initiated in Gabor [104], and then was developed by Andres et al. [12] (see also Andres and Pavlačková [14]). It is observed that the problems of differential equations and inclusions on noncompact intervals can be reformulated as fixed-point problems in Fréchet spaces which are inverse limits of Banach spaces that appear when we consider these differential problems on compact intervals.

Another important and interesting problem related to evolution inclusions is to study the stability of solutions. Since the question of uniqueness of solutions to evolution inclusions is no longer addressed, the Lyapunov theory for stability is not a suitable choice. Thanks to the theories of attractors for multivalued semiflows/processes given by Melnik and Valero [145, 146], Caraballo et al. [54, 55, 56], one can find a global attractor for semiflows/processes governed by solutions of evolution inclusions, which is a compact set attracting all solutions as the time goes to infinity in some contexts. Recently, Ke and Lan [132] proved the global solvability and the existence of a compact global attractor for the m-semiflow generated by evolution inclusions. Dac and Ke [77] obtained the existence of a pullback attractor for nonautonomous differential inclusions with infinite delays in Banach spaces by using measures of noncompactness.

This monograph is arranged and organized as follows:

In order to make the book self-contained, we devote chap. 1 to a description of general information on multivalued analysis, m-semiflow and attractor, evolution system, semigroups, weak compactness of sets and operators, and stochastic process.

Chapter 2 deals with a nonlinear delay differential inclusion of evolution type involving m-dissipative operator and source term of multivalued type in a Banach space. Under rather mild conditions, the R_δ-structure of C^0-solution set is studied on compact intervals, which is then used to obtain the R_δ-property on noncompact intervals. Secondly, the results about the structure are furthermore employed to show the existence of C^0-solutions for the inclusion (mentioned above) subject to nonlocal condition defined on right half line. No nonexpansive condition on nonlocal function is needed. As samples of applications, we consider a partial differential inclusion with time delay and then with nonlocal condition at the end of the chapter.

In Chap. 3, we study evolution inclusions involving a nondensely defined closed linear operator satisfying the Hille-Yosida condition. In the cases that the semigroup is noncompact, the topological structure of the set of solutions and the global solvability and the existence of a compact global attractor for the m-semiflow generated by evolution inclusions are investigated by using some techniques of measure of noncompactness.

In Chap. 4, we study different types of generalized solutions in general Banach spaces, called limit and weak solutions. Under appropriate assumptions, we show that the set of the limit solutions is a compact R_δ-set. When the right-hand side satisfies the one-sided Perron condition, a variant of the well-known lemma of Filippov-Pliś, as well as a relaxation theorem, is given. Next, we establish the relation between the solutions of the quasi-autonomous evolution inclusions and the solutions of the relaxed one. A variant of the well-known Filippov-Pliś lemma is also proved. Finally, we analyze the existence of a pullback attractor for functional evolution inclusions with infinite delays in Banach spaces by using measures of noncompactness. The results are applied to control systems driven by semilinear partial differential equations and multivalued feedbacks.

Chapter 5 is devoted to the investigation of the topological structure of solution sets of nonautonomous parabolic evolution inclusions with time delay, defined on noncompact intervals. The result restricted to compact intervals is then extended to nonautonomous parabolic control problems with time delay. Moreover, as the applications of the information about the structure, we establish the existence result of global mild solutions for nonautonomous Cauchy problems subject to nonlocal condition and prove the invariance of a reachability set for nonautonomous control problems under single-valued nonlinear perturbations.

In Chap. 6, we investigate the topological properties of the set of solutions for neutral functional evolution inclusions. It is shown that the solution set is nonempty, compact, and an R_δ-set when the semigroup is compact as well as noncompact.

Chapter 7 deals with the existence of mild solutions for impulsive differential inclusions in a reflexive Banach space. Weakly compact valued nonlinear terms are considered, combined with strongly continuous evolution operators generated by the linear part. A continuation principle or a fixed-point theorem is used, according to the various regularity and growth conditions assumed. Secondly, a topological structure of the set of solutions to impulsive functional differential inclusions on the half line is investigated. It is shown that the solution set is nonempty, compact, and, moreover, an R_δ-set. It is proved on compact intervals and then, using the inverse limit method, obtained on the half line.

Chapter 8 studies stochastic evolution inclusions in Hilbert spaces when the semigroup is compact as well as noncompact. It is shown that the solution set is a nonempty, compact, and R_δ-set.

The materials in this monograph are based on the research work carried out by the author and other excellent experts during the past four years. The results in Chap. 2 are taken from Chen, Wang, and Zhou [71]. The contents of Sect. 3.2 are new and due to Zhou and Peng. The results in Sect. 3.3 are taken from Ke and Lan [132]. The contents of Sect. 4.2 are adopted from Cârjă, Donchev, and Lazu [59]. The results in Sect. 4.3 are taken from Cârjă, Donchev, and Postolache [61]. The results in Sect. 4.4 are taken from Dac and Ke [77]. All the results in Chap. 5 are taken from Wang, Ma, and Zhou [196]. Chapter 6 is taken from Zhou and Peng [209]. Section 7.2 is the extension of Benedetti et al. [35], and the results in Sect. 7.3 are taken from Benedetti and Rubbioni [37], Gabor and Grudzka [106, 107]. The results in Chap. 8 are taken from Zhou, Peng, and Ahmad [208].

Keywords and Phrases: Evolution inclusions, multivalued mappings, one-sided Perron multifunctions, topological structure, relaxation results, multi-valued semiflows, global attractor, global solvability, pullback attractor, Cauchy problem, nonlocal problem, control problem, solution set, R_δ-set, mild solutions, integral solutions, limit solutions, weak solutions, C_0-semigroup, analytic semi-group, integrated semigroup, compact semigroup, noncompact semigroup, m-dis-spative operator, Hille-Yosida operator, weak topology, time delay, impulses, stochastic process, fixed-point theorem, measure of noncompactness, inverse limit method.

Chapter 1
Preliminaries

Abstract In this chapter, we introduce some basic facts on multivalued analysis, evolution system, semigroups, weak compactness of sets and operators, and stochastic process which are needed throughout this monograph.

1.1 Basic Facts and Notations

As usual $\mathbb{N}^+$ denotes the set of positive integer numbers and $\mathbb{N}_0$ the set of nonnegative integer numbers. $\mathbb{R}$ denotes the real line, $\mathbb{R}_+$ denotes the set of nonnegative reals, and $\mathbb{R}^+$ denotes the set of positive reals. Let $\mathbb{C}$ be the set of complex numbers.

We recall that a vector space X equipped with a norm $|\cdot|$ is called a normed vector space. A subset Ω of a normed space X is said to be bounded if there exists a number K such that $|u| \leq K$ for all $u \in \Omega$. A subset Ω of a normed vector space X is called convex if for any $u, v \in \Omega$, $au + (1-a)v \in \Omega$ for all $a \in [0, 1]$.

A sequence $\{u_n\}$ in a normed vector space X is said to converge to the vector u in X if and only if the sequence $|u_n - u|$ converges to zero as $n \to \infty$. A sequence $\{u_n\}$ in a normed vector space X is called a Cauchy sequence if for every $\varepsilon > 0$ there exists an $N = N(\varepsilon)$ such that for all $n, m \geq N(\varepsilon)$, $|u_n - u_m| < \varepsilon$. Clearly a convergent sequence is also a Cauchy sequence, but the converse may not be true. A space X where every Cauchy sequence of elements of X converges to an element of X is called a complete space. A complete normed vector space is said to be a Banach space.

Let Ω be a subset of a Banach space X. A point $u \in X$ is said to be a limit point of Ω if there exists a sequence of vectors in Ω which converges to u. We say a subset Ω is closed if Ω contains all of its limit points. The union of Ω and its limit points is called the closure of Ω and will be denoted by $\overline{\Omega}$. Let X, Y be normed vector spaces and Ω be a subset of X. An operator $\mathscr{T} : \Omega \to Y$ is continuous at a point $u \in \Omega$ if and only if for any $\varepsilon > 0$ there is a $\delta > 0$ such that $|\mathscr{T}u - \mathscr{T}v| < \varepsilon$ for all $v \in \Omega$ such that $|u - v| < \delta$. Further, $\mathscr{T}$ is continuous on Ω, or simply continuous, if it is continuous at all points of Ω.

Let $J = [a, b]$ $(-\infty < a < b < \infty)$ be a finite interval of $\mathbb{R}$. We assume that X is a Banach space with the norm $|\cdot|$. Denote $C(J, X)$ by the Banach space of all

© Springer Nature Singapore Pte Ltd. 2017
Y. Zhou et al., *Topological Structure of the Solution Set for Evolution Inclusions*,
Developments in Mathematics 51, https://doi.org/10.1007/978-981-10-6656-6_1

continuous functions from J into X with the norm

$$\|u\| = \sup_{t \in J} |u(t)|,$$

where $u \in C(J, X)$. $C^n(J, X)$ ($n \in \mathbb{N}_0$) denotes the set of mappings having n times continuously differentiable on J. Let $AC(J, X)$ be the space of functions which are absolutely continuous on J and $AC^n(J, X)$ ($n \in \mathbb{N}_0$) be the space of functions f such that $f \in C^{n-1}(J, X)$ and $f^{(n-1)} \in AC(J, X)$. In particular, $AC^1(J, X) = AC(J, X)$.

$\widetilde{C}([a, \infty), X)$ is the separated locally convex space consisting of all continuous functions u from $[a, \infty)$ to X endowed with the family of seminorms $\{\| \cdot \|_m : m \in \mathbb{N} \setminus \{0\}\}$, defined by

$$\|u\|_m = \sup_{t \in [a,m]} |u(t)|, \quad m \in \mathbb{N} \setminus \{0\},$$

and a metric

$$d(u_1, u_2) = \sum_{m=1}^{\infty} \frac{1}{2^m} \frac{\|u_1 - u_2\|_m}{1 + \|u_1 - u_2\|_m}.$$

Let us recall that by a Fréchet space we mean a locally convex space which is metrizable and complete. Thus, $\widetilde{C}([a, \infty), X)$ is a Fréchet space. Also, every Banach space is a Fréchet space.

$C_b([a, \infty), X)$ represents the set of all continuous and bounded functions from $[a, \infty)$ to X. Denote also by $\widetilde{C}_b([a, \infty), X)$ the subset of $\widetilde{C}([a, \infty), X)$ consisting of all functions $u \in C_b([a, \infty), X)$, subset equipped with the uniform convergence on compacta topology.

Let $1 \le p \le \infty$. We denote by $L^p(J, X)$ the set of those Lebesgue measurable functions $f : J \to X$ for which $\|f\|_{L^p(J,X)} < \infty$, where

$$\|f\|_{L^p(J,X)} = \begin{cases} \left(\int_J |f(t)|^p dt \right)^{\frac{1}{p}}, & 1 \le p < \infty, \\ \operatorname*{ess\,sup}_{t \in J} |f(t)|, & p = \infty. \end{cases}$$

In particular, $L^1(J, X)$ denotes the Banach space consisting of all measurable functions $f : J \to X$ with the norm

$$\|f\|_{L(J,X)} = \int_J |f(t)| dt,$$

and $L^\infty(J, X)$ denotes the Banach space consisting of all measurable functions $f : J \to X$ which are bounded, equipped with the norm

$$\|f\|_{L^\infty(J,X)} = \inf\{c > 0 : |f(t)| \le c \text{ a.e. } t \in J\}.$$

A subset K in $L^1(J, X)$ is called integrably bounded if there exists $l \in L^1(J, \mathbb{R}^+)$ such that

$$|f(t)| \le l(t) \text{ a.e. } t \in J$$

for each $f \in K$. We say that $K \subset L^1(J, X)$ is uniformly integrable if for each $\epsilon > 0$ there exists $\delta(\epsilon) > 0$ such that for each measurable subset $E \subset J$ with $\text{mes}(E) < \delta(\epsilon)$, we have

$$\int_E |f(t)|dt \le \epsilon \text{ uniformly for } f \in K.$$

It is easy to see that if $K \subset L^1(J, X)$ is integrably bounded, then K is uniformly integrable.

Lemma 1.1 ([162, Remark 2.3]) *If $K \subset L^1(J, X)$ is uniformly integrable, then it is norm bounded in $L^1(J, X)$.*

Lemma 1.2 (Hölder inequality) *Assume that $p, q \ge 1$ and $\frac{1}{p} + \frac{1}{q} = 1$. If $f \in L^p(J, X)$, $g \in L^q(J, X)$, then $fg \in L^1(J, X)$ and*

$$\|fg\|_{L(J,X)} \le \|f\|_{L^p(J,X)}\|g\|_{L^q(J,X)}.$$

We say that a subset Ω of a Banach space X is compact if every sequence of vectors in Ω contains a subsequence which converges to a vector in Ω. We say that Ω is relatively compact if every sequence of vectors in Ω contains a subsequence which converges to a vector in X, i.e., Ω is relatively compact if $\overline{\Omega}$ is compact.

Definition 1.1 A sequence $\{f_n\} \subset L^1(J, X)$ is said to be semicompact if it is integrably bounded and $\{f_n(t)\} \in K(t)$ for a.e. $t \in J$, where $K(t) \subset X$, $t \in J$, is a family of compact sets.

Lemma 1.3 (Arzela-Ascoli's theorem) *If a family $F = \{f(t)\}$ in $C(J, X)$ is uniformly bounded and equicontinuous on J, and for any $t^* \in J$, $\{f(t^*)\}$ is relatively compact, then F has a uniformly convergent subsequence $\{f_n(t)\}_{n=1}^\infty$.*

Remark 1.1 (a) If a family $F = \{f(t)\}$ in $C(J, \mathbb{R})$ is uniformly bounded and equicontinuous on J, then F has a uniformly convergent subsequence $\{f_n(t)\}_{n=1}^\infty$.
(b) Arzela-Ascoli's theorem is the key to the following result: A subset F in $C(J, \mathbb{R})$ is relatively compact if and only if it is uniformly bounded and equicontinuous on J.

Theorem 1.1 ([172, Theorem 3.20]) *The convex hull of a relatively compact set in a Banach space X is relatively compact.*

Theorem 1.2 (Lebesgue's dominated convergence theorem) *Let E be a measurable set and let $\{f_n\}$ be a sequence of measurable functions such that $\lim_{n\to\infty} f_n(t) = f(t)$*

a.e. in E, and for every $n \in \mathbb{N}^+$, $|f_n(t)| \le g(t)$ a.e. in E, where g is integrable on E. Then

$$\lim_{n \to \infty} \int_E f_n(t)dt = \int_E f(t)dt.$$

Theorem 1.3 (Bochner's theorem) *A measurable function $f : (a, b) \to X$ is Bochner integrable if $|f|$ is Lebesgue integrable .*

Let us state an elementary lemma of Gronwall's type and Halanay's inequality (see [119], also [193] for a generalized version).

Lemma 1.4 *Let $a, b \in \mathbb{R}$ with $a > b$. Suppose that u is a nonnegative continuous function defined on $[a - \tau, b]$ satisfying*

$$\begin{cases} u(t) \le \varphi(0) + \int_a^t g(s)u(s)ds + \int_a^t g(s) \sup_{\theta \in [s-\tau,s]} u(\theta)ds, & t \in [a, b], \\ u(t) = \varphi(t - a), & t \in [a - \tau, a], \end{cases}$$

where $g \in L^1([a, b], \mathbb{R}^+)$ and $\varphi \in C([-\tau, 0], \mathbb{R}^+)$ are given. Then,

$$u(t) \le \sup_{s \in [-\tau, 0]} \varphi(s) \exp\left(2 \int_a^t g(s)ds\right), \quad t \in [a, b].$$

Lemma 1.5 (Halanay's inequality) *Let the function $f : [t_0 - \tau, b) \to \mathbb{R}^+$, $0 \le t_0 < b < +\infty$ be continuous and satisfy the functional differential inequality*

$$f'(t) \le -\gamma f(t) + \nu \sup_{s \in [t-\tau,t]} f(s),$$

for $t \ge t_0$, where $\gamma > \nu > 0$. Then

$$f(t) \le \kappa e^{-l(t-t_0)}, \quad t \ge t_0,$$

where $\kappa = \sup_{s \in [t_0-\tau, t_0]} f(s)$ and l is the solution of the equation $\gamma = l + \nu e^{-l\tau}$.

Theorem 1.4 ([98, Theorem 4.1]) *Let X be an arbitrary metric space, D a closed subset of X, Y a locally convex linear space, and $f : D \to Y$ a continuous mapping. Then there exists an extension $F : X \to Y$ of f; furthermore, $F(X) \subset \text{co } f(A)$.*

Finally, we introduce some general concepts which will be used in the following chapters.

An inner product on a complex vector space X is a mapping $(\cdot, \cdot) : X \times X \to \mathbb{C}$ such that for all $u, v, w \in X$ and all $\lambda \in \mathbb{C}$:

(i) $(u, v) = \overline{(v, u)}$;
(ii) $(\lambda u, v) = \lambda(u, v)$;
(iii) $(u + v, w) = (u, w) + (v, w)$;
(iv) $(u, u) > 0$, when $u > 0$.

An inner product space is a pair $(X, (\cdot, \cdot))$, where X is a complex vector space and $(\cdot, \cdot)$ is an inner product on X. A Hilbert space is an inner product space which is a complete metric space with respect to the metric induced by its inner product.

Let X be a Banach space. By a cone $K \subset X$, we understand a closed convex subset K such that $\lambda K \subset K$ for all $\lambda \geq 0$ and $K \cap (-K) = \{0\}$. We define a partial ordering $\leq$ with respect to K by $u \leq v$ if and only if $u - v \in K$. Then

(i) K is called positive if the element $u \in K$ is positive;
(ii) K is regeneration if $K - K = X$, and total if $\overline{K - K} = X$;
(iii) K is called normal if $\inf\{|u + v| : u, v \in K \cap \partial B_1(0)\} > 0$.

1.2 Multivalued Analysis

1.2.1 Multivalued Mappings

Multivalued mappings play a significant role in the description of processes in control theory since the presence of controls provides an intrinsic multivalence in the evolution of the system. In this subsection, we introduce some general properties on multivalued mappings. The material in this subsection is taken from Kamenskii et al. [130].

Let Y and Z be metric spaces. $P(Y)$ stands for the collection of all nonempty subsets of Y. As usual, we denote

$P_b(Y) = \{D \in P(Y): D \text{ bounded}\}; \quad P_{cl}(Y) = \{D \in P(Y): D \text{ closed}\};$
$P_{cp}(Y) = \{D \in P(Y): D \text{ compact}\}; \quad P_{cl,cv}(Y) = \{D \in P_{cl}(Y): D \text{ convex}\};$
$P_{cp,cv}(Y) = \{D \in P_{cp}(Y): D \text{ convex}\}; \quad \text{co(D) (resp., } \overline{\text{co}}(D)) \text{ be the convex hull}$
(resp., convex closed hull in D) of a subset D.

A multivalued mapping φ of Y into Z is a correspondence which associates to every $y \in Y$ a nonempty subset $\varphi(y) \subseteq Z$, called the value of y. We write this correspondence as $\varphi : Y \to P(Z)$. If $D \subseteq Y$, then the set $\varphi(D) = \bigcup_{y \in D} \varphi(y)$ is called the image of D under φ. The set $\text{Gra}(\varphi) \subseteq Y \times Z$, defined by $\text{Gra}(\varphi) = \{(y, z) : y \in Y, z \in \varphi(y)\}$, is the graph of φ.

Let (J, τ, μ) be a finite measure space and X be a Banach space. We denote by Sel_φ the set of all X-valued Bochner integrable selections from $\varphi : J \to P(X)$, i.e.,

$$Sel_\varphi = \{f \in L^1(J, X) : f(t) \in \varphi(t) \text{ a.e. } t \in J\}.$$

Moreover, Aumann integral of the set-valued function $\varphi : J \to P(X)$ is defined as follows:

$$\int_J \varphi(t) d\mu(t) = \left\{ \int_J f(t) d\mu(t) : f \in Sel_\varphi \right\}.$$

The set-valued function $\varphi : J \to P(X)$ is said to be integrably bounded if there exists a map $h \in L^1(\mu, \mathbb{R})$ such that $\sup\{|f| : f \in \varphi(t)\} \leq h(t)$ μ-a.e.

Yannelis [201] extended some results of Aumann [22, Theorem 5] to separable Banach spaces.

Theorem 1.5 (Lebesgue-Aumann dominated convergence theorem) *Let (J, τ, μ) be a complete, finite measure space and X be a separable Banach space. Let $\varphi : [a, b] \to P(X)$ be a sequence of integrably bounded, nonempty valued correspondence having a measurable graph, such that:*

(i) *For all n $(n = 1, 2, \ldots)$, $\varphi_n(t) \subset K$ μ-a.e., where K is a compact, non-empty subset of X, and*

(i) *$\varphi_n(t) \to \varphi(t)$ μ-a.e.*

Then

$$\int_J \varphi_n(t) d\mu(t) \to \overline{\int_J \varphi(t) d\mu(t)}.$$

Moreover, if $\varphi(\cdot)$ is convex valued then

$$\int_J \varphi_n(t) d\mu(t) \to \int_J \varphi(t) d\mu(t).$$

Definition 1.2 Let $\varphi : Y \to P(Z)$ be a multivalued mapping and D be a subset of Z. The complete preimage $\varphi^{-1}(D)$ of a set D is the set

$$\varphi^{-1}(D) = \{y \in Y : \varphi(y) \cap D \neq \emptyset\}.$$

Definition 1.3 A multivalued mapping $\varphi : Y \to P(Z)$ is said to be

(i) closed if its graph $\text{Gra}(\varphi)$ is closed subset of the space $Y \times Z$;

(ii) upper semicontinuous (shortly, u.s.c.) if the set $\varphi^{-1}(D)$ is closed for every closed set $D \subset Z$.

The Hausdorff metric between the bounded sets $D, E \subset X$ is defined by

$$d_H(D, E) = \max\{\text{dist}_X(D, E), \text{dist}_X(E, D)\},$$

where $\text{dist}_X(D, E) = \sup_{a \in D} \inf_{b \in E} |a - b|$ is the Hausdorff semi-distance. For any bounded set $D \subset X$, the support function is given by $\sigma(l, D) = \sup_{a \in D} \langle l, a \rangle$, for every $l \in X^*$, and we denote $|D| = d_H(D, \{0\})$.

let $I \subset \mathbb{R}$ be a compact interval, μ be a Lebesgue measure on I and Y be a Banach space. A multivalued map $F : I \to P_{cp}(Y)$ is said to be measurable (resp. weakly measurable) if for every open (resp. closed) subset $V \subset Y$ the set $F^{-1}(V)$ is measurable.

A multivalued map $\varphi : I \to P_{cp}(Y)$ is said to be strongly measurable if there exists a sequence $\{\varphi_n\}_{n=1}^\infty$ of step multivalued maps such that $d_H(\varphi_n(t), \varphi(t)) \to 0$ as $n \to \infty$ for μ-a.e. $t \in I$.

Lemma 1.6 ([185]) *Let X, Y be a complete separable metric spaces and $\varphi :$ $[0, b] \times X \to P(Y)$ be a $\mathcal{L} \otimes \mathcal{B}_X$ measurable multifunction with closed values. Then for any continuous function $u : [0, b] \to X$ the multifunction $t \mapsto F(t, u(t))$ is measurable and has a strongly measurable selector.*

Definition 1.4 Let $\varphi : I \times X \to P(X)$ be a multifunction:

(i) $\varphi(\cdot, \cdot)$ is said to be lower semicontinuous (shortly, l.s.c.) if for any $(t, u) \in I \times X$, any $v \in \varphi(t, u)$, and any sequence $\{(t_n, u_n)\}_{n=1}^{\infty}$ with $t_n \to t$ and $u_n \to u$, there exists a sequence $\{v_n\}_{n=1}^{\infty}$ with $v_n \in F(t_n, u_n)$ such that $v_n \to v$;

(ii) $\varphi(\cdot, \cdot)$ is said to be continuous if it is continuous with respect to the Hausdorff metric;

(iii) $\varphi(\cdot, \cdot)$ is called upper hemicontinuous if the support function $\sigma(l, \varphi(\cdot, \cdot))$ is upper semicontinuous as a real valued function;

(iv) $\varphi(\cdot, \cdot)$ is said to be almost lower semicontinuous (almost continuous, with almost closed graph) if for any $\varepsilon > 0$ there exists a compact set $I_\varepsilon \subset I$ with mes$(I \setminus I_\varepsilon) < \varepsilon$ such that the restriction of $\varphi(\cdot, \cdot)$ to $I_\varepsilon \times X$ is lower semicontinuous (continuous, with closed graph).

Definition 1.5 Let D be nonempty subset of a Banach space Y and $\varphi : D \to P(Y)$ be a multivalued mapping:

(i) φ is said to have a weakly sequentially closed graph if for every sequence $\{u_n\} \subset D$ with $u_n \rightharpoonup u$ in D and for every sequence $\{u_n\}$ with $v_n \in \varphi(u_n)$, $\forall n \in \mathbb{N}$, $v_n \rightharpoonup v$ in Y implies $v \in \varphi(u)$;

(ii) φ is called weakly upper semicontinuous (shortly, weakly u.s.c.) if $\varphi^{-1}(A)$ is closed for all weakly closed $A \subset Y$;

(iii) β is ϵ-δ u.s.c. if for every $u_0 \in Y$ and $\epsilon > 0$ there exists $\delta > 0$ such that $\beta(v) \subset \beta(u_0) + B_\epsilon(0)$ for all $v \in B_\delta(u_0) \cap D$.

Lemma 1.7 ([45]) *Let $\varphi : D \subset Y \to P(Z)$ be a multivalued mapping with weakly compact values. Then*

(i) *φ is weakly u.s.c. if φ is ϵ-δ u.s.c., and*

(ii) *suppose further that φ has convex values. Then φ is weakly u.s.c. if and only if $\{u_n\} \subset D$ with $u_n \to u_0 \in D$ and $v_n \in \varphi(u_n)$ implies $v_n \rightharpoonup v_0 \in \varphi(u_0)$, up to a subsequence.*

Definition 1.6 A multivalued mapping $\varphi : Y \to P(Z)$ is

(i) compact if its range $\varphi(Y)$ is relatively compact in Z, i.e., $\overline{\varphi(Y)}$ is compact in Z;

(ii) locally compact if every point $v \in Y$ has a neighborhood $V(v)$ such that the restriction of φ to $V(v)$ is compact;

(iii) quasicompact if $\varphi(D)$ is relatively compact for each compact set $D \subset Y$.

It is clear that (i) $\Longrightarrow$ (ii) $\Longrightarrow$ (iii). The following facts will be used.

Lemma 1.8 *Let Y be a topological spaces, Z a regular topological space and φ : $Y \rightarrow P_{cl}(Z)$ an u.s.c. multivalued mapping. Then φ is closed.*

The inverse relation between u.s.c. mappings and closed ones is expressed in the following lemma.

Lemma 1.9 *Let Y and Z be metric spaces and φ : $Y \rightarrow P_{cp}(Z)$ a closed quasicompact multivalued mapping. Then φ is u.s.c.*

Let us consider some properties of closed and u.s.c. multivalued mapping.

Lemma 1.10 *Let φ : $Y \rightarrow P_{cl}(Z)$ be a closed multivalued mapping. If $D \subset Y$ is a compact set then its image $\varphi(D)$ is a closed subset of Z.*

Lemma 1.11 *Let φ : $Y \rightarrow P_{cp}(Z)$ be an u.s.c. multivalued mapping. If $D \subset Y$ is a compact set then its image $\varphi(D)$ is a compact subset of Z.*

Lemma 1.12 *Let Y and Z be Banach space, and let the multivalued mapping φ : $[0, b] \times Y \rightarrow P_{cp}(Z)$ be such that*

  (i) *for every $u \in Y$ the multifunction $\varphi(\cdot, u)$: $[0, b] \rightarrow P_{cp}(Z)$ has a strongly measurable selection;*
 (ii) *for a.e. $t \in [0, b]$ the multivalued mapping $\varphi(t, \cdot)$: $Y \rightarrow P_{cp}(Z)$ is u.s.c.*

Then for every strongly measurable function q : $[0, b] \rightarrow Y$ there exists a strongly measurable selection g : $[0, b] \rightarrow Z$ of the multifunction G : $[0, b] \rightarrow P_{cp}(Z)$, $G(t) = \varphi(t, q(t))$.

Theorem 1.6 *Let X be a Hausdorff locally convex topological vector space and $D \subset X$ be a nonempty compact convex subset. If the multivalued mapping φ : $D \rightarrow P(D)$ is u.s.c. with closed convex values, then φ has a fixed point.*

We propose now a continuation principle.

Theorem 1.7 *([35]) Let D be a closed, convex subset of a Banach space Y with nonempty interior and H : $D \times [0, 1] \rightarrow P(Y)$ be such that*

 (a) *H is nonempty convex valued and it has a closed graph;*
 (b) *H is compact;*
 (c) *$H(D, 0) \subset D$;*
 (d) *$H(\cdot, \lambda)$ is fixed points free on the boundary of D for all $\lambda \in [0, 1)$.*

Then there exists $u \in D$ such that $u \in H(u, 1)$.

1.2.2 Measure of Noncompactness

We recall here some definitions and properties of measure of noncompactness and condensing mappings. For more details, we refer the reader to Akhmerov et al. [9], Banaś and Goebel [28], Deimling [81], Heinz [122], Kamenskii et al. [130], and Lakshmikantham and Leela [139].

Definition 1.7 Let Y^+ be the positive cone of an order Banach space $(Y, \leq)$. A function α defined on the set of all bounded subsets of the Banach space X with values in Y^+ is called a measure of noncompactness (*MNC*) on X if $\alpha(\overline{co}\Omega) = \alpha(\Omega)$ for all bounded subsets $\Omega \subset X$.

The MNC α is said to be:

(i) *Monotone* if for all bounded subsets B_1, B_2 of X, $B_1 \subseteq B_2$ implies $\alpha(B_1) \leq \alpha(B_2)$;

(ii) *Nonsingular* if $\alpha(\{u\} \cup B) = \alpha(B)$ for every $u \in X$ and every nonempty subset $B \subseteq X$;

(iii) *Regular* $\alpha(B) = 0$ if and only if B is relatively compact in X.

One of the most important examples of MNC is Hausdorff MNC α defined on each bounded subset B of X by

$$\alpha(B) = \inf \left\{ \varepsilon > 0 : B \subset \bigcup_{j=1}^{m} B_\varepsilon(u_j), \text{ where } u_j \in X \right\},$$

where $B_\varepsilon(u_j)$ is a ball of radius $\leq \varepsilon$ centered at u_j, $j = 1, 2, \ldots, m$. Without confusion, Kuratowski MNC α_1 is defined on each bounded subset B of X by

$$\alpha_1(B) = \inf \left\{ \varepsilon > 0 : B \subset \bigcup_{j=1}^{m} M_j \text{ and } \text{diam}(M_j) \leq \varepsilon \right\},$$

where the diameter of M_j is defined by $\text{diam}(M_j) = \sup\{|u - v| : u, v \in M_j\}$, $j = 1, 2, \ldots, m$.

It is well known that Hausdorff MNC α and Kuratowski MNC α_1 enjoy the above properties (i)-(iii) and other properties.

(iv) $\alpha(B_1 + B_2) \leq \alpha(B_1) + \alpha(B_2)$, where $B_1 + B_2 = \{u + v : u \in B_1, v \in B_2\}$;

(v) $\alpha(B_1 \cup B_2) \leq \max\{\alpha(B_1), \alpha(B_2)\}$;

(vi) $\alpha(\lambda B) \leq |\lambda|\alpha(B)$ for any $\lambda \in \mathbb{R}$.

In particular, the relationship between Hausdorff MNC α and Kuratowski MNC α_1 is given by

(vii) $\alpha(B) \leq \alpha_1(B) \leq 2\alpha(B)$.

In the following, several examples of useful measures of noncompactness in spaces of continuous functions are presented.

Example 1.1 We consider general example of MNC in the space of continuous functions $C([a, b], X)$. For $\Omega \subset C([a, b], X)$ define

$$\phi(\Omega) = \sup_{t \in [a,b]} \alpha(\Omega(t)),$$

where α is Hausdorff MNC in X and $\Omega(t) = \{v(t) : v \in \Omega\}$.

Example 1.2 Consider another useful MNC in the space $C([a, b], X)$. For a bounded $\Omega \subset C([a, b], X)$, set

$$\nu(\Omega) = \left(\sup_{t \in [a,b]} \alpha(\Omega(t)), \text{mod}_C(\Omega) \right);$$

here, the modulus of equicontinuity of the set of functions $\Omega \subset C([a, b], X)$ has the following form:

$$\text{mod}_C(\Omega) = \lim_{\delta \to 0} \sup_{u \in \Omega} \max_{|t_1 - t_2| \le \delta} |u(t_1) - u(t_2)|. \tag{1.1}$$

Example 1.3 We consider one more MNC in the space $C([a, b], X)$. For a bounded $\Omega \subset C([a, b], X)$, set

$$\nu(\Omega) = \max_{D \in \Delta(\Omega)} \left(\sup_{t \in [a,b]} \exp^{-Lt} \alpha(D(t)), \text{mod}_C(D) \right),$$

where $\Delta(\Omega)$ is the collection of all denumerable subsets of Ω, L is a constant, and $\text{mod}_C(D)$ is given in formula (1.1).

Let $J = [0, b]$, $b \in \mathbb{R}^+$. For any $W \subset C(J, X)$, we define

$$\int_0^t W(s)ds = \left\{ \int_0^t u(s)ds : u \in W \right\} \quad \text{for } t \in [0, b],$$

where $W(s) = \{u(s) \in X : u \in W\}$.

We present here some useful properties.

Property 1.1 *If $W \subset C(J, X)$ is bounded and equicontinuous, then $\overline{co}W \subset C(J, X)$ is also bounded and equicontinuous.*

Property 1.2 *If $W \subset C(J, X)$ is bounded and equicontinuous, then $t \mapsto \alpha(W(t))$ is continuous on J, and*

$$\alpha(W) = \max_{t \in J} \alpha(W(t)), \quad \alpha\left(\int_0^t W(s)ds \right) \le \int_0^t \alpha(W(s))ds,$$

for $t \in [0, b]$.

Property 1.3 *Let $\{u_n\}_{n=1}^\infty$ be a sequence of Bochner integrable functions from J into X with $|u_n(t)| \le \tilde{m}(t)$ for almost all $t \in J$ and every $n \ge 1$, where $\tilde{m} \in L^1(J, \mathbb{R}^+)$, then the function $\psi(t) = \alpha(\{u_n(t)\}_{n=1}^\infty)$ belongs to $L^1(J, \mathbb{R}^+)$ and satisfies*

$$\alpha\left(\left\{ \int_0^t u_n(s)ds : n \ge 1 \right\} \right) \le 2 \int_0^t \psi(s)ds.$$

Property 1.4 *If W is bounded, then for each $\varepsilon > 0$, there is a sequence $\{u_n\}_{n=1}^{\infty} \subset W$ such that*

$$\alpha(W) \leq 2\alpha(\{u_n\}_{n=1}^{\infty}) + \varepsilon.$$

Using Properties 1.3 and 1.4, we have the following statement.

Property 1.5 *Let $D \subset L^1(J, X)$ be such that*

(i) $|\xi(t)| \leq \nu(t)$ *for all $\xi \in D$ and for a.e. $t \in J$,*
(ii) $\alpha(D(t)) \leq q(t)$ *for a.e. $t \in J$,*

where $\nu, q \in L^1(J, \mathbb{R})$. Then

$$\alpha\left(\int_0^t D(s)ds\right) \leq 4\int_0^t q(s)ds.$$

Consider an abstract operator $\mathscr{L} : L^1([0, b], X) \to C([0, b], X)$ satisfying the following conditions:
(L_1) there exists a constant $C > 0$ such that

$$|(\mathscr{L}g_1)(t) - (\mathscr{L}g_2)(t)| \leq C\int_0^t |g_1(s) - g_2(s)|ds$$

for all $g_1, g_2 \in L^1([0, b], X)$, $t \in [0, b]$;
(L_2) for each compact set $K \subset X$ and sequence $\{g_n\} \subset L^1([0, b], X)$ such that $\{g_n(t)\} \subset K$ for a.e. $t \in [0, b]$, the weak convergence $g_n \rightharpoonup g_0$ implies $\mathscr{L}(g_n) \to \mathscr{L}(g_0)$ strongly in $C([0, b], X)$.

Remark 1.2 A typical example for $\mathscr{L}$ is the Cauchy operator

$$(Gg)(t) = \int_0^t g(s)ds, \quad g \in L^1([0, b], X),$$

which satisfies conditions (L_1)-(L_2) with $C = 1$ (see [130]).

Property 1.6 *Let $\mathscr{L}$ satisfy (L_1)-(L_2). Then the following properties hold:*

(i) *if the sequence of functions $\{g_n\} \subset L^1([0, b], X)$ is integrably bounded for all $n = 1, 2, \ldots$ and*

$$\alpha(\{g_n(t)\}) \leq \varpi(t)$$

for a.e. $t \in [0, b]$, where $\varpi \in L^1([0, b], \mathbb{R}^+)$, then

$$\alpha(\{(\mathscr{L}g_n)(t)\}) \leq 2C\int_0^t \varpi(s)ds;$$

(ii) *for every semicompact sequence $\{g_n\} \subset L^1([0, b], X)$, the sequence $\{\mathscr{L}(g_n)\}$ is relatively compact in $C([0, b], X)$ and, moreover, if $g_n \rightharpoonup g_0$, then $\mathscr{L}(g_n) \to \mathscr{L}(g_0)$.*

The approximation estimate and α-estimate in L^p space are given as follows.

Lemma 1.13 *Let the sequence* $\{f_n\} \subset L^p([0, b], X)$ $(p \geq 1)$ *be integrably bounded:*

$$|f_n(t)| \leq \zeta(t) \text{ for a.e. } t \in [0, b],$$

where $\zeta \in L^p([0, b], \mathbb{R}^+)$. *Assume that*

$$\alpha(\{f_n(t)\}_{n=1}^\infty) \leq \eta(t)$$

for a.e. $t \in [0, b]$, *where* $\eta \in L^p([0, b], \mathbb{R}^+)$. *Then for every* $\delta > 0$, *there exist a compact set* $K_\delta \subset X$, *a set* $m_\delta \subset [0, b]$, $\mathrm{mes}(m_\delta) < \delta$ *and a set of functions* $G_\delta \subset L^p([0, b], X)$ *with values in* K_δ *such that for every* $n \geq 1$ *there exist* $g_n \in G_\delta$ *for which*

$$|f_n(t) - g_n(t)| \leq 2\eta(t) + \delta, \quad t \in [0, b] \setminus m_\delta.$$

Property 1.7 *Assume that* X *is a separable Banach space. Let* $F : [0, b] \to P(X)$ *be* $L^p(p \geq 1)$-*integrable bounded multifunction such that*

$$\alpha(F(t)) \leq q(t)$$

for a.e. $t \in [0, b]$; *here,* $q \in L^p([0, b], \mathbb{R}^+)$. *Then*

$$\alpha\left(\int_0^t F(\tau)d\tau\right) \leq \int_0^t q(\tau)d\tau$$

for a.e. $t \in [0, b]$. *In particular, if the multifunction* $F : [0, b] \to P_{cp}(X)$ *is measurable and* L^p-*integrably bounded, then the function* $\alpha(F(\cdot))$ *is integrable and, moreover,*

$$\alpha\left(\int_0^t F(\tau)d\tau\right) \leq \int_0^t \alpha(F(\tau))d\tau$$

for a.e. $t \in [0, b]$.

We also recall definition of condensing mappings and fixed point theorems via condensing mappings, see, e.g., Akhmerov et al. [9] and Kamenskii et al. [130].

Definition 1.8 A multivalued mapping $\varphi : X \to P_{cp}(X)$ is said to be condensing with respect to an MNC β (β-condensing) if for every bounded set $D \subset X$ that is not relatively compact, we have $\beta(\varphi(D)) \not\geq \beta(D)$.

Theorem 1.8 *Let* Ω *be a bounded convex closed subset of* X *and* $\mathscr{H} : \Omega \to \Omega$ *a* β-*condensing mapping. Then the fixed point set* $\mathrm{Fix}\mathscr{H} = \{u : u = \mathscr{H}(u)\}$ *is nonempty.*

Theorem 1.9 *Let* $\Omega \subset X$ *be a bounded open neighborhood of zero and* $\mathscr{H} : \overline{\Omega} \to X$ *a* β-*condensing mapping satisfying the boundary condition* $u \neq \hat{\lambda}\mathscr{H}(u)$ *for all* $u \in \partial\Omega$ *and* $\hat{\lambda} \in (0, 1]$. *Then* $\mathrm{Fix}\mathscr{H}$ *is a nonempty compact set.*

Theorem 1.10 *Let Ω be a compact convex subset of a Banach space X and $\varphi : \Omega \to P(\Omega)$ be a closed multimap with convex values. Then $\mathrm{Fix}\varphi$ is nonempty.*

Theorem 1.11 *Let Ω be a closed subset of a Banach space X, and $\varphi : \Omega \to P_{cp}(\Omega)$ β-condensing on every bounded set of Ω. If $\mathrm{Fix}\varphi$ is bounded, then it is compact.*

The next fixed point principle can be seen as a consequence of Theorems 1.10 and 1.11.

Theorem 1.12 *Let Ω be a bounded convex closed subset of a Banach space X, and $\varphi : \Omega \to P_{cp,cv}(\Omega)$ an u.s.c. β-condensing multivalued mapping. Then $\mathrm{Fix}\varphi = \{u : u \in \varphi(u)\}$ is a nonempty compact set.*

Theorem 1.13 *Let Ω be a bounded convex closed subset of a Banach space X. Let $\varphi_1 : \Omega \to X$ be a single-valued map and $\varphi_2 : \Omega \to P_{cp,cv}(X)$ be a multimap such that $\varphi_1(u) + \varphi_2(u) \in P(\Omega)$ for $u \in \Omega$. Suppose that*

(a) *φ_1 is a contraction with the contraction constant $k < \frac{1}{2}$, and*
(b) *φ_2 is u.s.c. and compact.*

Then the fixed point set $\mathrm{Fix}(\varphi_1 + \varphi_2) := \{u : u \in \varphi_1(u) + \varphi_2(u)\}$ is a nonempty compact set.

Proof Since φ_1 is a single-valued contraction, it is continuous on X. For $u \in \Omega$, $\varphi_1(u) + \varphi_2(u) \in P(\Omega)$. Therefore the multimap $\varphi : \Omega \to P(\Omega)$ defined by $\varphi(u) = \varphi_1(u) + \varphi_2(u)$ is u.s.c. Since φ_1 is a contraction with the contraction constant $k < \frac{1}{2}$, then we have that $\alpha(\varphi_1(S)) \le 2k\alpha(S)$ for any bounded subset S of X, that is, $\alpha(\varphi_1(u)) \le 2k\alpha(\{u\}) = 0$ for $u \in \Omega$. Obviously $\varphi_1 : \Omega \to P_{cp,cv}(\Omega)$. As a result, we have $\varphi : \Omega \to P_{cp,cv}(\Omega)$. Let S be a bounded subset of Ω. As φ_2 is compact, we have that $\alpha(\varphi_2(S)) = 0$. It follows

$$\begin{aligned}\alpha(\varphi(S)) &\le \alpha(\varphi_1(S) + \varphi_2(S)) \\ &\le \alpha(\varphi_1(S)) + \alpha(\varphi_2(S)) \\ &\le 2k\alpha(S) \\ &< \alpha(S),\end{aligned}$$

whenever $\alpha(S) > 0$. Hence, we have that $\alpha(\varphi(S)) < \alpha(S), \alpha(S) > 0$ for all bounded sets S in Ω. So $\varphi : \Omega \to P_{cp,cv}(\Omega)$ is a α-condensing multimap. By Theorem 1.12, the fixed point set $\mathrm{Fix}\varphi$ is a nonempty compact set. This completes the proof.

1.2.3 R_δ-Set

In the study of the topological structure of the solution set for differential equations and inclusions, an important aspect is the R_δ-property. Recall that a subset D of a metric space is an R_δ-set if there exists a decreasing sequence $\{D_n\}_{n=1}^{\infty}$ of compact and contractible sets such that $D = \bigcap_{n=1}^{\infty} D_n$ (see Definition 1.11 below). This means that an R_δ-set is acyclic (in particular, nonempty, compact, and connected) and may not be a singleton but, from the point of view of algebraic topology, it is equivalent to a point, in the sense that it has the same homology groups as one point space.

Definition 1.9 ([126])

(i) X is called an absolute retract (AR space) if for any metric space Y and any closed subset $D \subset Y$, there exists a continuous function $h : D \to X$ which can be extended to a continuous function $\tilde{h} : Y \to X$.
(ii) X is called an absolute neighborhood retract (ANR space) if for any metric space Y, closed subset $D \subset Y$ and continuous function $h : D \to X$, there exists a neighborhood $U \supset D$ and a continuous extension $\tilde{h} : U \to X$ of h.

Obviously, if X is an AR space then it is an ANR space. Furthermore, as in [98], if D is a convex set in a locally convex linear space then it is an AR space. This yields that each convex set of a Fréchet space is an AR space, since every Fréchet space is locally convex. In particular, every Banach space is an AR pace.

Definition 1.10 A nonempty subset D of a metric space Y is said to be contractible if there exists a point $v_0 \in D$ and a continuous function $h : D \times [0, 1] \to D$ such that $h(v, 1) = v_0$ and $h(v, 0) = v$ for every $v \in D$.

Definition 1.11 A subset D of a metric space is called an R_δ-set if there exists a decreasing sequence $\{D_n\}$ of compact and contractible sets such that

$$D = \bigcap_{n=1}^{\infty} D_n.$$

Remark 1.3 The R_δ-property of the solution set for differential equations and inclusions is also called the Aronszajn-type result.

Note that any R_δ-set is nonempty, compact, and connected. What followed is the hierarchy for nonempty subsets of a metric space:

$$\text{compact+convex} \subset \text{compact } AR \text{ space}$$
$$\subset \text{compact+ contractible} \subset R_\delta\text{-set}, \tag{1.2}$$

and all the above inclusions are proper. The characterization of compact absolute neighborhood contractible spaces is obtained in [129].

Lemma 1.14 (Hyman) *Let X be a complete metric space, and $D \subset X$ be compact and absolutely neighborhood contractible set, then there exists a decrease sequence of compact ANR spaces $\{D_n\}_{n \geq 1}$ such that*

(i) $D_1 = X$,
(ii) *each D_n is contractible in D_{n-1} and*

(iii) $D = \bigcap_{n=1}^{\infty} D_n$.

Proof Let $\{Y_k\}_{k\geq 1}$ be a decreasing sequence of compact ANR space such that

$$D = \bigcap_{k=1}^{\infty} Y_k,$$

where the existence of sequence $\{Y_k\}_{k\geq 1}$ follows from the methods of Borsuk [43]. We may assume that $Y_1 = X$. Take $D_1 = Y_1$ and recursively define $D_n = Y_{k(n)}$, where $k(n)$ is the smallest index greater than n such that $Y_{k(n)}$ is contractible in D_n. Such $Y_{k(n)}$ exists, for by the methods of Hyman [128] that some neighborhood V of D is contractible in D_n. Since $D = \bigcap_{k=1}^{\infty} Y_k$, some Y_k, $k > n$ lies in V; hence Y_k is contractible in D_n. $\{D_n\}_{n\geq 1}$ is the desired sequence.

The following characterization of R_δ-sets, which develops the well-known Hyman's theorem [129], was shown by Bothe.

Theorem 1.14 ([45]) *Let X be a complete metric space, α denote Hausdorff MNC in X, and let $\emptyset \neq D \subset X$. Then the following statements are equivalent:*

(i) *D is an R_δ-set;*

(ii) *D is an intersection of a decreasing sequence $\{D_n\}$ of closed contractible spaces with $\alpha(D_n) \to 0$;*

(iii) *D is compact and absolutely neighborhood contractible, i.e., D is contractible in each neighborhood in $Y \in ANR$.*

Proof Evidently (i) implies (ii), since every AR is contractible, and the last implication (iii) $\Rightarrow$ (i) follows from Lemma 1.14. To prove that (iii) follows from (ii), let Y be any ANR which contains D as a closed subset and let $V \subset Y$ be a neighborhood of D in Y. Since open subsets of ANR space are ANR space too, there is a continuous extension $f : U \to V$ of the identity $I : D \to V$ to some neighborhood U of D in X. Then $\alpha(D_n) \to 0$ implies $D_n \subset U$ for all large n. Fix such n, let $u_0 \subset D$, and $h : [0, 1] \times D \to D$ be continuous such that $h(0, u) = u_0$ and $h(1, u) = u$ on D_n. Then $h_0 := f \circ h|_{[0,1]\times D} : [0, 1] \times D \to V$ is continuous and satisfies

$$h_0(0, u) = f(u_0) \in V \quad \text{and} \quad h_0(1, u) = u \text{ on } D.$$

Hence D is absolutely neighborhood contractible and compactness of D is obvious.

Definition 1.12 A multivalued mapping $\varphi : Y \to P(Z)$ is an R_δ-mapping if φ is u.s.c. and $\varphi(v)$ is an R_δ-set for each $v \in Y$.

It is clear that every u.s.c. multivalued mapping with contractible values can be seen as an R_δ-mapping. In particular, every single-valued continuous mapping is an R_δ-mapping.

Recall that $\text{Liminf } R_n = \bigcup_{n=1}^{\infty} \bigcap_{k=n}^{\infty} R_k$, $\text{Limsup } R_n = \bigcap_{n=1}^{\infty} \bigcup_{k=n}^{\infty} R_k$ and, if $\text{Liminf } R_n = \text{Limsup } R_n$, then this set is called the set-theoretic limit Limes R_n of $\{R_n\}$.

A lower topological limit is the set

$$\mathrm{Li}\, R_n = \{u \in X : \text{there exist } \{u_n\} \text{ such that } u_n \in R_n \text{ and } u_n \to u\},$$

an upper topological limit of $\{R_n\}$ is the set

$$\mathrm{Ls}\, R_n = \{u \in X : \text{there exist } n_1 < n_2 < \cdots \text{ such that } u_{n_i} \in R_n \text{ and } u_{n_i} \to u\}$$

and, if $\mathrm{Li}\, R_n = \mathrm{Ls}\, R_n$, then this set is called a topological limit $\mathrm{Lim}\, R_n$ of $\{R_n\}$.

For the sake of completeness, we recall that the original version of Aronszajn's theorem [51, Lemma 5] can be formulated in terms of limit sets as follows.

Lemma 1.15 (Browder–Gupta) *Let Y be a given metric space. $\{R_n\}$ a sequence of absolute retracts in Y. Assume that $M \subset Y$ is such that the following conditions hold:*

(i)   *$M \subset R_n$ for every n;*
(ii) *M is the set-theoretic limit of the sequence $\{R_n\}$ of absolute retracts;*
(iii) *For each neighborhood V of M in Y there is an infinite subsequence $\{R_{n_k}\}$ of $\{R_n\}$ such that $\{R_{n_k}\}$ is contained in V for every n_k.*

Then M is an R_δ-set.

Proof Let h_n denote the continuous retraction of Y onto R_n for each n. We shall first construct an infinite subsequence $\{\widetilde{R}_n\}$ of $\{R_n\}$ inductively such that for each $k < n$ and every $u \in \widetilde{R}_n$, we have $d(u, \psi_k^n(u)) < \frac{1}{n}$, where ψ_k^n denotes the continuous mapping $\psi_k^n = \widetilde{h}_k \widetilde{h}_{k+1} \cdots \widetilde{h}_{n-1}$ of Y into $\widetilde{R}_k$. We shall use this subsequence $\widetilde{R}_n$ of the sequence R_n to construct a decreasing sequence of absolute retracts in the product metric space $\prod_{k=1}^{\infty} Y$ and show that the intersection of this decreasing sequence of absolute retracts in $\prod_{k=1}^{\infty} Y$ is homeomorphic to M. (The metric on the product metric space $\prod_{k=1}^{\infty} Y$ is given by

$$\rho(u, v) = \sum_{k=1}^{\infty} \frac{1}{2^k} \frac{d(u_k, v_k)}{1 + d(u_k, v_k)},$$

where $u = (u_1, \ldots, u_k, \ldots)$ and $v = (v_1, \ldots, v_k, \ldots)$ are points of $\prod_{k=1}^{\infty} Y$).

Let $\widetilde{R}_1 = R_1$, and suppose that $\widetilde{R}_1, \ldots, \widetilde{R}_n$ have been chosen. Then for each $k < n + 1$ the continuous mapping $\psi_k^{n+1} = \widetilde{h}_k \cdots \widetilde{h}_n$ of Y into $\widetilde{R}_k$ is defined and for every u in M we have $\psi_k^{n+1}(u) = u$. It follows that there is a neighborhood V of M such that $d(u, \psi_k^{n+1}(u)) < \frac{1}{n+1}$ for every u in V. and $k < n + 1$. Now choose $\widetilde{R}_{n+1}$ to be the first R_n, after $\widetilde{R}_1, \cdots \widetilde{R}_n$ contained in V, which exists because of condition (iii). This completes the inductive step of the construction.

For each n, set $Q_n = \{u : u \in \prod_{k=1}^{\infty} \widetilde{R}_k, u_k = \psi_k^n(u_n), i < n\}$. It follows easily that the natural mapping of Q_n onto $\prod_{k=n}^{\infty} \widetilde{R}_k$ is a homeomorphism. Hence Q_n is an absolute retract for $\prod_{k=n}^{\infty} \widetilde{R}_k$ being the Cartesian product of absolute retracts is certainly an absolute retract. The sequence $\{Q_n\}$ is clearly decreasing and contains the diagonal set of $\prod_{k=1}^{\infty} M$. We shall show that $\mathrm{diag}(\prod_{k=1}^{\infty} M) = \bigcap_{n=1}^{\infty} Q_n$. Indeed, $\mathrm{diag}(\prod_{k=1}^{\infty} M) \subset \bigcap_{n=1}^{\infty} Q_n$ and if $(u_1, \ldots, u_n, \ldots)$ is a point in $\bigcap_{n=1}^{\infty} Q_n$ we have for every n and each $k < n$ that

$$d(u_n, u_k) = d(u_n, \psi_k^{n+1}(u_n)) < \frac{1}{n}.$$

This shows that sequence $\{u_n\}$ converges to u_k for every k and so $u_1 = u_2 = \cdots = u_n = \cdots = u$ (say). It follows that u belongs to $\widetilde{R}_n$ for every n and hence u is in the limit M of $\{\widetilde{R}_n\}$. This shows that $\bigcap_{n=1}^{\infty} Q_n$ is contained in $\mathrm{diag}(\prod_{k=1}^{\infty} M)$. This proves the assertion. Since now, M is homeomorphic to the diagonal of $\prod_{k=1}^{\infty} M$, we get that M is an R_δ-set. This completes the proof.

Among several remarks on Browder–Gupta results the following should be added.

Remark 1.4 If R_n is a compact absolute retract for every n in Lemma 1.15, then M is a compact R_δ-set.

Remark 1.5 Following the proof of Lemma 1.15, assumption (i) and the fact that the set-theoretic upper limit of any sequence of sets is contained in the topological upper limit of it, we can assume in (ii) that M is a topological limit of $\{R_n\}$.

Example 1.4 It is easy to see that the intersection M ($\emptyset \neq M \subset Y$) of a decreasing sequence of closed subsets of Y is its topological limit.

Lemma 1.16 ([103]) *Let Y be a given metric space. $\{R_n\}$ a sequence of compact R_δ-sets in Y. Assume that $M \subset Y$ is such that the following conditions hold:*

(i) *$M \subset R_n$ for every n;*
(ii) *M is the topological limit of the sequence $\{R_n\}$;*
(iii) *For each open neighborhood V of M in Y there is an infinite subsequence $\{R_{n_k}\}$ of $\{R_n\}$ such that $\{R_{n_k}\}$ is contained in V for every n_k.*

Then M is an R_δ-set.

Proof Let $R_n = \bigcap_{i=1}^{\infty} R_n^i$ for every $n \geq 1$, where each R_n^i is a compact absolute retract and $R_n^{i+1} \subset R_n^i$ for all $n, i \geq 1$. For every n, by the compactness of R_n^i, there exists $W_n = R_n^{i_n}$ such that $R_n^{i_n} \subset N_{\frac{1}{n}}(R_n)$, where $N_{\frac{1}{n}}(R_n)$ denotes the $\frac{1}{n}$-neighbourhood of the set R_n.

Notice that

(i)' $M \subset R_n \subset R_n^{i_n}$ for every n.
(ii)' From (i) it follows that $M \subset \mathrm{Li} W_n$. Let $u \in \mathrm{Li} W_n$ be an arbitrary point. Then there are a subsequence $n_1 < n_2 < \cdots$ and points $u_{n_k} \in W_{n_k}$ such that $u = \lim_{k \to \infty} u_{n_k}$. By the definition of W_{n_k}, for every $k \geq 1$ there exists $w_{n_k} \in R_{n_k}$ such that $d(u_{n_k}, w_{n_k}) < \frac{1}{n_k}$. Thus $w_{n_k} \to u$, which implies that $u \in \mathrm{Ls} R_n$. From assumption (ii) we obtain $\mathrm{Ls} W_n \subset M$ which gives $M = \mathrm{Lim} W_n$.

(iii)' Let U be an arbitrary open neighbourhood of M in Y. By the compactness of M, there is $n_0 \geq 1$ such that $N_{\frac{2}{n_0}}(M) \subset U$. By assumption (iii), we can find a subsequence $\{R_{n_k}\}$, $n_0 < n_1 < n_2 < \cdots$, such that $R_{n_k} \subset N_{\frac{1}{n_k}}(M)$, for every $k \geq 1$. Since $W_{n_k} \subset N_{\frac{1}{n_k}}(R_{n_k})$, one can easily obtain that $W_{n_k} \subset N_{\frac{2}{n_0}}(M) \subset U$.

The Browder–Gupta Lemma 1.15 and Remark 1.5 end the proof.

The following result on the topological structure of the solution set of nonlinear functional equations is due to Aronszajn [19] and further developed by Browder and Gupta in [51], which is also called Aronszajn-type result.

Theorem 1.15 *Let Y be a metric space and E a Banach space. Suppose that $\mathscr{F}$: $Y \to E$ is a proper mapping, i.e., $\mathscr{F}$ is continuous and $\mathscr{F}^{-1}(K)$ is compact for each compact set $K \subset E$. In addition, if there exists a sequence $\{\mathscr{F}_n\}$ of mappings from Y into E such that*

(i) *$\mathscr{F}_n$ is proper and $\{\mathscr{F}_n\}$ converges to $\mathscr{F}$ uniformly on Y, and*
(ii) *for a given point $v_0 \in E$ and for all v in a neighborhood $U(v_0)$ of v_0 in E, there exists exactly one solution u_n of the equation $\mathscr{F}_n(u) = v$,*

then $\mathscr{F}^{-1}(v_0)$ is an R_δ-set.

Proof Let $\rho > 0$ be such that the open ball $B_\rho(v_0)$ of radius ρ and center v_0 is contained in $U(v_0)$. Let $\{\varepsilon_n\}$ be a sequence of positive real numbers such that $\varepsilon_n < \rho$ for every n and $\varepsilon_n \to 0$ as $n \to \infty$. Since the sequence $\{\mathscr{F}_n\}$ of continuous mappings of Y into E converges uniformly to the mapping $\mathscr{F}$ of Y into E we may suppose (by choosing a subsequence of $\{\mathscr{F}_n\}$ if necessary) that $|\mathscr{F}_n(u) - \mathscr{F}(u)| < \varepsilon_n$ for every $u \in Y$ and every n.

Let K denote the compact set $\mathscr{F}^{-1}(v_0)$. For every n, the compact set $\mathscr{F}_n(K)$ is contained in $B_{\varepsilon_n}(v_0)$ since for every $u \in K$,

$$|\mathscr{F}_n(u) - v_0| = |\mathscr{F}_n(u) - \mathscr{F}(u)| < \varepsilon_n.$$

Let Q_n denote the convex closure of the compact set $\mathscr{F}_n(K)$ in E. Then Q_n is a compact convex subset of E and is accordingly an absolute retract for each n. Further Q_n is contained in $B_{\varepsilon_n}(v_0)$ for every n. The mapping $\mathscr{F}_n$ is a continuous one-to-one mapping of the compact set $R_n = \mathscr{F}_n^{-1}(Q_n)$ onto Q_n for every n. (This follows easily from our assumptions.) Accordingly $\mathscr{F}_n$ is a homeomorphism of R_n onto Q_n and so R_n is a compact absolute retract for every n. We shall show that the sequence $\{R_n\}$ of compact absolute retracts satisfies the conditions (i)-(iii) of Lemma 1.15 to conclude that K is a compact R_δ-set.

Clearly K is contained in R_n for every n and so K is contained in the inferior set-theoretic limit of the sequence $\{R_n\}$. Let u be a point in the superior set-theoretic limit of the sequence $\{R_n\}$ so that u is in R_{n_k} for every n_k, for some subsequence $\{R_{n_k}\}$ of $\{R_n\}$. It follows that $|\mathscr{F}_{n_k}(u) - \mathscr{F}(u)| < \varepsilon_{n_k}$ and $|\mathscr{F}_{n_k}(u) - v_0| < \varepsilon_{n_k}$ for every n_k, so that $|\mathscr{F}_n(u) - v_0| < 2\varepsilon_{n_k}$ for every n_k. Hence $f(u) = v_0$, which implies that u is in K. Thus the superior set theoretic limit of $\{R_n\}$ is contained in K. Hence K is the set-theoretic limit of $\{R_n\}$. To verify condition (iii) of Lemma 1.15 it suffices to show that each neighborhood V of K contains at least one member of the sequence $\{R_n\}$, as the set-theoretic limit remains unchanged if finitely many members of $\{R_n\}$ are omitted. Suppose V is a neighborhood of K such that $\{R_n\}$ is not contained in V for any n. So there exists a sequence u_n in Y such that u_n belongs to R_n, for every n and u_n does not belong to V for any n. Now u_n belongs to R_n for every n gives that $|\mathscr{F}_n(u_n) - v_0| < \varepsilon_n$. Also $|\mathscr{F}_n(u_n) - \mathscr{F}(u_n)| < \varepsilon_n$ for every n. Hence $|\mathscr{F}(u_n) - v_0| < 2\varepsilon_n$ for every n and so the sequence $\{\mathscr{F}(u_n)\}$ converges to v_0 in E. Since $\mathscr{F}$ is proper, there is a subsequence $\{u_{n_k}\}$ of $\{u_n\}$ and an u in Y such that $\{u_{n_k}\}$ converges to u in Y. It follows that the subsequence $\{\mathscr{F}(u_{n_k})\}$ of $\{\mathscr{F}(u_n)\}$ converges to $\mathscr{F}(u)$ as well as to v_0 in E. Hence $\mathscr{F}(u) = v_0$. So the subsequence $\{u_{n_k}\}$ of $\{u_n\}$ converges to the point u in K, which is a contradiction, for u_n does not belong to the neighborhood V of K for any n. This proves that each neighborhood V of K contains at least one member of the sequence. Hence the conditions of Lemma 1.15 are verified. Thus K is a compact R_δ-set by Remark 1.4.

We need the following fixed point theorem which is due to [114].

Theorem 1.16 *Let Y be an ANR space. Assume that $\varphi : Y \to P(Y)$ can be factorized as*

$$\varphi = \varphi_n \circ \varphi_{n-1} \circ \cdots \circ \varphi_1,$$

where $\varphi_i : Y_{i-1} \to P(Y_i)$, $i = 1, \ldots, n$, are R_δ-mappings and Y_i, $i = 1, \ldots, n-1$, are ANR spaces and $Y_0 = Y_n = Y$ are AR spaces. If there exists a compact subset $K \subset Y$ satisfying $\varphi(Y) \subset K$, then φ has a fixed point in Y.

We also need the following result, which can be seen from the inclusion relation (1.2) and Theorem 1.16.

Theorem 1.17 *Let X be a Banach space and $D \subset X$ be a nonempty compact convex subset. If the multivalued mapping $\varphi : D \to P(D)$ is u.s.c. with contractible values, then φ has a fixed point.*

1.2.4 Inverse Limit Method

To study the Cauchy problem defined on right half-line, we shall use the inverse limit method. Let us sketch the necessary abstract framework.

Let Σ be a directed set ordered by a relation $\leq$, X_α a metric space for every $\alpha \in \Sigma$ and $\pi_\alpha^\beta : X_\beta \to X_\alpha$ a continuous mapping for each two elements α, β with $\alpha \leq \beta$. Moreover, for each $\alpha \leq \beta \leq \gamma$,

$$\pi_\alpha^\alpha = id_{X_\alpha} \text{ and } \pi_\alpha^\beta \circ \pi_\beta^\gamma = \pi_\alpha^\gamma.$$

By an inverse system we mean a family $S = \{X_\alpha, \pi_\alpha^\beta, \Sigma\}$. The limit of inverse system S, denoted by $\lim_{\leftarrow} S$, is defined by

$$\lim_{\leftarrow} S = \left\{ (u_\alpha) \in \prod_{\alpha \in \Sigma} X_\alpha : \pi_\alpha^\beta(u_\beta) = u_\alpha \text{ for all } \alpha \leq \beta \right\}.$$

If we denote by $\pi_\alpha : \lim_{\leftarrow} S \to X_\alpha$ a restriction of the projection $p_\alpha : \Pi_{\alpha \in \Sigma} X_\alpha \to X_\alpha$ onto the α-th axis, then it holds $\pi_\alpha = \pi_\alpha^\beta \circ \pi_\beta$ for each $\alpha \leq \beta$.

Remark 1.6 It is noted that every Fréchet space is a limit of an inverse system of Banach spaces (cf. e.g., [13]).

Consider two inverse systems $S = \{X_\alpha, \pi_\alpha^\beta, \Sigma\}$ and $S' = \{X_{\alpha'}, \pi_{\alpha'}^{\beta'}, \Sigma'\}$. By a multivalued mapping of $S = \{X_\alpha, \pi_\alpha^\beta, \Sigma\}$ into $S' = \{X_{\alpha'}, \pi_{\alpha'}^{\beta'}, \Sigma'\}$ we means a family $\{\sigma, \varphi_{\sigma(\alpha')}\}$ consisting of a monotone mapping $\sigma : \Sigma' \to \Sigma$ and multivalued mappings $\varphi_{\sigma(\alpha')} : X_{\sigma(\alpha')} \to 2^{X_{\alpha'}}$ such that

$$\pi_{\alpha'}^{\beta'} \circ \varphi_{\sigma(\beta')} = \varphi_{\sigma(\alpha')} \circ \pi_{\sigma(\alpha')}^{\sigma(\beta')} \text{ for each } \alpha' \leq \beta'.$$

A family $\{\sigma, \varphi_{\sigma(\alpha')}\}$ induces a limit mapping $\varphi : \lim_{\leftarrow} S \to 2^{\lim_{\leftarrow} S'}$ defined by

$$\varphi(u) = \prod_{\alpha' \in \Sigma'} \varphi_{\sigma(\alpha')}(u_{\sigma(\alpha')}) \cap \lim_{\leftarrow} S'.$$

In other words, a limit mapping is that $\pi_{\alpha'} \circ \varphi = \varphi_{\sigma(\alpha')} \circ \pi_{\sigma(\alpha')}$ for every $\alpha' \in \Sigma'$.
    We will make use of the following results.

Theorem 1.18 ([101]) *Let $S = \{X_\alpha, \pi_\alpha^\beta, \Sigma\}$ be an inverse system. If for each $\alpha \in \Sigma$, X_α is nonempty and compact, then $\lim_{\leftarrow} S$ is also nonempty and compact.*

Lemma 1.17 ([103]) *Let $S = \{X_m, \pi_m^p, \mathbb{N} \setminus \{0\}\}$ be an inverse system. If each X_m is a compact R_δ-set, then $\lim_{\leftarrow} S$ is an R_δ-set, too.*

Proof The assertion follows from Example 1.4 and Lemma 1.16. Indeed, define

$$Q_n = \left\{ (u_i) \in \prod_{i=1}^\infty X_i : u_i = \pi_i^n(u_n) \text{ for all } i \leq n \right\}.$$

It is easy to see that each Q_n is homeomorphic to the R_δ-set $\prod_{i=n}^\infty X_i$. Notice that

$$\bigcap_{i=n}^\infty Q_n = \left\{ (u_i) \in \prod_{i=1}^\infty X_n : u_i = \pi_i^n(u_n) \text{ for every } n \geq 1 \text{ and } i \leq n \right\} = \lim_{\leftarrow} S.$$

This implies (comp. Example 1.4) that $\lim \lim_{\leftarrow} S = \mathrm{Lim}\, Q_n$, and by Lemma 1.16, it is an R_δ-set, as required.

Theorem 1.19 ([104]) *Let $S = \{X_m, \pi_m^p, \mathbb{N} \setminus \{0\}\}$ be an inverse system and $\varphi : \lim_{\leftarrow} S \to 2^{\lim_{\leftarrow} S}$ a limit mapping induced by a family $\{id, \varphi_m\}$, where φ_m is a multivalued mapping from X_m into itself. If $\mathrm{Fix}(\varphi_m)$ is an R_δ-set for each $m \in \mathbb{N} \setminus \{0\}$, then $\mathrm{Fix}(\varphi)$ is an R_δ-set.*

Proof We show that $\pi_m^p(\mathrm{Fix}(\varphi_p)) \subset \mathrm{Fix}(\varphi_m)$. Let $u_p \in \mathrm{Fix}(\varphi_p)$. Then $u_p \in \varphi_p(u_p)$ and $\pi_m^p(u_p) \in \pi_m^p \varphi_p(u_p) \subset \varphi_m \pi_m^p(u_p)$, which implies that $\pi_m^p(u_p) \in \mathrm{Fix}(\varphi_m)$.

Similarly we show that $\pi_m(\mathrm{Fix}(\varphi)) \subset \mathrm{Fix}(\varphi_m)$. Denote by $\overline{\pi}_m^p : \mathrm{Fix}(\varphi_p) \to \mathrm{Fix}(\varphi_m)$ the restriction of π_m^p. One can see that $\overline{S} = \{\mathrm{Fix}(\varphi_m), \overline{\pi}_m^p, \mathbb{N} \setminus \{0\}\}$ is an inverse system.

By Lemma 1.17, the set $\mathrm{Fix}(\varphi)$ is an R_δ-set and the proof is complete.

For more details about the inverse limit method, we refer reader to [104, 105] and the references therein.

In the following, two examples of useful inverse systems are presented.

Example 1.5 Let $\tau \in \mathbb{R}$ and $\mathbb{N}(\tau) = \{m \in \mathbb{N} \setminus \{0\} : m > \tau\}$. For each $p, m \in \mathbb{N}(\tau)$ with $p \geq m$, let us consider a projection $\pi_{\tau,m}^p : C([\tau, p], X) \to C([\tau, m], X)$, defined by

$$\pi_{\tau,m}^p(u) = u|_{[\tau,m]}, \quad u \in C([\tau, p], X). \tag{1.3}$$

It is readily checked that $\{C([\tau, m], X), \pi_{\tau,m}^p, \mathbb{N}(\tau)\}$ is an inverse system and its limit is isometrically homeomorphic to $\widetilde{C}([\tau, \infty), X)$, so for convenience we write

$$\widetilde{C}([\tau, \infty), X) = \lim_{\leftarrow}\{C([\tau, m], X), \pi_{\tau,m}^p, \mathbb{N}(\tau)\}.$$

Example 1.6 Let $L_{loc}^1(\mathbb{R}^+, X)$ be the separated locally convex space consisting of all locally Bocher integrable functions from $\mathbb{R}^+$ to X endowed with a family of seminorms $\{\|\cdot\|_m^1 : m \in \mathbb{N} \setminus \{0\}\}$, defined by

$$\|u\|_m^1 = \int_0^m |u(t)|dt, \quad m \in \mathbb{N} \setminus \{0\}.$$

In a similar manner as above, we also obtain that $\{L^1([0, m], X), \dot{\pi}_m^p, \mathbb{N} \setminus \{0\}\}$ is an inverse system, where $p \geq m$ and

$$\dot{\pi}_m^p(f) = f|_{[0,m]}, \quad f \in L^1([0, p], X).$$

Moreover, it is clear that

$$L_{loc}^1(\mathbb{R}^+, X) = \lim_{\leftarrow}\{L^1([0, m], X), \dot{\pi}_m^p, \mathbb{N} \setminus \{0\}\}.$$

At the end of this subsection, we establish the following compactness characterization in Fréchet spaces.

Lemma 1.18 *Let $\tau \in \mathbb{R}$. For each $m \in \mathbb{N}(\tau)$, define a projection $\pi_{\tau,m} : \tilde{C}([\tau, \infty), X) \to C([\tau, m], X)$ by*

$$\pi_{\tau,m}(u) = u|_{[\tau,m]} \quad for \ each \ u \in \tilde{C}([\tau, \infty), X).$$

Then a nonempty set $D \subset \tilde{C}([\tau, \infty), X)$ is relatively compact (resp. compact) if $\pi_{\tau,m}(D)$ is relatively compact (resp. compact) in $C([\tau, m], X)$ for each $m \in \mathbb{N}(\tau)$.

Proof Let $D \subset \tilde{C}([\tau, \infty), X)$ and let $\pi_{\tau,m}(D)$ be relatively compact in $C([\tau, m], X)$ for each $m \in \mathbb{N}(\tau)$. Then it is readily checked that $\pi_{\tau,m}(\overline{D}) = \overline{\pi_{\tau,m}(D)}$ for each $m \in \mathbb{N}(\tau)$ and

$$\overline{D} = \lim_{\leftarrow}\{\pi_{\tau,m}(\overline{D}), \pi_{\tau,m}^p, \mathbb{N}(\tau)\},$$

where $\pi_{\tau,m}^p$ is the mapping defined by (1.3). Then, applying Theorem 1.18 yields the compactness of $\overline{D}$ in $\tilde{C}([\tau, \infty), X)$, as desired. The rest is clear.

1.3 Multivalued Semiflows

Let X be a complete metric space with metric $\rho(\cdot, \cdot)$. Let Γ be a nontrivial subgroup of the additive group of real numbers $\mathbb{R}$ and $\Gamma_+ = \Gamma \cap [0, \infty)$. For some formulations regarding multivalued semiflows and global attractors, we can see [145, 207].

Definition 1.13 The multimap $G : \Gamma \times X \to P(X)$ is called a multivalued flow (m-flow) if the next conditions are satisfied:

(i) $G(0, \cdot) = I$ is the identity map;
(ii) $G(t_1 + t_2, x) \subset G(t_1, G(t_2, x))$ for all $t_1, t_2 \in \Gamma, x \in X$,

where $G(t, B) = \bigcup_{x \in B} G(t, x), \ B \subset E$.

Remark 1.7 The multimap $G : \Gamma \times X \to P(X)$ is called an m-semiflow if conditions (i) and (ii) of Definition 1.13 hold for any $t_1, t_2 \in \Gamma_+$.

The m-semiflow G is called strict if $G(t_1 + t_2, x) = G(t_1, G(t_2, x))$ for all $x \in X$ and $t_1, t_2 \in \Gamma_+$.

Definition 1.14 The map $x(\cdot) : \Gamma_+ \to X$ is said to be a trajectory of the m-semiflow G corresponding to the initial condition x_0 if $x(t + \tau) \in G(t, x(\tau))$ for all $t, \tau \in \Gamma_+$, $x(0) = x_0$.

G is said to be eventually bounded if for each bounded set $B \subset X$, there is a number $T(B) > 0$ such that $\gamma_{T(B)}^+(B)$ is bounded. Here $\gamma_{T(B)}^+(B)$ is the orbit after time $T(B) : \gamma_{T(B)}^+(B) = \bigcup_{t \geq T(B)} G(t, B)$.

Definition 1.15 A bounded set $B_1 \subset X$ is called an absorbing set for m-semiflow G if for any bounded set $B \subset X$, there exists $\tau = \tau(B) \geq 0$ such that $\gamma_{\tau(B)}^+(B) \subset B_1$.

Definition 1.16 The subset $A \subset X$ is called a global attractor of the m-semiflow G if it satisfies the following conditions:

(i) A attracts any bounded set B, i.e., $\text{dist}_X(G(t, B), A) \to 0$ as $t \to \infty$, for all bounded set $B \subset X$, where $\text{dist}_X(\cdot, \cdot)$ is the Hausdorff semi-distance of two subsets in X;

(ii) A is negatively semi-invariant, i.e., $A \subset G(t, A)$ for all $t \in \Gamma_+$. It is called invariant if $A = G(t, A)$ for all $t \in \Gamma_+$.

Lemma 1.19 *Let bounded set M be a negatively semiinvariant set with respect to the m-semiflow G, which has an attracting set Z. Then $M \subset \overline{Z}$.*

Definition 1.17 The m-semiflow G is called asymptotically upper semicompact if B is a bounded set in X such that for some $T(B) \in \Gamma_+$, $\gamma_{\tau(B)}^+(B)$ is bounded, any sequence $\xi_n \in G(t_n, B)$ with $t_n \to \infty$ is precompact in X.

The next propositions are useful to check in applications that an m-semiflow is asymptotically upper semicompact.

Lemma 1.20 *Let the map $G(t, \cdot) : X \to P(X)$ be compact for some $t \in \Gamma_+ \setminus \{0\}$, that is, for any bounded set B, $G(t, B)$ is precompact in X. Then the m-semiflow G is asymptotically upper semicompact.*

Definition 1.18 The m-semiflow G is called pointwise dissipative if there exists $K > 0$ such that for $x \in X$, $u(t) \in G(t, x)$, one has $|u(t)| \leq K$ for $t \geq t_0(|x|)$.

Consider now some theorems which state the existence of compact attractors for m-semiflows.

Theorem 1.20 *Let G be a pointwise dissipative and asymptotically upper semicompact m-semiflow. Suppose that $G(t, \cdot) : X \to P(X)$ is upper semicontinuous for any $t \in \Gamma_+$. If G is eventually bounded; then it possesses a compact global attractor $\mathscr{A}$ in X. Moreover, if G is a strict m-semiflow, then $\mathscr{A}$ is invariant, that is, $\mathscr{A} = G(t, \mathscr{A})$ for any $t \in \Gamma_+$.*

Theorem 1.21 *Let $G(t, \cdot) : X \to P(X)$ be an upper semicontinuous map. If there exists a compact set $K \subset X$ such that for any bounded set B*

$$\text{dist}(G(t, B), K) \to 0 \ \text{ as } t \to \infty,$$

the m-semiflow G has the global compact attractor $\mathscr{A} \subset K$. It is the minimal closed set attracting each bounded set B.

1.4 Pullback Attractor

We now recall some formulations regarding multivalued autonomous dynamical system and pullback attractors. For more detail, we refer to [55, 77].

Definition 1.19 A multivalued map $\mathcal{U} : \mathbb{R}_d^2 \times X \to P_c(X)$, where $\mathbb{R}_d^2$ is called a multivalued non-autonomous dynamical system (MNDS) on X iff

(i) $\mathcal{U}(t, t, x) = \{x\}$ for all $t \in \mathbb{R}$, $x \in X$;
(ii) $\mathcal{U}(t, \tau, x) \subset \mathcal{U}(t, s, \mathcal{U}(s, \tau, x))$ for all $\tau \leq s \leq t$, $x \in X$.

The MNDS $\mathcal{U}$ is said to be strict if $\mathcal{U}(t, \tau, x) = \mathcal{U}(t, s, \mathcal{U}(s, \tau, x))$ for all $\tau \leq s \leq t$, and $x \in X$.

The MNDS $\mathcal{U}$ is said to be strict if $\mathcal{U}(t, \tau, x) = \mathcal{U}(t, s, \mathcal{U}(s, \tau, x))$ for all $\tau \leq s \leq t$, and $x \in X$.

A multivalued map $D : \mathbb{R} \to P(X)$ is called a multifunction. Let $\mathcal{D}$ be a family of multifunctions taking values in $P_b(X)$ and having the inclusion-closed property: if $D \in \mathcal{D}$ and D' is a multifunction such that $D'(t) \subset D(t)$ for all $t \in \mathbb{R}$, then $D' \in \mathcal{D}$. The family $\mathcal{D}$ is called a universe.

Definition 1.20 A multifunction $B \in \mathcal{D}$ is said to be pullback $\mathcal{D}$-absorbing (with respect to the MNDS $\mathcal{U}$) if for every $D \in \mathcal{D}$, there exists $T = T(t, D) > 0$ such that

$$\mathcal{U}(t, t - s, D(t - s)) \subset B(t) \quad \text{for all } s \geq T.$$

We say that a multifunction $B \in \mathcal{D}$ is pullback $\mathcal{D}$-attracting if for every $D \in \mathcal{D}$

$$\lim_{s \to +\infty} \text{dist}_X(\mathcal{U}(t, t - s, D(t - s)), B(t)) = 0,$$

for all $t \in \mathbb{R}$. Here $\text{dist}_X(\cdot, \cdot)$ is the Hausdorff semi-distance between two subsets of X.

Definition 1.21 A multifunction $A \in \mathcal{D}$ is said to be a global pullback $\mathcal{D}$-attractor for the MNDS $\mathcal{U}$ if it satisfies

(i) $A(t)$ is compact for any $t \in \mathbb{R}$;
(ii) A is pullback $\mathcal{D}$-attracting;
(iii) A is negatively invariant, that is $A(t) \subset \mathcal{U}(t, \tau, A(\tau))$ for all $(t, \tau) \in \mathbb{R}_d^2$.

The pullback $\mathcal{D}$-attractor A is called strict if the invariance property in the third item is strict.

For a multifunction D, we define the pullback ω-limit set of D as a t-dependent set

$$\Lambda(t, D) = \bigcap_{\tau \geq 0} \overline{\bigcup_{s \geq \tau} \mathcal{U}(t, t - s, D(t - s))}.$$

Lemma 1.21 ([55]) *Let $\mathcal{U}$ be a u.s.c. MNDS on X, i.e., $\mathcal{U}(t, \tau, \cdot)$ is u.s.c. for $(t, \tau) \in \mathbb{R}_d^2$. Assume that B is a multifunction such that $\mathcal{U}$ is asymptotically compact with respect to B, i.e., for every sequence $s_n \to +\infty$, $t \in \mathbb{R}$, every sequence $y_n \in \mathcal{U}(t, t - s_n, B(t - s_n))$ is relatively compact. Then, for $t \in \mathbb{R}$, the pullback ω-limit set $\Lambda(t, B)$ is nonempty, compact, and*

$$\lim_{s \to +\infty} \mathrm{dist}_X(\mathcal{U}(t, t - s, B(t - s)), \Lambda(t, B)) = 0,$$

$$\Lambda(t, B) \subset \mathcal{U}(t, s, \Lambda(s, B)) \text{ for all } (t, s) \in \mathbb{R}_d^2.$$

Theorem 1.22 ([55]) *Let $\mathcal{U}$ be a u.s.c MNDS on X, and $B \in \mathcal{D}$ be a pullback $\mathcal{D}$-absorbing set for $\mathcal{U}$ such that $\mathcal{U}$ is asymptotically compact with respect to B. Then the multifunction A given by*

$$A(t) = \Lambda(t, B)$$

is a pullback $\mathcal{D}$-attractor for $\mathcal{U}$, and A is the unique element in $\mathcal{D}$ with this property. Moreover, if $\mathcal{U}$ is a strict MNDS then A is strictly invariant.

Let α be the Hausdorff MNC in Banach space X.

Definition 1.22 The MNDS $\mathcal{U}$ is said to be α-contracting if there exists a continuous function $k : \mathbb{R} \times \mathbb{R} \to \mathbb{R}^+$ such that $k(t, \tau) \to 0$ as $\tau \to -\infty$ for each t fixed and the inequality

$$\alpha(\mathcal{U}(t, \tau, B(\tau))) \le k(t, \tau)\alpha(B(t))$$

holds for all $B \in \mathcal{D}$ such that $B(\tau) \in P_b(X)$ for any $\tau \in \mathbb{R}$.

We now give a sufficient condition for the existence of global pullback $\mathcal{D}$-attractor.

Theorem 1.23 ([77]) *Let $\mathcal{U}$ be a u.s.c MNDS on X. If there is a monotone pullback $\mathcal{D}$-absorbing B, i.e., $B(t_1) \subset B(t_2)$ whenever $t_1 \le t_2$, and $\mathcal{U}$ is α-contracting, then the MNDS $\mathcal{U}$ admits a global pullback $\mathcal{D}$-attractor.*

Proof By Theorem 1.22, it is sufficient to prove that $\mathcal{U}$ is asymptotically compact with respect to B. For arbitrary $\varepsilon > 0$ and $t \in \mathbb{R}$ fixed, there exists $s_0 > 0$ such that

$$k(t, t - s)\alpha(B(t)) < \varepsilon, \quad \forall s \ge s_0.$$

Since B is pullback $\mathcal{D}$-absorbing, there is a positive number $C(t - s_0, B)$ such that

$$\mathcal{U}(t - s_0, t - s_0 - \tau, B(t - s_0 - \tau)) \subset B(t - s_0), \quad \forall \tau \ge C(t - s_0, B). \quad (1.4)$$

Let $t_n \to +\infty$ and $\xi_n \in \mathcal{U}(t, t - t_n, B(t - t_n))$. Choose $N_0 \in \mathbb{N}$ such that $t_n \ge s_0 + C(t - s_0, B)$ for all $n \ge N_0$. Then for each $n \ge N_0$, one has

$$\mathcal{U}(t, t - t_n, B(t - t_n)) \subset \mathcal{U}(t, t - s_0, \mathcal{U}(t - s_0, B(t - t_n)))$$
$$= \mathcal{U}(t, t - s_0, \mathcal{U}(t - s_0, t - s_0 - (t_n - s_0), B(t - t_n)))$$
$$\subset \mathcal{U}(t, t - s_0, B(t - s_0))$$

in view of (1.4). It follows that

$$\bigcup_{n\geq N_0} \mathcal{U}(t, t - t_n, B(t - t_n)) \subset \mathcal{U}(t, t - s_0, B(t - s_0)).$$

Therefore

$$
\begin{aligned}
\alpha(\{\xi_n\}_{n\geq 1}) &= \alpha(\{\xi_n : n \leq N_0\}) + \alpha(\{\xi_n : n \geq N_0\}) \\
&= \alpha(\{\xi_n : n \geq N_0\}) \\
&\leq \alpha\left(\bigcup_{n\geq N_0} \mathcal{U}(t, t - t_n, B(t - t_n))\right) \\
&\leq \alpha(\mathcal{U}(t, t - s_0, B(t - s_0))) \\
&\leq k(t, t - s_0)\alpha(B(t - s_0)) \\
&\leq k(t, t - s_0)\alpha(B(t)) \\
&< \varepsilon.
\end{aligned}
$$

Since $\varepsilon > 0$ is arbitrary, we get $\alpha(\{\xi_n\}_{n\geq 1}) = 0$. Consequently, $\{\xi_n\}$ is relatively compact in X. The proof is complete.

Remark 1.8 There are two concepts of pullback attractors for non-autonomous dynamical systems. In the first one, it requires the attraction of autonomous bounded sets, i.e., $\mathrm{dist}_X(\mathcal{U}(t, t - s, B), A(t)) \to 0$ as $s \to +\infty$ for any $B \in P_b(X)$. The second one involves a universe $\mathcal{D}$ as described and used in this note and the attraction means

$$\lim_{s\to+\infty} \mathrm{dist}_X(\mathcal{U}(t, t - s, B(t - s)), A(t)) = 0 \quad \text{for any } B \in \mathcal{D}$$

that is, A attracts non-autonomous bounded sets. A comparison of these concepts for single-valued non-autonomous dynamical systems was given in [143]. In [206], Zelati and Kalita have improved the framework for pullback attractors of MNDSs using the first concept, so that the continuity of MNDS is relaxed. In addition, the asymptotic compactness of MNDS is characterized by the ω-limit compact property, namely,

$$\lim_{\tau\to-\infty} \alpha\left(\bigcup_{s\leq\tau} \mathcal{U}(t, s, B)\right) = 0 \quad \text{for any } B \in P_b(X).$$

In the framework using second concept, it is not known that the u.s.c. condition on $\mathcal{U}$ can be removed, but one can make use of the following form of ω-limit compactness

$$\lim_{\tau\to+\infty} \alpha\left(\bigcup_{s\geq\tau} \mathcal{U}(t, t - s, B(t - s))\right) = 0 \quad \text{for any } B \in \mathcal{D}$$

to obtain the asymptotic compactness for $\mathcal{U}$. In fact, this condition is not easy to testify directly in applications. In a concrete model formed by partial differential equations, one usually test the flattening condition (see [206]) as an alternative, but in turn the testing flattening condition is impractical if the phase space X is not a separable Hilbert space.

1.5 Evolution System

In what follows, let $\mathcal{L}(X)$ be the Banach space of linear bounded operators with the norm $\|\cdot\|$, and let $A(t)$ for $t \in \mathbb{R}$ be closed linear operators on X with domain $D(A(t))$ (possibly not densely defined) satisfying the following hypotheses:

(i) for $t \geq 0$, the resolvent $R(\lambda, A(t)) = (\lambda I - A(t))^{-1}$ exists for all λ with $Re(\lambda) \leq 0$, and there is a constant M independent of λ and t such that

$$\|R(\lambda, A(t))\| \leq \frac{M}{1 + |\lambda|} \quad \text{for } Re(\lambda) \leq 0;$$

(ii) there exist constants $L > 0$ and $0 < \alpha \leq 1$ such that

$$\left\|(A(t) - A(s))A^{-1}(r)\right\| \leq L|t - s|^{\alpha} \quad \text{for } t, s, r \in \mathbb{R}.$$

Among other things, it ensures that there exists a unique evolution family U on X such that:

(i) $U(t, r)U(r, s) = U(t, s)$ and $U(s, s) = I$ for $0 \leq s \leq r \leq t < \infty$;
(ii) $(t, s) \to U(t, s) \in \mathcal{L}(X)$ is continuous for $t > s$;
(iii) $U(\cdot, s) \in C^1((s, \infty), \mathcal{L}(X))$, $\frac{\partial U}{\partial t}(t, s) = A(t)U(t, s)$ and

$$\|A^k(t)U(t, s)\| \leq L(t - s)^{-k} \quad \text{for } 0 < t - s \leq 1, \ k = 0, 1;$$

(iv) $\partial_s^+ U(t, s)u = -U(t, s)A(s)u$ for $t > s$ and $u \in D(A(s))$ with $A(s)u \in D(A(s))$.

Let us recall the definition of evolution family below, sometimes also called an evolution process or evolution system. For a detailed account and bibliographic references see, e.g., the survey by Acquistapace [2, 3], Engel and Nagel [100, Sect. 6.9.] and Goldstein [112].

Definition 1.23 A two parameter family $U = \{U(t, s)\}_{0 \leq s \leq t < \infty}$ of bounded linear operators on X is called an evolution family if

(i) $U(t, r)U(r, s) = U(t, s)$ and $U(s, s) = I$ for $0 \leq s \leq r \leq t < \infty$;
(ii) the mapping $(t, s) \mapsto U(t, s)u$ is strongly continuous for $0 \leq s \leq t < \infty$, $u \in X$.

Definition 1.24 The evolution family U is called compact if $U(t, s)$ is compact for each $t > s$, that is, $U(t, s)$ mappings bounded subsets of X into precompact subsets of X for all $t > s \geq 0$.

For information on the nonlinear version of evolution family we refer the reader to, e.g., Carvalho and Langa [67, 68] and to the references therein.

1.6   Semigroups

1.6.1 C_0-Semigroup

Let us recall the definitions and properties of operator semigroups, for details see Banasiak et al. [29] and Pazy [167].

Let X be a Banach space and $\mathcal{L}(X)$ be the Banach space of linear bounded operators with the norm $\| \cdot \|$.

Definition 1.25 A one parameter family $\{T(t)\}_{t \geq 0} \subset \mathcal{L}(X)$ is a semigroup of bounded linear operators on X if

(i) $T(t)T(s) = T(t + s)$, for $t, s \geq 0$;
(ii) $T(0) = I$; here, I denotes the identity operator in X.

Definition 1.26 A semigroup of bounded linear operators $\{T(t)\}_{t \geq 0}$ is uniformly continuous if

$$\lim_{t \to 0^+} \|T(t) - I\| = 0.$$

From the definition it is clear that if $\{T(t)\}_{t \geq 0}$ is a uniformly continuous semigroup of bounded linear operators, then

$$\lim_{s \to t} \|T(s) - T(t)\| = 0.$$

Definition 1.27 We say that the semigroup $\{T(t)\}_{t \geq 0}$ is strongly continuous (or a C_0-semigroup) if the mapping $t \mapsto T(t)u$ is strongly continuous, for each $u \in X$, i.e.,

$$\lim_{t \to 0^+} T(t)u = u, \quad \forall u \in X.$$

Definition 1.28 Let $\{T(t)\}_{t \geq 0}$ be a C_0-semigroup defined on X. The linear operator A is the infinitesimal generator of $\{T(t)\}_{t \geq 0}$ defined by

$$Au = \lim_{t \to 0^+} \frac{T(t)u - u}{t} \quad \text{for } u \in D(A),$$

where $D(A) = \left\{ u \in X : \lim_{t \to 0^+} \frac{T(t)u - u}{t} \text{ exists in } X \right\}$.

If there are $M \geq 0$ and $v \in \mathbb{R}$ such that $\|T(t)\| \leq Me^{vt}$, then

$$(\lambda I - A)^{-1}u = \int_0^\infty e^{-\lambda t} T(t)u dt, \quad \text{Re}(\lambda) > v, \ u \in X. \tag{1.5}$$

A C_0-semigroup $\{T(t)\}_{t \geq 0}$ is called exponentially stable if there exist constants $M > 0$ and $\delta > 0$ such that

$$\|T(t)\| \leq Me^{-\delta t}, \quad t \geq 0. \tag{1.6}$$

The growth bound v_0 of $\{T(t)\}_{t \geq 0}$ is defined by

$$v_0 = \inf\{\delta \in \mathbb{R} : \text{there exists } M_\delta > 0 \text{ such that } \|T(t)\| \leq M_\delta e^{\delta t}, \ \forall t \geq 0\}. \tag{1.7}$$

Furthermore, v_0 can also be obtained by the following formula:

$$v_0 = \limsup_{t \to +\infty} \frac{\ln \|T(t)\|}{t}. \tag{1.8}$$

Definition 1.29 A C_0-semigroup $\{T(t)\}_{t \geq 0}$ is called uniformly bounded if there exists a constant $M > 0$ such that

$$\|T(t)\| \leq M, \quad t \geq 0. \tag{1.9}$$

Definition 1.30 A C_0-semigroup $\{T(t)\}_{t \geq 0}$ is called compact if $T(t)$ is compact for $t > 0$.

Property 1.8 *If $\{T(t)\}_{t \geq 0}$ is compact, then $\{T(t)\}$ is equicontinuous for $t > 0$.*

Definition 1.31 A C_0-semigroup $\{T(t)\}_{t \geq 0}$ is called positive if $T(t)u \geq \theta$ for all $u \geq \theta$ and $t \geq 0$.

1.6.2 Analytic Semigroup

Definition 1.32 Let $\quad \Delta := \{z : \varphi_1 < arg z < \varphi_2, \varphi_1 < 0 < \varphi_2\}$. The family $\{T(z)\}_{z \in \Delta} \subset \mathcal{L}(X)$ is called an analytic semigroup in Δ if

(i) $z \mapsto T(z)$ is analytic in Δ;
(ii) $T(0) = I$ and $\lim_{z \in \Delta, z \to 0} T(z)x = x$ for every $x \in X$;
(iii) $T(z_1 + z_2) = T(z_1)T(z_2)$ for $z_1, z_2 \in \Delta$.

A semigroup $T(t)$ will be called analytic if it is analytic in some sector Δ containing the nonnegative real axis.

Theorem 1.24 ([167]) *Let $\{T(t)\}_{t\geq 0}$ be a uniformly bounded C_0-semigroup. Let A be the infinitesimal generator of $\{T(t)\}_{t\geq 0}$ and assume $0 \in \rho(A)$. The following statements are equivalent:*

(i) *$T(t)$ can be extended to an analytic semigroup in a sector $\Sigma_\delta := \{z \in \mathbb{C} : |argz| < \delta\}$ and $\|T(z)\|$ is uniformly bounded in every closed subsector $\overline{\Sigma}_{\delta'}$, $\delta' < \delta$ of Σ_δ;*

(ii) *there exists a constant C such that for every $\sigma > 0$, $\tau \neq 0$,*

$$\|R(\sigma + i\tau, A)\| \leq \frac{C}{|\tau|};$$

(iii) *there exist $0 < \delta < \frac{\pi}{2}$ and $M > 0$ such that*

$$\rho(A) \supset \Sigma := \{\lambda \in \mathbb{C} : |\arg \lambda| < \frac{\pi}{2} + \delta\} \setminus \{0\}$$

and

$$\|R(\lambda, A)\| \leq \frac{M}{|\lambda|} \ \text{for} \ \lambda \in \Sigma, \ \lambda \neq 0;$$

(iv) *$T(t)$ is differentiable for $t > 0$ and there is a constant C such that*

$$\|AT(t)\| \leq \frac{C}{|t|} \ \text{for} \ t \geq 0.$$

1.6.3 Integrated Semigroup

Definition 1.33 Let X be a Banach space. An integrated semigroup is a family of bounded linear operators $\{S(t)\}_{t\geq 0}$ on X with the following properties:

(i) $S(0)=0$;

(ii) $t \mapsto S(t)$ is strongly continuous;

(iii) $S(s)S(t) = \int_0^s \big(S(t+r) - S(r)\big)dr$ for all $t, s \geq 0$.

Definition 1.34 An operator A is called a generator of an integrated semigroup if there exists $\omega \in \mathbb{R}$ such that $(\omega, \infty) \subset \rho(A)$ ($\rho(A)$ is the resolvent set of A) and there exists a strongly continuous exponentially bounded family $\{S(t)\}_{t\geq 0}$ of bounded operators such that $S(0) = 0$ and

$$R(\lambda, A) := (\lambda I - A)^{-1} = \lambda \int_0^\infty e^{-\lambda t} S(t)dt \ \text{exists for all } \lambda \text{ with } \lambda > \omega.$$

Property 1.9 ([134]) *Let A be the generator of an integrated semigroup $\{S(t)\}_{t\geq 0}$. Then for all $u \in X$ and $t \geq 0$,*

$$\int_0^t S(s)ds \in D(A) \quad and \quad S(t)u = A \int_0^t S(s)uds + tu.$$

Definition 1.35 (i) An integrated semigroup $\{S(t)\}_{t\geq 0}$ is called locally Lipschitz continuous if for all $\tau > 0$ there exists a constant L such that

$$\|S(t) - S(s)\| \leq L|t - s|, \quad t, s \in [0, \tau].$$

(ii) An integrated semigroup $\{S(t)\}_{t\geq 0}$ is called nondegenerate if $S(t)u = 0$ for all $t \geq 0$ implies that $u = 0$.

Definition 1.36 We say that the linear operator $A : D(A) \subset X \to X$ satisfies the Hille-Yosida condition if there exist two constant $\omega \in \mathbb{R}$ and $M > 0$ such that $(\omega, +\infty) \subset \rho(A)$ and

$$\|(\lambda I - A)^{-k}\| \leq \frac{M}{(\lambda - \omega)^k} \quad \text{for all } \lambda > \omega, \ k \geq 1.$$

Theorem 1.25 ([17]) *The following assertions are equivalent:*

(i) *A is the generator of a nondegenerate, locally Lipschitz continuous integrated semigroup;*
(ii) *A satisfies the Hille-Yosida condition.*

If A is the generator of an integrated semigroup $\{S(t)\}_{t\geq 0}$ which is locally Lipschitz, then from [134], $S(t)u$ is continuously differentiable if and only if $u \in \overline{D(A)}$ and $\{S'(t)\}_{t\geq 0}$ is a C_0 semigroup on $\overline{D(A)}$.

1.7 Weak Compactness of Sets and Operators

In this section we consider a situation where a set is given a topology which is natural under the circumstances. This is the "weak topology". Let us briefly describe the starting point for the introduction of this topology. So let X be our set. For the weak topology the situation is the following. We are given a family $\{Y_i, f_i\}_{i \in I}$ of pairs, each consisting of a topological space Y_i and a map $f_i : X \to Y_i$. Any topology of X that makes all the f_i is continuous, is said to be admissible. Evidently, the set of admissible topologies on X is nonempty, since the discrete topology is such a topology. We will see that there exists a topology w on X such that every admissible topology is stronger or equal to w.

Definition 1.37 Let $\{Y_i, f_i\}_{i \in I}$ (I is an arbitrary index set) be a family of pairs where each Y_i is a topological space and $f_i : X \to Y_i$ is a map. The "weak topology" or "initial topology" on X, generated by the family $\{f_i\}_{i \in I}$, is the weakest topology on X that makes all the functions f_i continuous. So it is the topology generated by

$$\mathcal{F} = \{f_i^{-1}(V) : i \in I, \ V \subset Y_i \text{ is open}\} \ (\text{i.e., } \mathcal{F} = \bigcup_{i \in I} f_i^{-1}(\tau_{Y_i})).$$

This is a subbase for the weak topology. In fact we can economize in the definition of the subbase for the weak topology and take

$$\mathcal{F} = \{f_i^{-1}(V) : i \in I, \ V \subset Y_i \text{ is subbasic open}\}.$$

This also is a subbase for the weak topology. We denote the weak topology on X generated by the family $\{f_i\}_{i \in I}$ by $w(X, \{f_i\}_{i \in I})$ or simply by w if no confusion is possible. Of course a base for the weak topology is given by all the sets of the form $\bigcap_{i=1}^n f_i^{-1}(V_i)$ with $V_i \subset \tau_{Y_i}$ and $n \geq 1$ an arbitrary integer.

We noted that X endowed with the weak topology, denoted by X_w, is a locally convex topological vector space, (see [64, 83]).

Lemma 1.22 *If a set X is furnished with the weak topology $w(X, \{f_i\}_{i \in I})$, then $u_n \to u$ if and only if for all $i \in I$ we have $f_i(u_n) \to f_i(u)$ in Y_i.*

Let X be a real Banach space with norm $|\cdot|$ and X^* be its topological dual. A subset A of a Banach space X is called weakly closed if it is closed in the weak topology. The symbol $\overline{D}^w$ denotes the weak closure of D.

We will say that $\{u_n\} \subset X$ converges weakly to $u_0 \in X$, and we write $u_n \rightharpoonup u_0$, if for each $f \in X^*$, $f(u) \to f(u_0)$. We recall (see [41]) that a sequence $\{u_n\} \subset C([0, b], X)$ weakly converges to an element $u \in C([0, b], X)$ if and only if

(i) there exists $N > 0$ such that, for every $n \in \mathbb{N}^+$ and $t \in [0, b]$, $|u_n(t)| \leq N$;
(ii) for every $t \in [0, b]$, $u_n(t) \rightharpoonup u(t)$.

Definition 1.38 (i) A subset A of a normed space X is said to be (relatively) weakly compact if (the weak closure of) A is compact in the weak topology of X.
(ii) A subset A of a Banach space X is weakly sequentially compact if any sequence in A has a subsequence which converges weakly to an element of X.

Definition 1.39 Suppose that X and Y are Banach spaces. A linear operator T from X into Y is weakly compact if $T(B)$ is a relatively weakly compact subset of Y whenever B is a bounded subset of X.

We mention also two results that are contained in the so-called Eberlein-Šmulian theory.

Theorem 1.26 ([97]) *Let Ω be a subset of a Banach space X. The following statements are equivalent:*

(i) *Ω is relatively weakly compact;*
(ii) *Ω is relatively weakly sequentially compact.*

Corollary 1.1 ([97]) *Let Ω be a subset of a Banach space X. The following statements are equivalent:*

(i) Ω *is weakly compact*;

(ii) Ω *is weakly sequentially compact.*

We recall Krein-Šmulian theorem and Pettis measurability theorem.

Theorem 1.27 ([97]) *The convex hull of a weakly compact set in a Banach space X is weakly compact.*

Theorem 1.28 ([168]) *Let (E, Σ) be a measure space, X be a separable Banach space. Then a function $f : E \to X$ is measurable if and only if for every $u^* \in X^*$ the function $u^* \circ f : E \to \mathbb{R}$ is measurable with respect to Σ and the Borel σ-algebra in $\mathbb{R}$.*

We recall that a bounded subset in a reflexive Banach space is relatively weakly compact.

Lemma 1.23 ([198]) *Let X be reflexive and $1 < p < \infty$. A subset $K \subset L^p([0, b], X)$ is relatively weakly sequentially compact in $L^p([0, b], X)$ if and only if K is bounded in $L^p([0, b], X)$.*

The following compactness criterion in $L^1([0, b], X)$ is an easy consequence of Dunford theorem. See Diestel and Uhl [85, Theorem 1 (Dunford), p. 101] (cf. also Diestel et al. [84] or Kamenskii et al. [130]).

Lemma 1.24 *Let $K \subset L^1([0, b], X)$. If K is uniformly integrable and X is reflexive, then K is relatively weakly compact in $L^1([0, b], X)$.*

Lemma 1.25 *Assume that $\{f_n\} \subset L^1([0, b], X)$ is integrably bounded, and $f_n(t)$ are relatively compact for a.e. $t \in [0, b]$. Then $\{f_n\}$ is weakly compact in $L^1([0, b], X)$.*

Theorem 1.29 ([158]) *Let X be a metrizable locally convex linear topological space and let D be a weakly compact, convex subset of X. Suppose $\varphi : D \to P_{cl,cv}(D)$ has a weakly sequentially closed graph. Then φ has a fixed point.*

1.8 Stochastic Process

We present some important concepts and results of stochastic process in this section. The material is taken from Arnold [18], Gawarecki et al. [111], and Prato et al. [171].

1.8.1 Random Variables

Let Ω be a sample space and $\mathscr{F}$ a σ-algebra of the subset of Ω. A function $\mathbb{P}(\cdot)$ defined on $\mathscr{F}$ and taking values in the unit interval $[0, 1]$ is called a probability measure, if

(i) $\mathbb{P}(\Omega) = 1$;

(ii) $\mathbb{P}(A) \geq 0$ for all $A \in \mathscr{F}$;

(iii) for an at most countable family $\{A_n : n \geq 1\}$ of mutually disjoint event, we
      have

$$\mathbb{P}\left\{\bigcup_{n \geq 1} A_n\right\} = \sum_{n \geq 1} \mathbb{P}(A_n).$$

The triple $(\Omega, \mathscr{F}, \mathbb{P})$ is a probability space.

$\mathbb{F} = \{\mathscr{F}_t\}_{t \geq 0}$ is a family of sub-σ-algebras $\mathscr{F}_t$ of σ-algebra $\mathscr{F}$ such that $\mathscr{F}_s \subset \mathscr{F}_t$
for $0 \leq s < t < \infty$. $\mathbb{P}_{\mathbb{F}} = (\Omega, \mathscr{F}, \mathbb{F}, \mathbb{P})$ is said to be a filtered probability space. We
say that a filtration $\mathscr{F}_t$ satisfies the usual conditions if $\mathscr{F}_0$ contains all $\mathbb{P}$-null sets
of $\mathscr{F}$ and $\mathscr{F}_t = \bigcap_{\varepsilon > 0} \mathscr{F}_{t+\varepsilon}$ for every $t \geq 0$. If the last condition is satisfied, we say
that a filtration $\mathscr{F}$ is right continuous.

Let $(X, \mathscr{B}_X)$ be measurable space, we mean an $(\mathscr{F}, \mathscr{B}_X)$-measurable mapping
$u : \Omega \to X$, i.e., such that $u^{-1}(A) \in \mathscr{F}$ for every $A \in \mathscr{B}_X$, where as usual, $\mathscr{B}_X$
denotes the Borel σ-algebra on X and $u^{-1}(A) = \{\omega \in \Omega : u(\omega) \in A\}$. We shall also
say that u is a random variable on Ω with values at X.

The integral of an integrable random variable u is called its mean value or expec-
tation and is denoted by

$$\mathbb{E}(u) = \int u(w)d\mathbb{P}.$$

Let K and H be separable Hilbert spaces, and Q be either a symmetric nonnegative
definite trace-class operator on K or $Q = I_K$, the identity operator on K. In case Q is
trace-class, we will always assume that its all eigenvalues $\lambda_j > 0, \ldots$; otherwise we
can start with the Hilbert space $\ker(Q)^{\perp}$ instead of K. The associated eigenvectors
forming an orthonormal basis (ONB) in K will be denoted by e_j.

Denote $\mathcal{L}(K, H)$ by all bounded linear operators from K to H. Then the space
of Hilbert–Schmidt operators from K to H is defined as

$$\mathcal{L}_2(K, H) = \left\{\Phi \in \mathcal{L}(K, H) : \sum_{i=1}^{\infty} |\Phi e_i|_H^2 < \infty\right\}.$$

It is well known (see [175]) that $\mathcal{L}_2(K, H)$ equipped with the norm

$$\|\Phi\|_{\mathcal{L}_2(K,H)} = \sum_{i=1}^{\infty} |\Phi e_i|_H^2$$

is a Hilbert space.

On the other hand, the space $Q^{\frac{1}{2}} K$ equipped with the scalar product

$$\langle u, v \rangle_{Q^{\frac{1}{2}} K} = \sum_{j=1}^{\infty} \frac{1}{\lambda_j}(u, e_j)_K (v, e_j)_K$$

is a separable Hilbert space with an ONB $\{\lambda_j^{\frac{1}{2}} e_j\}_{j=1}^{\infty}$.

Consider $\mathcal{L}_2^0 = \mathcal{L}_2(Q^{\frac{1}{2}} K, H)$, the space of Hilbert–Schmidt operators from $Q^{\frac{1}{2}} K$ to H. If $\{\tilde{e}_j\}_{j=1}^{\infty}$ is an ONB in H, then the Hilbert–Schmidt norm of an operator $\Phi \in \mathcal{L}_2^0$ is given by

$$\|\Phi\|_{\mathcal{L}_2^0} = \sum_{i,j=1}^{\infty} (\Phi(\lambda_j^{\frac{1}{2}} e_j), \tilde{e}_i)_H^2 = \sum_{i,j=1}^{\infty} (\Phi(Q^{\frac{1}{2}} e_j), \tilde{e}_i)_H^2$$

$$= \|\Phi Q^{\frac{1}{2}}\|_{\mathcal{L}_2(K,H)}^2 = \mathrm{tr}(\Phi Q^{\frac{1}{2}})(\Phi Q^{\frac{1}{2}})^*.$$

1.8.2 Stochastic Calculus

An X-valued stochastic process (briefly, an X-valued process) indexed by a set I is a family of X-valued random variables $\{X(i) : i \in I\}$ defined on some underlying probability space $(\Omega, \mathcal{F}, \mathbb{P})$.

Definition 1.40 An X-valued process $\{X(i) : i \in I\}$ is called Gaussian, if for all $N > 1$ and $i_1, \ldots, i_N \in I$ the X^N-valued random variable $(X(i_1), \ldots, X(i_N))$ is Gaussian.

Definition 1.41 A real-valued process $\{W(t) : t \in [0, T]\}$ is called a Brownian motion, if it enjoys the following properties:

 (i) $W(0) = 0$;
 (ii) $W(t) - W(s)$ is independent of $\{W(r) : r \in [0, s]\}$ for $0 \le s \le t \le T$;
 (iii) $W(t) - W(s)$ is Gaussian with variance $(t - s)$.

Definition 1.42

(a) For an $\mathcal{L}(H, X)$-valued step function of the form $\Phi(t, \omega) = \phi_1(\omega) I_{[t_0, t_1]}(t) + \sum_{i=2}^{n} \phi_i(\omega) I_{(t_{i-1}, t_i]}(t)$, where $0 = t_0 < t_1 < \cdots < t_n = T$ and ϕ_i, $i = 1, \ldots, n$, are, respectively, $\mathcal{F}_0$-measurable and $\mathcal{F}_{t_i}$-measurable $\mathcal{L}_2(K, H)$-valued random variables such that $\phi_i(\omega) \in \mathcal{L}(K, H), i = 1, \ldots, n$. We define the stochastic integral process $\int_0^t \Phi(s) dW(s), 0 \le t \le T$, by

$$\int_0^t \Phi(s) dW(s) = \sum_{i=1}^{n} \phi_i(W(t_i) - W(t_{i-1})).$$

(b) A function $\Phi : [0, T] \to \mathcal{L}(H, X)$ is said to be stochastically integrable with respect to the H-cylindrical Brownian motion W if there exists a sequence of finite rank step functions $\Phi_n : [0, T] \to \mathcal{L}(H, X)$ such that:

 (i) for all $h \in H$, we have $\lim_{n \to \infty} \Phi_n h = \Phi h$ in measure;

(ii) there exists an X-valued random variable u such that

$$\lim_{n\to\infty} \int_0^t \Phi_n(s)dW(s) = u$$

in probability.

The stochastic integral of a stochastically integrable function $u : [0, T] \to \mathcal{L}(H, X)$ is then defined as the limit in probability

$$\int_0^t \Phi(s)dW(s) = \lim_{n\to\infty} \int_0^t \Phi_n(s)dW(s).$$

The relationship

$$u(t, \omega) = \int_0^t \Phi(s, \omega)dW(s, \omega)$$

can also be written as

$$du(t) = \Phi(t)dW(t).$$

This is a special so-called stochastic differential. Let us look at a somewhat more general stochastic process of the form

$$u(t, \omega) = u(0, \omega) + \int_0^t f(s, \omega)ds + \int_0^t \Phi(s, \omega)dW(s, \omega); \qquad (1.10)$$

here, $\int_0^t f(s, \omega)ds$ is the usual Lebesgue or possibly Riemann integral.

Definition 1.43 We shall say that a stochastic process $u(t)$ defined by equation (1.10) possesses the stochastic differential $f(t)dt + \Phi(t)dW(t)$ and we shall write

$$\begin{aligned} du(t) &= f(t)dt + \Phi(t)dW(t) \\ &= fdt + \Phi dW. \end{aligned}$$

Chapter 2
Evolution Inclusions with m-Dissipative Operator

Abstract This chapter deals with a nonlinear delay differential inclusion of evolution type involving m-dissipative operator and source term of multivalued type in a Banach space. Under rather mild conditions, the R_δ-structure of C^0-solution set is studied on compact intervals, which is then used to obtain the R_δ-property on noncompact intervals. Secondly, the result about the structure is furthermore employed to show the existence of C^0-solutions for the inclusion (mentioned above) subject to nonlocal condition defined on right half-line. No nonexpansive condition on nonlocal function is needed. As samples of applications, we consider a partial differential inclusion with time delay and then with nonlocal condition at the end of the chapter.

2.1 Introduction

It is worth mentioning that for differential inclusions on noncompact intervals, governed by a nonlinear multivalued operator (specially, an m-dissipative operator), the research of topological structure of solution sets is much more delicate and the related results are still very rare. Furthermore, much of the previous research on differential inclusions in infinite dimensional spaces was done provided the nonlinearity (a multivalued function), with compact values, is upper semicontinuous with respect to solution variable. This condition turns out to be restrictive to some extent and is not satisfied usually in practical applications (see, e.g., Vrabie [192, Example 5.1, Example 5.2] and [189]). To make things more applicable, an appropriate alterative is that the nonlinearity, with closed and convex values, is weakly upper semicontinuous with respect to solution variable.

Throughout this section, X is a real Banach space with norm $|\cdot|$, X^* denotes the topological dual of Banach space X. Denote by $|\cdot|_0$ the sup-norm of $C([-\tau, 0], X)$. Note that $X \times C([-\tau, 0], X)$, endowed with the norm

$$|(x, v)|_\tau := \max\{|x|, |v|_0\}, \quad (x, v) \in X \times C([-\tau, 0], X),$$

is a Banach space.

© Springer Nature Singapore Pte Ltd. 2017
Y. Zhou et al., *Topological Structure of the Solution Set for Evolution Inclusions*,
Developments in Mathematics 51, https://doi.org/10.1007/978-981-10-6656-6_2

We consider the Cauchy problem of nonlinear delay differential inclusion of evolution type

$$\begin{cases} u'(t) \in Au(t) + f(t), & t \in \mathbb{R}^+, \\ f(t) \in F(t, u(t), u_t), & t \in \mathbb{R}^+, \\ u(t) = \phi(t), & t \in [-\tau, 0]. \end{cases} \tag{2.1}$$

Here $A : D(A) \subset X \to P(X)$ is an m-dissipative operator (possible multivalued and/or nonlinear), the forcing source $F : \mathbb{R}^+ \times \overline{D(A)} \times C([-\tau, 0], \overline{D(A)}) \to P(X)$ is a multivalued function with convex, closed values, and $\phi \in C([-\tau, 0], \overline{D(A)})$. $u_t \in C([-\tau, 0], \overline{D(A)})$ is defined by $u_t(s) = u(t + s)$ ($s \in [-\tau, 0]$) for every $u \in \widetilde{C}([-\tau, \infty), \overline{D(A)})$ and $t \in \mathbb{R}^+$.

Here, we are interested in studying the topological characterization of the solution set for the Cauchy problem (2.1) in some Fréchet spaces. We first investigate the existence of C^0-solutions and R_δ-structure of the solution set for the Cauchy problem (2.1) on compact intervals, then proceed to study the R_δ-structure of the solution set for the Cauchy problem (2.1). In the proof of the latter result, the key tool is the inverse limit method.

As an application of the information about the structure, we shall deal with the C^0-solutions for the nonlocal Cauchy problem of nonlinear delay evolution inclusion of the form

$$\begin{cases} u'(t) \in Au(t) + f(t), & t \in \mathbb{R}^+, \\ f(t) \in F(t, u(t), u_t), & t \in \mathbb{R}^+, \\ u(t) = g(u)(t), & t \in [-\tau, 0], \end{cases} \tag{2.2}$$

where A and F are defined the same as those in the problem (2.1), and

$$g : \widetilde{C}_b([-\tau, \infty), \overline{D(A)}) \to C([-\tau, 0], \overline{D(A)})$$

is a function to be specified later. As can be seen, g constitutes a nonlocal condition. It is also noted that the nonlocal function g depends on history states, that is, it takes history values. We emphasize that in the proof of our main result, no nonexpansive condition on nonlocal function g will be required.

The consideration for nonlocal initial condition g is stimulated by the observation that this type of conditions is more realistic than usual ones in treating physical problems, see, e.g., [5, 8, 82, 110, 192, 194, 197] for more detailed information about the importance of nonlocal initial conditions in applications. Some typical examples for g are

- $g(u)(t) = u(t + \omega)$ for each $t \in [-\tau, 0]$ (Periodicity condition);
- $g(u)(t) = -u(t + \omega)$ for each $t \in [-\tau, 0]$ (Anti-periodicity condition);
- $g(u)(t) = \displaystyle\int_\tau^\infty k(\theta) u^{\frac{1}{3}}(t + \theta) d\theta$ for each $t \in [-\tau, 0]$ with $k \in L^1(\mathbb{R}^+, \mathbb{R}^+)$ and

$\displaystyle\int_\tau^\infty k(\theta) d\theta = 1$ (Mean value condition);

- $g(u)(t) = \sum_{i=1}^{n} \alpha_i u^{\frac{1}{3}}(t_i + t)$ for each $t \in [-\tau, 0]$, where $\sum_{i=1}^{n} |\alpha_i| \leq 1$ and $\tau < t_1 < t_2 < \cdots < t_n < \infty$ are constants   (Multi-point discrete mean condition).

Remark 2.1 The final case on g above can be seen as a generalization of the nonlocal function introduced in Deng [82], where the nonlocal function is used to describe the diffusion phenomenon of a small amount of gas in a transparent tube.

It is noted that by using an interplay of compactness arguments and invariance techniques, Vrabie [192] obtained an existence result of C^0-solutions to the nonlocal Cauchy problem (2.2). Similar arguments are also used to solve other nonlocal problems, we refer the reader to Paicu and Vrabie [162], Vrabie [191], Wang and Zhu [197] and references therein. However, there exists a limitation among these results, that is, it is assumed that the nonlocal function is nonexpansive. Thus, there naturally arises a question: "Is there any chance to solve this problem without this condition?". The results in Sect. 2.4 in fact gives an affirmative answer to this question and close this gap.

Remark 2.2 Let us mention that the lack of nonexpansive condition on nonlocal function prevents us from using the well-known tools such as Banach and Schauder fixed point theorems to show the existence of C^0-solutions to the nonlocal Cauchy problem (2.2). This difficulty leads us to study the topological structure of the solution set to the Cauchy problem (2.1), before applying a fixed point theorem for multivalued mappings with non-convex values.

This chapter is organized as follows. Section 2.2 gives some properties of m-dissipative operators and the definition of C^0-Solutions. Section 2.3 is devoted to the existence of C^0-solutions and R_δ-structure of the solution set for the Cauchy problem (2.1) on compact intervals. In Sect. 2.3.2, we obtain the R_δ-structure of the solution set for the Cauchy problem (2.1) on noncompact intervals by the inverse limit method. Section 2.4 is concerned with the existence of C^0-solutions to the nonlocal Cauchy problem (2.2) defined on right half-line. Finally, as an illustration of the developed theory, we apply it to the examples of partial differential inclusions defined on right half-line.

The results in this chapter are taken from Chen, Wang and Zhou [71].

2.2 The m-Dissipative Operators and C^0-Solution

Given a multivalued operator $A : D(A) \subset X \to P(X)$ with the domain $D(A)$, we let $R(A) = \bigcup_{x \in D(A)} Ax$ stand for the range of A.

Let $x, y \in X$ and $h \in \mathbb{R} \setminus \{0\}$. We put

$$[x, y]_h = \frac{|x + hy| - |x|}{h}$$

and then note that there exists the limit

$$[x, y]_+ = \lim_{h \to 0^+} [x, y]_h.$$

Furthermore, for each $x, y \in X$ and $\alpha > 0$,

$$[\alpha x, y]_+ = [x, y]_+, \quad |[x, y]_+| \le |y|.$$

Recall that $A : D(A) \subset X \to P(X)$ is m-dissipative if $R(I - \lambda A) = X$ for all $\lambda > 0$ and A is dissipative, i.e.,

$$[x_1 - x_2, y_2 - y_1]_+ \ge 0 \quad \text{for all } (x_i, y_i) \in \text{Gra}(A), \ i = 1, 2.$$

Consider the following evolution inclusion

$$u'(t) \in Au(t) + f(t), \tag{2.3}$$

where A is m-dissipative. By a C^0-solution of (2.3) on $[a, b]$, it will be understood an element $u \in C([a, b], X)$, $u(t) \in \overline{D(A)}$ for each $t \in [a, b]$ and u verifies

$$|u(t) - x| \le |u(s) - x| + \int_s^t [u(\sigma) - x, f(\sigma) - y]_+ d\sigma$$

for each $(x, y) \in \text{Gra}(A)$ and $a \le s \le t \le b$.

From [139, Theorems 3.5.1 and 3.6.1] it follows that for each $x \in \overline{D(A)}$ and $f \in L^1([a, b], X)$, there exists an unique C^0-solution to (2.3) on $[a, b]$ which satisfies $u(a) = x$. Moreover, as proved in [31, Theorem 2.1], if $f, g \in L^1([a, b], X)$ and u, v are two C^0-solutions to (2.3) corresponding to f and g, respectively, then

$$|u(t) - v(t)| \le |u(s) - v(s)| + \int_s^t [u(\sigma) - v(\sigma), f(\sigma) - g(\sigma)]_+ d\sigma$$

for all $a \le s \le t \le b$. In particular, we see

$$|u(t) - v(t)| \le |u(s) - v(s)| + \int_s^t |f(\sigma) - g(\sigma)| d\sigma$$

for all $a \le s \le t \le b$.

Let $x \in \overline{D(A)}$, $c \in [a, b)$ and $f \in L^1([a, b], X)$. We denote by $u(\cdot, c, x, f)$ the unique C^0-solution $v : [c, b] \to \overline{D(A)}$ of (2.3) on $[c, b]$ which satisfies $v(c) = x$. Define

$$S(t) : \overline{D(A)} \to \overline{D(A)} \quad \text{with } S(t)x = u(t, 0, x, 0) \quad \text{for each } t \ge 0, \ x \in \overline{D(A)}.$$

Then it follows readily that $\{S(t)\}_{t\geq 0}$ is a semigroup of contractions on $\overline{D(A)}$ (see, e.g., Barbu [31] for more details). We say that this semigroup is generated by A.

The semigroup $\{S(t)\}_{t\geq 0}$ is called compact if $S(t)$ is a compact operator for each $t > 0$.

Definition 2.1 An m-dissipative operator $A : D(A) \subset X \to P(X)$ is called of compact type if for each $a < b$ and each sequence $\{(f_n, u_n)\}$ in $L^1([a, b], X) \times C([a, b], X)$ such that u_n is a C^0-solution on $[a, b]$ of the evolution inclusion

$$u'_n(t) \in Au_n(t) + f_n(t), \quad n = 1, 2, \ldots,$$

$f_n \rightharpoonup f$ in $L^1([a, b], X)$ and $u_n \to u$ in $C([a, b], X)$, then it follows that u is a C^0-solution on $[a, b]$ of the limit problem

$$u'(t) \in Au(t) + f(t).$$

Lemma 2.1 [189, Corollary 2.3.1]) *Let X^* be uniformly convex and A an m-dissipative operator generating a compact semigroup. Then A is of compact type.*

The following compactness result is due to Baras [30]. See also Vrabie [189, Theorem 2.3.3].

Lemma 2.2 *Let A be an m-dissipative operator generating a compact semigroup. Suppose in addition that B is a bounded set in $\overline{D(A)}$ and $\mathscr{F}$ is uniformly integrable in $L^1([a, b], X)$. Then for each $c \in (a, b)$, the C^0-solution set*

$$\{u(\cdot, a, x, f) : x \in B, f \in \mathscr{F}\}$$

is relatively compact in $C([c, b], X)$. If, in addition, B is relatively compact, then the C^0-solution set is relatively compact in $C([a, b], X)$.

Next, for each $\phi \in C([-\tau, 0], \overline{D(A)})$ and $f \in L^1([0, b], X)$, we define the mapping $S_{\phi,b} : L^1([0, b], X) \to C([-\tau, b], \overline{D(A)})$ by setting

$$S_{\phi,b}(f)(t) = \begin{cases} \phi(t), & t \in [-\tau, 0), \\ u(t, 0, \phi(0), f), & t \in [0, b]. \end{cases}$$

Clearly, $S_{\phi,b}(f)$ is the unique C^0-solution for the evolution inclusion with time delay of the form

$$\begin{cases} u'(t) \in Au(t) + f(t), & t \in [0, b], \\ u(t) = \phi(t), & t \in [-\tau, 0]. \end{cases}$$

As an immediate consequence of Lemmas 2.1 and 2.2, we obtain the following result.

Lemma 2.3 *Let X^* be uniformly convex and A an m-dissipative operator generating a compact semigroup. Then the following results hold:*

(i) *if $\mathscr{F}$ is uniformly integrable in $L^1([0, b], X)$ and $\mathcal{B} \subset C([-\tau, 0], \overline{D(A)})$ is relatively compact, then $S_{\mathcal{B},b}(\mathscr{F})$ is relatively compact in $C([-\tau, b], X)$;*
(ii) *for each sequence $\{(f_n, u_n)\}$ in $L^1([0, b], X) \times C([-\tau, b], X)$ such that $u_n = S_{\phi,b}(f_n)$, $n \geq 1$, f_n converges weakly to f and u_n converges to u, it follows that $u = S_{\phi,b}(f)$.*

2.3 Topological Structure of Solution Set

We introduce the following assumptions:

(H_0) $A : D(A) \subset X \to P(X)$ is an m-dissipative operator with $0 \in A0$ and A generates a compact semigroup. In addition, $\overline{D(A)}$ is convex and X^* is uniformly convex.

(H_1) $F : \mathbb{R}^+ \times \overline{D(A)} \times C([-\tau, 0], \overline{D(A)}) \to P_{cl,cv}(X)$ is a multivalued function for which $F(t, \cdot, \cdot)$ is weakly u.s.c. for a.e. $t \in \mathbb{R}^+$ and $F(\cdot, x, v)$ has a strongly measurable selection for each $(x, v) \in \overline{D(A)} \times C([-\tau, 0], \overline{D(A)})$.

(H_2) There exists $L \in L^1_{loc}(\mathbb{R}^+, \mathbb{R}^+)$ such that

$$|F(t, x, v)| = \sup\{|f| : f \in F(t, x, v)\} \leq L(t)(1 + |x| + |v|_0)$$

for a.e $t \in \mathbb{R}^+$ and each $(x, v) \in \overline{D(A)} \times C([-\tau, 0], \overline{D(A)})$.

Define a multivalued mapping $Sel_F : C([-\tau, \infty), \overline{D(A)}) \to P(L^1_{loc}(\mathbb{R}^+, X))$ by setting

$$Sel_F(u) = \{f \in L^1_{loc}(\mathbb{R}^+, X) : f(t) \in F(t, u(t), u_t) \text{ for a.e. } t \in \mathbb{R}^+\}$$

for each $u \in C([-\tau, \infty), \overline{D(A)})$.

Remark 2.3 Let us note that if $u \in C([-\tau, T], \overline{D(A)})$, then Sel_F will be seen as a multivalued mapping from $C([-\tau, T], \overline{D(A)})$ into $L^1([0, T], X)$.

2.3.1 Compact Intervals Case

For the sake of convenience, put $J_\tau = [-\tau, 0] \cup J$ with $J = [0, T]$. Let us consider the Cauchy problem

$$\begin{cases} u'(t) \in Au(t) + f(t), & t \in J, \\ f(t) \in F(t, u(t), u_t), & t \in J, \\ u(t) = \phi(t), & t \in [-\tau, 0]. \end{cases} \quad (2.4)$$

The following lemma provides an useful property of Sel_F.

Lemma 2.4 *Let* (H_1) *and* (H_2) *be satisfied and let* X *be reflexive. Then* Sel_F *is weakly u.s.c. with nonempty, convex and weakly compact values.*

Proof Let us first show that $Sel_F(u) \neq \emptyset$ for each $u \in C(J_\tau, \overline{D(A)})$. For this purpose we assume that $u \in C(J_\tau, \overline{D(A)})$ and $\{(u_n, v_n)\}$ is a sequence of step functions from J to $\overline{D(A)} \times C([-\tau, 0], \overline{D(A)})$ such that

$$\sup_{t \in J} |u_n(t) - u(t)| \to 0, \quad \sup_{t \in J} |v_n(t) - u_t|_0 \to 0 \text{ as } n \to \infty.$$

By (H_1) we see readily that for each n, $F(\cdot, u_n(\cdot), v_n(\cdot))$ admits a strongly measurable selection $f_n(\cdot)$. Furthermore, it follows from (H_2) that $\{f_n\}$ is integrably bounded in $L^1(J, X)$. Making use of Lemma 1.24 we then see that $\{f_n\}$ is relatively weakly compact in $L^1(J, X)$. Hence, we may assume, by passing to a subsequence if necessary, that $f_n \rightharpoonup f$ in $L^1(J, X)$. An application of Mazur's theorem enables us to find that there exists a sequence $\{\tilde{f}_n\} \subset L^1(J, X)$ such that $\tilde{f}_n \in \text{co}\{f_k : k \geq n\}$ for each $n \geq 1$ and $\tilde{f}_n \to f$ in $L^1(J, X)$. Hence, $\tilde{f}_{n_k}(t) \to f(t)$ in X for a.e. $t \in J$ with some subsequence $\{\tilde{f}_{n_k}\}$ of $\{\tilde{f}_n\}$.

Denote by E the set of all $t \in J$ such that $\tilde{f}_{n_k}(t) \to f(t)$ in X and $f_n(t) \in F(t, u_n(t), v_n(t))$ for all $n \geq 1$. Let $x^* \in X^*$, $\varepsilon > 0$, and $t \in E$ be fixed. From (H_1), it follows immediately that $(x^* \circ F)(t, \cdot, \cdot) : X \to P(\mathbb{R})$ is u.s.c. with compact convex values, so $\varepsilon - \delta$ u.s.c. with compact convex values. Accordingly, we have

$$x^*(\tilde{f}_{n_k}(t)) \in \text{co}\{x^*(f_k(t)) : k \geq n\} \subset x^*(F(t, u_n(t), v_n(t)))$$
$$\subset x^*(F(t, u(t), u_t)) + (-\varepsilon, \varepsilon)$$

with k large enough. Therefore, we obtain that $x^*(\tilde{f}(t)) \in x^*(F(t, u(t), u_t))$ for each $x^* \in X^*$ and $t \in E$. Since F has convex and closed values, we conclude that $f(t) \in F(t, u(t), u_t)$ for each $t \in E$, which implies that $f \in Sel_F(u)$.

In the sequel, let $\{u_n\}$ be a sequence converging to $u \in C(J_\tau, \overline{D(A)})$ and $f_n \in Sel_F(u_n), n \geq 1$. Using the same argument as above, we obtain that $\{f_n\}$ is relatively weakly compact, and there exists a subsequence $\{f_{n_k}\}$ of $\{f_n\}$ and $f \in Sel_F(u)$ such that $f_{n_k} \rightharpoonup f$ in $L^1(J, X)$. This, together with Lemma 1.7 (ii), shows that Sel_F is weakly u.s.c. Also, from the arguments above it is easy to see that Sel_F has weakly compact values. Moreover, it is readily checked that Sel_F has convex values. The proof is complete.

In order to study the topological structure of solution set for the Cauchy problem (2.4), we first establish the following existence result.

Theorem 2.1 *Let* (H_0)–(H_2) *be satisfied. Then the Cauchy problem (2.4) has at least one* C^0-*solution for each* $\phi \in C([-\tau, 0], \overline{D(A)})$.

Proof Let $\phi \in C([-\tau, 0], \overline{D(A)})$. Consider the set

$K_T = \{u \in C(J_\tau, \overline{D(A)}) : u(t) = \phi(t) \text{ for } t \in [-\tau, 0] \text{ and } |u(t)| \le x_\phi(t) \text{ for all } t \in J\}$,

where $x_\phi \in C(J, \mathbb{R}^+)$ is the unique continuous solution of the integral equation in the form

$$x_\phi(t) = |\phi|_0 + \int_0^t L(\sigma)\left(1 + 2x_\phi(\sigma)\right) d\sigma, \quad t \in J. \tag{2.5}$$

We seek for solutions in K_T. To the end, let us define a multivalued mapping W^ϕ on K_T by setting

$$W^\phi(u) = S_{\phi,T}(Sel_F(u)), \quad u \in K_T.$$

It is clear that we obtain the result if we show that the map W^ϕ admits a fixed point in K_T. Below, we shall omit the subscript "T" and write only S_ϕ instead of $S_{\phi,T}$ if there is no danger of confusion.

Observe that for every $u \in K_T$, $Sel_F(u) \ne \emptyset$ due to Lemma 2.4 and hence $W^\phi(u) \subset C(J_\tau, \overline{D(A)})$. Also, $\{v|_{[-\tau,0]} : v \in W^\phi(u)\} = \{\phi\}$ for all $u \in K_T$. Moreover, taking $f \in Sel_F(u)$ with $u \in K_T$, it follows from (H_2) that for every $t \in J$,

$$
\begin{aligned}
|S_\phi(f)(t)| &\le |\phi(0)| + \int_0^t |f(\sigma)| d\sigma \\
&\le |\phi(0)| + \int_0^t L(\sigma)\left(1 + |u(\sigma)| + |u_\sigma|_0\right) d\sigma \\
&\le |\phi|_0 + \int_0^t L(\sigma)\left(1 + 2x_\phi(\sigma)\right) d\sigma \\
&= x_\phi(t).
\end{aligned}
$$

Here, we have tacitly used the condition $0 \in A0$ and the fact $|u_t|_0 \le x_\phi(t)$ for every $t \in J$ and $u \in K_T$. Hence, it is proved that $W^\phi(u) \subset K_T$ for every $u \in K_T$.

We process to verify that W^ϕ is u.s.c. on K_T. Due to Lemma 1.9, it suffices to prove that W^ϕ is quasi-compact and closed. By (H_2) we obtain that for all $f \in \mathscr{F} := Sel_F(K_T)$,

$$|f(t)| \le L(t)(1 + 2x_\phi(T)) \quad \text{for a.e. } t \in J, \tag{2.6}$$

which implies that $\mathscr{F}$ is integrably bounded and thus uniformly integrable. From this and Lemma 2.3 (i) we see that $W^\phi(K_T)(= S_\phi(\mathscr{F}))$ is relatively compact in $C(J_\tau, X)$. This in particular implies that W^ϕ is quasi-compact.

Let $\{(u_n, v_n)\}$ be a sequence in $\mathrm{Gra}(W^\phi)$ such that $(u_n, v_n) \to (u, v)$ in $C(J_\tau, X) \times C(J_\tau, X)$. Since $v_n \in W^\phi(u_n)$, there exists a sequence $\{f_n\} \subset L^1(J, X)$ satisfying $f_n \in Sel_F(u_n)$ and $v_n = S_\phi(f_n)$. Therefore, noticing that Sel_F is weakly u.s.c. with convex, weakly compact values due to Lemma 2.4, an application of Lemma 1.7 (ii) yields that there exists $f \in Sel_F(u)$ and a subsequence of $\{f_n\}$, still denoted by $\{f_n\}$, such that $f_n \rightharpoonup f$ in $L^1(J, X)$. From this and Lemma 2.3 (ii) we see that $v = S_\phi(f)$ and then $v \in W^\phi(u)$. It follows that W^ϕ is closed.

Consider the set

$$\mathscr{K}_T = \overline{\mathrm{co}}(\, W^\phi(K_T)),$$

the closed convex hull of $W^\phi(K_T)$. Clearly, $\mathscr{K}_T$ is a compact, convex set in $C(J_\tau, X)$ and $W^\phi(\mathscr{K}_T) \subset \mathscr{K}_T$.

Below, we shall prove that W^ϕ has a fixed point in $\mathscr{K}_T$. Due to Theorem 1.17, it suffices to show that W^ϕ has compact, contractible values. Given $u \in \mathscr{K}_T$, it is easy to see that $W^\phi(u)$ is compact because of the closedness and qusi-compactness of W^ϕ. Fix $f^* \in Sel_F(u)$ and put $u^* = S_\phi(f^*)$. Define a function $H : [0, 1] \times W^\phi(u) \to W^\phi(u)$ by setting

$$H(\lambda, v)(t) = \begin{cases} v(t), & t \in [-\tau, \lambda T], \\ u(t, \lambda T, v(\lambda T), f^*), & t \in (\lambda T, T] \end{cases}$$

for each $(\lambda, v) \in [0, 1] \times W^\phi(u)$, where $u(\cdot, \lambda T, v(\lambda T), f^*)$, as prescribed in Sect. 2.2, is the unique C^0-solution of the evolution inclusion in the form

$$\begin{cases} u'(t) \in Au(t) + f^*(t), & t \in [\lambda T, T], \\ u(\lambda T) = v(\lambda T). \end{cases}$$

What followed is to show that $H(\lambda, v) \in W^\phi(u)$ for each $(\lambda, v) \in [0, 1] \times W^\phi(u)$. Note that for each $v \in W^\phi(u)$, there exists $\tilde{f} \in Sel_F(u)$ such that $v = S_\phi(\tilde{f})$. Put

$$\hat{f}(t) = \tilde{f}(t)\chi_{[0,\lambda T]}(t) + f^*(t)\chi_{(\lambda T, T]}(t) \text{ for each } t \in J.$$

It is clear that $\hat{f} \in Sel_F(u)$. Also, it is readily checked that $S_\phi(\hat{f})(t) = v(t)$ for all $t \in [-\tau, \lambda T]$ and $S_\phi(\hat{f})(t) = u(t, \lambda T, v(\lambda T), f^*)$ for all $t \in (\lambda T, T]$, which gives $S_\phi(\hat{f}) = H(\lambda, v)$ and hence $H(\lambda, v) \in W^\phi(u)$.

To show that $W^\phi(u)$ is contractible, we first note that

$$H(0, v) = u^* \text{ and } H(1, v) = v \text{ for every } v \in W^\phi(u).$$

It remains to show that H is continuous. Given $(\lambda_i, v_i) \in [0, 1] \times W^\phi(u)$, $i = 1, 2$, with $\lambda_1 \le \lambda_2$, we can choose $f_i \in Sel_F(u)$ such that $H(\lambda_i, v_i) = S_\phi(f_i)$ and $f_i(t) = f^*(t)$ for all $t \in [\lambda_i T, T]$. Then, we have that for $-\tau \le s \le t \le T$,

$$|H(\lambda_1, v_1)(t) - H(\lambda_2, v_2)(t)| \le |H(\lambda_1, v_1)(s) - H(\lambda_2, v_2)(s)|$$
$$+ \int_s^t |f_1(\sigma) - f_2(\sigma)| d\sigma.$$

Noticing (2.6) and the fact $f_1(t) = f_2(t)$ for $t \in [\lambda_2 T, T]$, we see that for all $t \in [\lambda_1 T, T]$,

$$|H(\lambda_1, v_1)(t) - H(\lambda_2, v_2)(t)|$$

$$\leq |H(\lambda_1, v_1)(\lambda_1 T) - H(\lambda_2, v_2)(\lambda_1 T)| + \int_{\lambda_1 T}^{\lambda_2 T} |f_1(\sigma) - f_2(\sigma)| d\sigma$$

$$\leq |H(\lambda_1, v_1)(\lambda_1 T) - H(\lambda_2, v_2)(\lambda_1 T)| + (2 + 4x_\phi(T)) \int_{\lambda_1 T}^{\lambda_2 T} L(\sigma) d\sigma,$$

which combining with the fact that $H(\lambda_i, v_i)(t) = v_i(t)$ for all $t \in [-\tau, \lambda_i T]$ yields

$$\sup_{t \in J_\tau} |H(\lambda_1, v_1)(t) - H(\lambda_2, v_2)(t)| \leq \|v_1 - v_2\| + (2 + 4x_\phi(T)) \int_{\lambda_1 T}^{\lambda_2 T} L(\sigma) d\sigma.$$

The continuity of H follows immediately.

Finally, an application of Theorem 1.17 yields that W^ϕ has at least one fixed point, which is a C^0-solution of the Cauchy problem (2.4). This completes the proof.

In the sequel, we denote by $\Sigma_{\phi,T}^F$ the solution set of the Cauchy problem (2.4), i.e.,

$$\Sigma_{\phi,T}^F = \{u \in C(J_\tau, \overline{D(A)}) : u \text{ is the } C^0\text{-solution of (2.4)}$$
$$\text{satisfying } u(t) = \phi(t) \text{ for } t \in [-\tau, 0]\},$$

and, by $\hat{K}_T$ the set

$$\hat{K}_T = \{u \in C(J_\tau, \overline{D(A)}) : u(t) = \phi(t), \ t \in [-\tau, 0]\}.$$

Let $\mathrm{Fix}(W^\phi)$ be the fixed point set of W^ϕ acting on K_T, where K_T and W^ϕ were introduced in Theorem 2.1. We present the following characterization.

Lemma 2.5 *Let the hypotheses in Theorem 2.1 hold. Then $\Sigma_{\phi,T}^F = \mathrm{Fix}(W^\phi)$ and $\Sigma_{\phi,T}^F$ is compact in $C(J_\tau, X)$ for each $\phi \in C([-\tau, 0], \overline{D(A)})$.*

Proof Let $\phi \in C([-\tau, 0], \overline{D(A)})$ and let x_ϕ be the unique continuous solution of (2.5). Along the same line with the proof of Theorem 2.1, we define a mapping $\hat{W}^\phi$ on $\hat{K}_T$ by

$$\hat{W}^\phi(u) = S_\phi(Sel_F(u)), \ u \in \hat{K}_T,$$

which is regarded as an extension of W^ϕ. Observe that $\Sigma_{\phi,T}^F = \mathrm{Fix}(\hat{W}^\phi)$. Below, it will be sufficient to show that $u \in K_T$ whenever $u \in \mathrm{Fix}(\hat{W}^\phi)$. Taking $u \in \mathrm{Fix}(\hat{W}^\phi)$, it follows that there exists $f \in Sel_F(u)$ such that $u = S_\phi(f)$. Then, noticing (H_2) and the condition $0 \in A0$ and using the same arguments as in the proof of Theorem 2.1 one can show

$$|u_t|_0 \leq |\phi|_0 + \int_0^t L(\sigma)(1 + 2|u_\sigma|_0) d\sigma, \ t \in J.$$

With the aid of the generalized Gronwall-Bellman's inequality we obtain that for each $t \in J$,

$$
\begin{aligned}
|u_t|_0 \leq & |\phi|_0 + \int_0^t L(\sigma)d\sigma \\
& + 2\int_0^t L(s)\left(|\phi|_0 + \int_0^s L(\sigma)d\sigma\right)\exp\left(2\int_s^t L(\sigma)d\sigma\right)ds \\
= & x_\phi(t),
\end{aligned}
$$

which implies that $u \in K_T$. Based on the considerations above, we have $\Sigma_{\phi,T}^F =$ Fix(W^ϕ).

Moreover, as in the proof of Theorem 2.1, $\mathscr{K}_T$ is compact in $C(J_\tau, X)$ and W^ϕ is closed, from this we see that Fix(W^ϕ) is a compact set in $\mathscr{K}_T$, so is $\Sigma_{\phi,T}^F$. The proof is complete.

We present the following approximation result.

Lemma 2.6 *Put $\mathscr{D} = \overline{D(A)} \times C([-\tau, 0], \overline{D(A)})$. Suppose that F satisfies the hypotheses (H_1) and (H_2). Then there exists a sequence of multivalued functions $\{F_n\}$ with $F_n : J \times \mathscr{D} \to P_{cl,cv}(X)$ such that*

(i) *$F(t, x, v) \subset F_{n+1}(t, x, v) \subset F_n(t, x, v) \subset \overline{co}(F(t, B_{3^{1-n}}(x, v) \cap \mathscr{D}))$, $n \geq 1$, for each $t \in J$, $(x, v) \in \mathscr{D}$;*

(ii) *$|F_n(t, x, v)| \leq L(t)(3 + |x| + |v|_0)$, $n \geq 1$, for a.e. $t \in J$ and each $(x, v) \in \mathscr{D}$;*

(iii) *there exists $\mathscr{T} \subset J$ with $\mathrm{mes}(\mathscr{T}) = 0$ such that for each $x^* \in X^*$, $\varepsilon > 0$ and $(t, x, v) \in J \backslash \mathscr{T} \times \mathscr{D}$, there exists $N > 0$ such that for all $n \geq N$,*

$$
x^*(F_n(t, x, v)) \subset x^*(F(t, x, v)) + (-\varepsilon, \varepsilon);
$$

(iv) *$F_n(t, \cdot) : \mathscr{D} \to P_{cl,cv}(X)$ is continuous for a.e. $t \in J$ with respect to Hausdorff metric for each $n \geq 1$;*

(v) *for each $n \geq 1$, there exists a selection $G_n : J \times \mathscr{D} \to X$ of F_n such that $G_n(\cdot, x, v)$ is strongly measurable for each $(x, v) \in \mathscr{D}$ and for any compact subset $\mathscr{D}' \subset \mathscr{D}$ there exist constants $C_V > 0$ and $\delta > 0$ for which the estimate*

$$
|G_n(t, x_1, v_1) - G_n(t, x_2, v_2)| \leq C_V L(t)(|x_1 - x_2| + |v_1 - v_2|_0) \quad (2.7)
$$

holds for a.e. $t \in J$ and each $(x_1, v_1), (x_2, v_2) \in V$ with $V := (\mathscr{D}' + B_\delta(0)) \cap \mathscr{D}$;

(vi) *F_n verifies the condition (H_1) with F_n instead of F for each $n \geq 1$, provided that X is reflexive.*

Proof Put $r_n = 3^{-n}, n \geq 1$. For each $n \geq 1$, let $\{B_{r_n}(x, v)\}_{(x,v)\in\mathscr{D}}$ be an open cover of $\mathscr{D}$. Therefore, there exists a locally finite refinement $\{V_{j,n}\}_{j\in I_n}$ of $\{B_{r_n}(x, v)\}_{(x,v)\in\mathscr{D}}$. For each $j \in I_n$, we can choose $y_{j,n} := (x_{j,n}, v_{j,n}) \in \mathscr{D}$ such that $V_{j,n} \subset B_{r_n}(y_{j,n})$.

Now let $\{p_{j,n}(x, v)\}_{j \in I_n}$ be a locally Lipschitz partition of unity subordinated to the open cover $\{V_{j,n}\}_{j \in I_n}$. For each $n \geq 1$, define

$$F_n(t, x, v) = \sum_{j \in I_n} p_{j,n}(x, v)\overline{\mathrm{co}}(F(t, B_{2r_n}(y_{j,n}) \cap \mathscr{D})), \quad (t, x, v) \in J \times \mathscr{D},$$

and

$$G_n(t, x, v) = \sum_{j \in I_n} p_{j,n}(x, v)g_{j,n}(t), \quad (t, x, v) \in J \times \mathscr{D},$$

where $g_{j,n}(\cdot)$ is a strongly measurable selection of $F(\cdot, y_{j,n})$ for each $j \in I_n$.

With the preparation above at hand, the assertions (i), (iv) and (v) can be proved by the same kind of manipulations as in [106, Theorem 3.5] (see also [80, Lemma 2.2]). The assertion (ii) is an immediate consequence of (i) and (H_2).

We process to prove the assertion (iii). Let $\mathscr{T}$ be the set of all $t \in J$ such that both $F(t, \cdot, \cdot) : \mathscr{D} \to P_{cl,cv}(X)$ is weakly u.s.c. and $F(t, x, v)$ verifies the condition (H_2) for all (t, x, v) with $(x, v) \in \mathscr{D}$. Given $y = (x, v) \in \mathscr{D}$, we put $I_n^y = \{j \in I_n : p_{j,n}(y) > 0\}$, which is a finite set due to the local finiteness of the cover $\{V_{j,n}\}_{j \in I_n}$. It is readily checked that

$$j \in I_n^y \text{ implies } y \in B_{r_n}(y_{j,n}), \quad F_n(t, y) = \sum_{j \in I_n^y} p_{j,n}(y)\overline{\mathrm{co}}(F(t, B_{2r_n}(y_{j,n}) \cap \mathscr{D}))$$
(2.8)

and hence $|z - y|_\tau < 3r_n$ for each $j \in I_n^y$ and $z \in B_{2r_n}(y_{j,n})$, which gives $B_{2r_n}(y_{j,n}) \subset B_{3r_n}(y)$.

Let $x^* \in X^*$, $\varepsilon > 0$ and $t \in \mathscr{T}$ be fixed. From (H_1) it follows immediately that $(x^* \circ F)(t, \cdot, \cdot) : \mathscr{D} \to 2^{\mathbb{R}}$ is u.s.c. and thus ε-δ u.s.c. That is, there exists $\delta > 0$ such that for all $z \in B_\delta(y) \cap \mathscr{D}$,

$$x^*(F(t, z)) \subset x^*(F(t, y)) + \left(-\frac{\varepsilon}{3}, \frac{\varepsilon}{3}\right). \tag{2.9}$$

Selecting N large enough so that $n \geq N$ implies $3r_n \leq \delta$, we conclude from (2.9) that

$$x^*(F(t, B_{2r_n}(y_{j,n}) \cap \mathscr{D})) \subset x^*(F(t, y)) + \left(-\frac{\varepsilon}{3}, \frac{\varepsilon}{3}\right) \text{ for each } n \geq N \text{ and } j \in I_n^y. \tag{2.10}$$

On the other hand, since $x^*(F(t, y))$ is convex due to (H_1), we obtain

$$\overline{\mathrm{co}}\left(x^*(F(t, y)) + \left(-\frac{\varepsilon}{3}, \frac{\varepsilon}{3}\right)\right) = \overline{\mathrm{co}\,(x^*(F(t, y)))} + \left(-\frac{\varepsilon}{3}, \frac{\varepsilon}{3}\right)$$
$$\subset x^*(F(t, y)) + \left(-\frac{2\varepsilon}{3}, \frac{2\varepsilon}{3}\right),$$

whence (2.10) gives

$$x^* \left(\overline{co}(F(t, B_{2r_n}(y_{j,n}) \cap \mathcal{D})) \right) = \overline{co}(x^*(F(t, B_{2r_n}(y_{j,n}) \cap \mathcal{D})))$$
$$\subset x^*(F(t, y)) + \left(-\frac{2\varepsilon}{3}, \frac{2\varepsilon}{3} \right)$$

for each $n \geq N$ and $j \in I_n^y$. We thus use (2.8) to obtain that for all $n \geq N$,

$$x^*(F_n(t, y)) \subset \overline{co}\left(x^*(F(t, y)) + \left(-\frac{2\varepsilon}{3}, \frac{2\varepsilon}{3} \right) \right) \subset x^*(F(t, y)) + (-\varepsilon, \varepsilon).$$

This proves the assertion (iii).

It remains to verify the assertion (vi). Let $n \geq 1$ be fixed and $\mathcal{T}'$ the set of all $t \in J$ such that both $F_n(t, \cdot, \cdot) : \mathcal{D} \to P_{cl,cv}(X)$ is continuous with respect to Hausdorff metric and $F_n(t, x, v)$ verifies the inequality in the assertion (ii) for all (t, x, v) with $(x, v) \in \mathcal{D}$. Clearly, $J \setminus \mathcal{T}'$ has null measure and $F_n(t, \cdot, \cdot)$ is ε-δ u.s.c. for each $t \in \mathcal{T}'$. From the reflexivity of X it follows that $F_n(t, \cdot, \cdot)$ has weakly compact values for each $t \in \mathcal{T}'$. Therefore, we conclude from Lemma 1.7 (i) that $F_n(t, \cdot, \cdot)$ is weakly u.s.c. for a.e $t \in J$. Also, it is clear that $F_n(\cdot, x, v)$ has a strongly measurable selection $G_n(\cdot, x, v)$ for each $(x, v) \in \mathcal{D}$, and thereby the assertion is established.

Remark 2.4 It is assumed in Lemma 2.6 that for a.e. $t \in J$, $F(t, \cdot, \cdot)$ is weakly u.s.c. rather than u.s.c. Such condition is more easily verified usually in practical applications (see Sect. 5 below and [192, Sect. 5]). The latter condition can be found in some situations of previous research such as [1, 14, 130].

The following result is the main result in this subsection.

Theorem 2.2 *Let the hypotheses in Theorem 2.1 be satisfied. Then $\Sigma_{\phi,T}^F$ is an R_δ-set for each $\phi \in C([-\tau, 0], \overline{D(A)})$.*

Proof Assume that $\{F_n\}$ is the approximate sequence established in Lemma 2.6. For each $n \geq 1$, consider the approximate problem of the form

$$\begin{cases} u'(t) \in Au(t) + f(t), & t \in J, \\ f(t) \in F_n(t, u(t), u_t), & t \in J, \\ u(t) = \phi(t), & t \in [-\tau, 0], \end{cases} \tag{2.11}$$

where $\phi \in C([-\tau, 0], \overline{D(A)})$. Let $\Sigma_{\phi,T}^{F_n}$ be the solution set of (2.11).

Noticing Lemma 2.6 (ii) and (vi) and performing similar arguments as in Theorem 2.1 and Lemma 2.5, we infer that $\Sigma_{\phi,T}^{F_n}$ is nonempty and compact in $C(J_\tau, X)$. Moreover, by Lemma 2.6 (i) we have

$$\Sigma_{\phi,T}^F \subset \cdots \subset \Sigma_{\phi,T}^{F_n} \cdots \subset \Sigma_{\phi,T}^{F_2} \subset \Sigma_{\phi,T}^{F_1}.$$

We claim that $\Sigma_{\phi,T}^F = \bigcap_{n \geq 1} \Sigma_{\phi,T}^{F_n}$. Note first that $\Sigma_{\phi,T}^F \subset \bigcap_{n=1} \Sigma_{\phi,T}^{F_n}$. To prove the reverse inclusion, we take $u \in \bigcap_{n=1} \Sigma_{\phi,T}^{F_n}$. Therefore, there exists a sequence $\{f_n\} \subset L^1(J, X)$ such that $f_n \in Sel_{F_n}(u)$, $u = S_\phi(f_n)$, and for all $n \geq 1$,

$$|f_n(t)| \leq L(t)(3 + 2|u_t|_0) \text{ for a.e. } t \in J$$

in view of Lemma 2.6 (ii). From which together with the fact that X is reflexive it follows that $\{f_n\}$ is relatively weakly compact in $L^1(J, X)$ due to Lemma 1.24. Thus, there exists a subsequence of $\{f_n\}$, still denoted by $\{f_n\}$, such that f_n converges weakly to $f \in L^1(J, X)$. An application of Mazur's theorem yields that there exists a sequence $\{\tilde{f}_n\} \subset L^1(J, X)$ such that $\tilde{f}_n \in \text{co}\{f_k : k \geq n\}$ for each $n \geq 1$ and $\tilde{f}_n \to f$ in $L^1(J, X)$ as $n \to \infty$. Passing to a subsequence if necessary, we may assume that for a.e. $t \in J$, $\tilde{f}_n(t) \to f(t)$ in X. Denote by $\mathscr{T}_c$ the set of all $t \in J$ such that $\tilde{f}_n(t) \to f(t)$ in X and $f_n(t) \in F_n(t, u(t), u_t)$ for all $n \geq 1$. Clearly, $J \backslash \mathscr{T}_c$ has null measure.

Now by Lemma 2.6 (iii) we have that there exists $E \subset J$ with $\text{mes}(E) = 0$ such that for each $t \in (J \backslash E) \cap \mathscr{T}_c$, $\varepsilon > 0$ and $x^* \in X^*$,

$$x^*(\tilde{f}_n(t)) \in \text{co}\{x^*(f_k(t)) : k \geq n\} \subset x^*(F_n(t, u(t), u_t) \subset x^*(F(t, u(t), u_t)) + (-\varepsilon, \varepsilon)$$

with n large enough. Here we use Lemma 2.6 (i) and the result that F_n has convex values for each $n \geq 1$. Passing to the limit in the inclusion above for $n \to \infty$ and taking into account the arbitrariness of ε, we get that $x^*(f(t)) \in x^*(F(t, u(t), u_t))$ for each $x^* \in X^*$ and $t \in (J \backslash E) \cap \mathscr{T}_c$. Since x^* is arbitrary and F has convex and closed values, we conclude that $f(t) \in F(t, u(t), u_t)$ for each $t \in (J \backslash E) \cap \mathscr{T}_c$, which implies that $f \in Sel_F(u)$. Moreover, noticing $f_n \rightharpoonup f$ in $L^1(J, X)$, we deduce, in view of Lemma 2.3 (ii), that $S_\phi(f) = u$. This proves that $u \in \Sigma_{\phi,T}^F$, as desired.

Finally, in order to show that $\Sigma_{\phi,T}^F$ is an R_δ-set, it suffices to verify that $\Sigma_{\phi,T}^{F_n}$ is contractible for each $n \geq 1$. Let G_n be the selection of F_n which is established in Lemma 2.6 (v). Observe, thanks to Lemma 2.6 (v), that $G_n(t, \cdot, \cdot)$ is continuous for a.e. $t \in J$. Also, $\mathcal{D}_n := \{(u(t), u_t) : t \in J, u \in \Sigma_{\phi,T}^{F_n}\}$ is a relatively compact set in $X \times C([-\tau, 0], X)$, since $\Sigma_{\phi,T}^{F_n}$ is compact in $C(J_\tau, X)$. Therefore, we conclude, again by Lemma 2.6 (v), that there exists a neighborhood U of $\overline{\mathcal{D}}_n$ and a constant $C_U > 0$ such that (2.7) is satisfied with C_V replaced by C_U. Furthermore, it is easy to see that G_n verifies

$$|G_n(t, x, v)| \leq L(t)(3 + |x| + |v|_0) \tag{2.12}$$

for a.e. $t \in J$ and each $(x, v) \in \overline{D(A)} \times C([-\tau, 0], \overline{D(A)})$.

Now, performing a trivial variant of an argument from Theorem 2.1, we obtain the existence of C^0-solutions of the Cauchy problem of the form

$$\begin{cases} v'(t) \in Av(t) + G_n(t, v, v_t), & t \in [s, T], \\ v(s + \theta) = \varphi(\theta), & \theta \in [-\tau, 0] \end{cases} \quad (2.13)$$

for each $s \in J$ and $\varphi \in C([-\tau, 0], \overline{D(A)})$. Moreover, we point out that the C^0-solution to (2.13) is unique. Indeed, if v_1 and v_2 are two solutions of (2.13) corresponding to $s \in J$ and $\varphi \in C([-\tau, 0], \overline{D(A)})$, then there exists a neighborhood U' related to v_1 and v_2 and $w := v_1 - v_2$ satisfies

$$|w(t)| \le \int_s^t |G_n(t, v_1(\sigma), v_{1\sigma}) - G_n(t, v_2(\sigma), v_{2\sigma})| d\sigma$$

$$\le C_{U'} \int_s^t L(\sigma)(|w(\sigma)| + |w_\sigma|_0) d\sigma$$

for every $t \in [s, T]$. We here used the result (v) of Lemma 2.6. Therefore, by Lemma 1.4 we see that $v_1 \equiv v_2$, as desired.

We denote by $v(\cdot, s, \varphi)$ the unique C^0-solution of (2.13) corresponding to $s \in J$ and $\varphi \in C([-\tau, 0], \overline{D(A)})$. Define a function $\hat{H} : [0, 1] \times \Sigma_{\phi,T}^{F_n} \to \Sigma_{\phi,T}^{F_n}$ by setting

$$\hat{H}(\lambda, u)(t) = \begin{cases} u(t), & t \in [-\tau, \lambda T], \\ v(t, \lambda T, u_{\lambda T}), & t \in (\lambda T, T] \end{cases}$$

for each $(\lambda, u) \in [0, 1] \times \Sigma_{\phi,T}^{F_n}$. In a manner similar to the proof in Theorem 2.1 we can show that $\hat{H}(\lambda, u) \in \Sigma_{\phi,T}^{F_n}$ for each $(\lambda, u) \in [0, 1] \times \Sigma_{\phi,T}^{F_n}$, and $\hat{H}(0, u) = v(\cdot, 0, \phi)$ and $\hat{H}(1, u) = u$ for each $u \in \Sigma_{\phi,T}^{F_n}$.

Below is to show that $\hat{H}$ is continuous. Let us consider a sequence $\{(\lambda_k, u_k)\} \subset [0, 1] \times \Sigma_{\phi,T}^{F_n}$ with $(\lambda_k, u_k) \to (\lambda, u)$ in $[0, 1] \times C(J_\tau, X)$ as $k \to \infty$. Set

$$\rho_k(t) = |\hat{H}(\lambda, u)(t) - \hat{H}(\lambda_k, u_k)(t)| \text{ for } t \in J_\tau.$$

We are going to show that $\sup_{t \in J_\tau} \rho_k(t) \to 0$ as $k \to \infty$. Without loss of generality we assume that $\lambda_k \le \lambda$ for all $k \ge 1$, since the remaining cases can be treated in a similar way. For simplicity in presentation, we put $\hat{v}_k = \hat{H}(\lambda_k, u_k), k \ge 1$, and $\hat{v} = \hat{H}(\lambda, u)$. From Lemma 2.6 (v) it follows that for each $t \in [\lambda T, T]$,

$$\rho_k(t) = |\hat{v}(t) - \hat{v}_k(t)|$$

$$\le |\hat{v}(\lambda T) - \hat{v}_k(\lambda T)| + \int_{\lambda T}^t |G_n(\sigma, \hat{v}(\sigma), \hat{v}_\sigma) - G_n(\sigma, \hat{v}_k(\sigma), \hat{v}_{k\sigma})| d\sigma$$

$$\le \rho_k(\lambda T) + C_U \int_{\lambda T}^t L(\sigma) \left(\rho_k(\sigma) + \sup_{\theta \in [\sigma-\tau, \sigma]} \rho_k(\theta) \right) d\sigma.$$

Then an application of Lemma 1.4 yields

$$\rho_k(t) \leq \sup_{\theta \in [\lambda T - \tau, \lambda T]} \rho_k(\theta) \exp\left(2C_U \int_{\lambda T}^t L(\sigma)d\sigma\right), \quad \lambda T \leq t \leq T. \qquad (2.14)$$

Also, noticing that $\Sigma_{\phi,T}^{F_n}$ is compact, we can find a constant $M > 0$ for which the estimates

$$|u(t)| \leq M \quad \text{and} \quad |u_t|_0 \leq M$$

hold for all $u \in \Sigma_{\phi,T}^{F_n}$ and $t \in J$, which together with (2.12) imply that for every $\lambda_k T \leq t \leq \lambda T$,

$$\rho_k(t) \leq |\hat{v}(\lambda_k T) - \hat{v}_k(\lambda_k T)| + \int_{\lambda_k T}^t |G_n(\sigma, \hat{v}(\sigma), \hat{v}_\sigma) - G_n(\sigma, \hat{v}_k(\sigma), \hat{v}_{k\sigma})| d\sigma$$

$$\leq |u(\lambda_k T) - u_k(\lambda_k T)| + (6 + 4M) \int_{\lambda_k T}^t L(\sigma)d\sigma.$$

$$(2.15)$$

Then, note that

$$\rho_k(t) = |u(t) - u_k(t)| \quad \text{for } t \in [-\tau, \lambda_k T], \qquad (2.16)$$

which, together with (2.15), yields

$$\sup_{\theta \in [\lambda T - \tau, \lambda T]} \rho_k(\theta) \leq \|u - u_k\| + (6 + 4M) \int_{\lambda_k T}^{\lambda T} L(\sigma)d\sigma. \qquad (2.17)$$

Recalling (2.14)–(2.17), we end up with

$$\rho_k(t) \leq 2\|u - u_k\| + (6 + 4M) \int_{\lambda_k T}^{\lambda T} L(\sigma)d\sigma$$

$$+ \left(\|u - u_k\| + (6 + 4M) \int_{\lambda_k T}^{\lambda T} L(\sigma)d\sigma\right) \exp\left(2C_U \int_{\lambda T}^T L(\sigma)d\sigma\right)$$

for every $t \in J_\tau$. The right-hand side of the inequality above can be made small when k is large independently of $t \in J_\tau$. Accordingly, our result follows. Therefore, we conclude that $\Sigma_{\phi,T}^{F_n}$ is contractible, and thus $\Sigma_{\phi,T}^F$ is an R_δ-set. This proof is complete.

2.3.2 Noncompact Intervals Case

Throughout this subsection, let $\tilde{J}_\tau = [-\tau, 0] \cup \mathbb{R}^+$. We first present the following result.

Lemma 2.7 *Let X be reflexive. Suppose further that F satisfies the hypotheses* (H_1) *and* (H_2). *Then* $Sel_F(u) \neq \emptyset$ *for each* $u \in \widetilde{C}(\tilde{J}_\tau, \overline{D(A)})$.

Proof Let $u \in \widetilde{C}(\tilde{J}_\tau, \overline{D(A)})$. By Lemma 2.4, one can choose $f_m \in Sel_{F|_{[0,m]}}(u|_{[-\tau,m]})$ for each $m \in \mathbb{N} \setminus \{0\}$, where $F|_{[0,m]}$ is the restriction of F to $[0, m]$, it is to say

$$F|_{[0,m]}(t, x, v) = F(t, x, v) \text{ on } [0, m] \times \overline{D(A)} \times C([-\tau, 0], \overline{D(A)}).$$

Consider the function $f : \mathbb{R}^+ \to X$ defined as

$$f(t) = \sum_{m=1}^{\infty} \chi_{[m-1,m)}(t) f_m(t), \quad t \in \mathbb{R}^+,$$

where $\chi_{[m-1,m)}$ denotes the characteristic function of interval $[m - 1, m)$. It is not difficult to see that $f \in Sel_F(u)$ and it is locally integrable. This gives desired result.

Assume that $\{C([a, m], X), \pi_{a,m}^P, \mathbb{N}(a)\}$ and $\{L^1([0, m], X), \dot{\pi}_m^P, \mathbb{N} \setminus \{0\}\}$ are the inverse systems established in Sect. 1.2.4. Given $\phi \in C([-\tau, 0], \overline{D(A)})$, we have that the family $\{id, S_{\phi,m}\}$ is a mapping from $\{L^1([0, m], X), \dot{\pi}_m^P, \mathbb{N} \setminus \{0\}\}$ into $\{C([-\tau, m], X), \pi_{-\tau,m}^P, \mathbb{N} \setminus \{0\}\}$. Indeed, this can be seen from the observation

$$\pi_{-\tau,m}^P(S_{\phi,p}(f)) = S_{\phi,m}(\dot{\pi}_m^P(f)) \text{ for all } f \in L^1(0, p, X) \text{ and } m \leq p.$$

So the family $\{id, S_{\phi,m}\}$ induces a limit mapping $S_{\phi,\infty} : L^1_{loc}(\mathbb{R}^+, X) \to \widetilde{C}(\tilde{J}_\tau, X)$ such that $S_{\phi,\infty}(f)|_{[-\tau,m]} = S_{\phi,m}(f|_{[0,m]})$ for each $f \in L^1_{loc}(\mathbb{R}^+, X)$ and $m \in \mathbb{N} \setminus \{0\}$.

In this subsection, by a C^0-solution of the Cauchy problem (2.1), we mean a continuous function $u : \tilde{J}_\tau \to \overline{D(A)}$ which satisfies $u(t) = \phi(t)$ for all $t \in [-\tau, 0]$ and is a C^0-solution in the sense of *Benilan* to $u'(t) + Au(t) \ni f(t)$, where $f \in L^1_{loc}(\mathbb{R}^+, X)$ and $f(t) \in F(t, u(t), u_t)$ for a.e. $t \in \mathbb{R}^+$.

Let $\Sigma_{\phi,\infty}^F$ stand for the set of all C^0-solutions to the Cauchy problem (2.1). We are in the position to present our main result in this subsection.

Theorem 2.3 *Assume that the hypotheses* (H_0)-(H_2) *are satisfied. Then* $\Sigma_{\phi,\infty}^F$ *is an* R_δ-*set for each* $\phi \in C([-\tau, 0], \overline{D(A)})$.

Proof Assume that $\phi \in C([-\tau, 0], \overline{D(A)})$. For every $m \in \mathbb{N} \setminus \{0\}$, let $W_m^\phi : \hat{K}_m \to 2^{\hat{K}_m}$ be a multivalued mapping defined by

$$W_m^\phi(u) = S_{\phi,m}(Sel_{F|_{[0,m]}}(u)) \text{ for each } u \in \hat{K}_m,$$

where

$$\hat{K}_m := \{u \in C([-\tau, m], \overline{D(A)}) : u(t) = \phi(t), \ t \in [-\tau, 0]\}.$$

Applying Theorem 2.1 and Lemma 2.5 to $F|_{[0,m]}$ we obtain that $\text{Fix}(W_m^\phi) = \Sigma_{\phi,m}^{F|_{[0,m]}}$ and $\text{Fix}(W_m^\phi)$ is nonempty and compact. Also, it is seen, thanks to Theorem 2.2, that

Fix(W_m^ϕ) is an R_δ-set. Moreover, one finds that $\{\hat{K}_m, \pi_{-\tau,m}^p, \mathbb{N} \setminus \{0\}\}$ is an inverse system and

$$\widetilde{K} := \{u \in \widetilde{C}(\tilde{J}_\tau, \overline{D(A)}) : u(t) = \phi(t) \text{ for all } t \in [-\tau, 0]\}$$
$$= \lim_{\leftarrow} \{\hat{K}_m, \pi_{-\tau,m}^p, \mathbb{N} \setminus \{0\}\}.$$

In order to apply Theorem 1.19, we first show that the family $\{id, W_m^\phi\}$ is a mapping from $\{\hat{K}_m, \pi_{-\tau,m}^p, \mathbb{N} \setminus \{0\}\}$ into itself. Let $p, m \in \mathbb{N}$ with $p \geq m$ and $u \in \hat{K}_p$. We claim that

$$Sel_{F|_{[0,m]}}(u|_{[-\tau,m]}) = \{f|_{[0,m]} : f \in Sel_{F|_{[0,p]}}(u)\}. \tag{2.18}$$

The case $p = m$ is obvious. For the case $p > m$ it is readily checked that $\{f|_{[0,m]} : f \in Sel_{F|_{[0,p]}}(u)\} \subset Sel_{F|_{[0,m]}}(u|_{[-\tau,m]})$. It remains to prove the reverse inclusion. Let $f \in Sel_{F|_{[0,m]}}(u|_{[-\tau,m]})$. Choose $g \in Sel_{F|_{[0,p]}}(u)$ and put

$$\hat{f}(t) = f(t)\chi_{[0,m]}(t) + g(t)\chi_{(m,p]}(t), \quad t \in [0, p].$$

We then obtain that $\hat{f} \in Sel_{F|_{[0,p]}}(u)$, which gives $Sel_{F|_{[0,m]}}(u|_{[-\tau,m]}) \subset \{f|_{[0,m]} : f \in Sel_{F|_{[0,p]}}(u)\}$, as desired.

Now, by using (2.18) and the fact $\pi_{-\tau,m}^p(S_{\phi,p}(f)) = S_{\phi,m}(\dot{\pi}_m^p(f))$ for every $f \in L^1(0, p, X)$, we have

$$\begin{aligned}
\pi_{-\tau,m}^p(W_p^\phi(u)) &= \pi_{-\tau,m}^p(S_{\phi,p}(Sel_{F|_{[0,p]}}(u))) \\
&= \{S_{\phi,m}(\dot{\pi}_m^p(f)) : f \in Sel_{F|_{[0,p]}}(u)\} \\
&= \{S_{\phi,m}(f) : f \in Sel_{F|_{[0,m]}}(u|_{[-\tau,m]})\} \\
&= W_m^\phi(\pi_{-\tau,m}^p(u)).
\end{aligned}$$

Hence, $\{id, W_m^\phi\}$ induces a limit mapping $W_\infty^\phi : \widetilde{K} \to 2^{\widetilde{K}}$, defined by

$$W_\infty^\phi(u) = \{w \in \widetilde{K} : w|_{[-\tau,m]} = S_{\phi,m}(f|_{[0,m]}) \text{ for every } m \in \mathbb{N} \setminus \{0\},$$
$$f \in L_{loc}^1(\mathbb{R}^+, X) \text{ and } f(t) \in F(t, u(t), u_t) \text{ for a.e. } t \in \mathbb{R}^+\}$$

for each $u \in \widetilde{K}$. Here we used Lemma 2.7. Moreover, it readily follows that

$$W_\infty^\phi(u) = S_{\phi,\infty}(Sel_F(u)) \text{ for every } u \in \widetilde{K}.$$

Now, applying Theorem 1.19 yields that Fix(W_∞^ϕ) is an R_δ-set, which together with relation $\Sigma_{\phi,\infty}^F = \text{Fix}(W_\infty^\phi)$ implies that $\Sigma_{\phi,\infty}^F$ is an R_δ-set. Thus, the proof is complete.

2.4 Nonlocal Cauchy Problem

We are concerned with the existence of C^0-solutions to the nonlocal Cauchy problem (2.2) defined on right half-line.

The next lemma, which gives the convergence property of Sel_F in the case when $J = \mathbb{R}^+$, plays an important role in the sequel.

Lemma 2.8 *Let X be reflexive and F verify the hypotheses (H_1) and (H_2). If $\{u_n\} \subset \widetilde{C}(\tilde{J}_\tau, \overline{D(A)})$ with $u_n \to u_0$ in $\widetilde{C}(\tilde{J}_\tau, X)$ and $f_n \in Sel_F(u_n)$, then there exists $f \in Sel_F(u_0)$ and a subsequence $\{f_{n'}\}$ of $\{f_n\}$ such that $f_{n'} \rightharpoonup f$ in $L^1([0, m], X)$ for each $m \in \mathbb{N} \setminus \{0\}$.*

Proof Observe that $u_n \to u_0$ in $C([-\tau, m], X)$ for each $m \in \mathbb{N} \setminus \{0\}$. Also, from Lemma 2.4 it follows that $Sel_{F|_{[0,1]}}$ is weakly u.s.c. with convex and weakly compact values. Since $f_n|_{[0,1]} \in Sel_{F|_{[0,1]}}(u_n|_{[-\tau,1]})$, we see, in view of Lemma 1.7 (ii), that there exists a subsequence of $\{f_n\}$, say $\{f_{n,1}\}$, and $\hat{f}_1 \in Sel_{F|_{[0,1]}}(u_0|_{[-\tau,1]})$ such that $f_{n,1}|_{[0,1]}$ converges weakly to $\hat{f}_1$ in $L^1([0, 1], X)$. Similarly, we can select a subsequence $\{f_{n,2}\}$ of $\{f_{n,1}\}$ and $\hat{f}_2 \in Sel_{F|_{[0,2]}}(u_0|_{[-\tau,2]})$ such that $f_{n,2}|_{[0,2]} \rightharpoonup \hat{f}_2$ in $L^1([0, 2], X)$. Proceeding in this manner, we can choose a family of subsequences $\{f_{n,m}\}$, $m \geq 1$, of $\{f_n\}$ and a sequence $\{\hat{f}_m\}$ such that $f_{n,m}|_{[0,m]} \rightharpoonup \hat{f}_m$ in $L^1([0, m], X)$. Note that $\hat{f}_m \in Sel_{F|_{[0,m]}}(u_0|_{[-\tau,m]})$. Write

$$\hat{f}(t) = \sum_{m=1}^{\infty} \chi_{[m-1,m)}(t)\hat{f}_m(t), \quad t \in \mathbb{R}^+.$$

It is clear that $\hat{f} \in Sel_F(u_0)$. Moreover, we see that the diagonal sequence $\{f_{n,n}\}$, as a subsequence of $\{f_n\}$, verifies $f_{n,n}|_{[0,m]} \rightharpoonup \hat{f}|_{[0,m]}$ in $L^1([0, m], X)$ for each $m \in \mathbb{N} \setminus \{0\}$. The lemma is proved.

To present our main result, we also need the following conditions.
(H_3) There exists $r > 0$ such that $[x, f]_+ \leq 0$ for each $x \in \overline{D(A)}$ with $|x| = r$, $t \in \mathbb{R}^+, v \in C([-\tau, 0], \overline{D(A)})$ with $|v|_0 \leq r$ and $f \in F(t, x, v)$.
(H_4) $g : \widetilde{C}_b(\tilde{J}_\tau, \overline{D(A)}) \to C([-\tau, 0], \overline{D(A)})$ verifies

(i) the restriction of g to Ω_r is continuous and $|g(u)|_0 \leq r$ for each $u \in \Omega_r$, where $\Omega_r = \{u \in \widetilde{C}_b(\tilde{J}_\tau, \overline{D(A)}) : |u(t)| \leq r$ for all $t \in \tilde{J}_\tau\}$, and
(ii) for each subset $\mathscr{U} \subset \Omega_r$ which restricted to $[\delta, \infty)$ is relatively compact in $\widetilde{C}([\delta, \infty), X)$ for each $\delta \in (0, \infty)$, $g(\mathscr{U})$ is relatively compact in $C([-\tau, 0], X)$, where r is given by (H_3).

Remark 2.5 (a) Let us mention that the condition (ii) above on g is quite general. In particular, we claim that the condition (ii) is satisfied when the condition (i) above and the following condition are fulfilled: (H_g) There exists $\delta' \in (0, \infty)$ such that for every $u, w \in \Omega_r$ satisfying $u(t) = w(t)$ $(t \in [\delta', \infty))$, $g(u) = g(w)$. To illustrate it, let us define a linear operator $\Lambda : \widetilde{C}([\delta', \infty), X) \to \widetilde{C}(\tilde{J}_\tau, X)$ by

$$\Lambda(u) = \begin{cases} u(t), & t \in (\delta', \infty), \\ u(\delta'), & t \in [-\tau, \delta']. \end{cases}$$

Then it is clear that Λ is bounded and hence $\hat{g} := g \circ \Lambda$ is a continuous function from $\Omega_r|_{[\delta',\infty)}$ to $C([-\tau, 0], \overline{D(A)})$. Moreover, if $\mathscr{U} \subset \Omega_r$ and $\mathscr{U}_{[\delta,\infty)}$ is relatively compact in $\widetilde{C}([\delta, \infty), X)$ for each $\delta \in (0, \infty)$, then we see that $\hat{g}(\mathscr{U}|_{[\delta',\infty)})$ is compact in $C([-\tau, 0], X)$. From this and (H_g) it follows that $g(\mathscr{U})(= \hat{g}(\mathscr{U}|_{[\delta',\infty)}) \subset \hat{g}(\overline{\mathscr{U}|_{[\delta',\infty)}}))$ is relatively compact in $C([-\tau, 0], X)$.

(b) Note that the condition (H_g), which was used in some situations of previous research (cf. e.g., Wang [194] et al. and references therein), covers the multipoint discrete mean condition mentioned in the Introduction.

For some $\widetilde{r} > 0$, denote $Q_{\widetilde{r}} := \{w \in C([-\tau, 0], \overline{D(A)}) : |w|_0 \leq \widetilde{r}\}$ below.

Lemma 2.9 *Let $\widetilde{r} > 0$ be fixed. Under the hypotheses (H_0)-(H_2), the multivalued mapping $\Gamma : Q_{\widetilde{r}} \to P(\widetilde{C}(\widetilde{J}_\tau, X))$, defined by $\Gamma(\phi) = \Sigma_{\phi,\infty}^F$ for each $\phi \in Q_{\widetilde{r}}$, is an R_δ-mapping.*

Proof As proved in Theorem 2.3, $\Gamma(\phi)$ is an R_δ-set for each $\phi \in Q_r$. It suffices to verify the upper semi-continuity of Γ.

We first show that Γ is quasi-compact. Let $\mathcal{A} \subset Q_{\widetilde{r}}$ be a compact set and

$$\mathscr{F}_{\widetilde{r}} = \{f \in L_{loc}^1(\mathbb{R}^+, X) : |f(t)| \leq L(t)(1 + 2x_{\widetilde{r}}(t)) \text{ for a.e. } t \in \mathbb{R}^+\}, \quad (2.19)$$

where $x_{\widetilde{r}}$ is the unique continuous solution of

$$x_{\widetilde{r}}(t) = \widetilde{r} + \int_0^t L(\sigma)(1 + 2x_{\widetilde{r}}(\sigma)) d\sigma, \quad t \in \mathbb{R}^+.$$

An argument similar to that in Lemma 2.5 enables us to obtain that $|v_t|_0 \leq x_{\widetilde{r}}(t)$ for each $t \in \mathbb{R}^+$ and $v \in \Gamma(Q_{\widetilde{r}})$. From this and the fact that

$$\Gamma(\phi) \subset S_{\phi,\infty}(Sel_F(\Gamma(\phi))) \text{ for each } \phi \in Q_{\widetilde{r}},$$

we deduce, thanks to (H_2), that $\Gamma(\phi) \in S_{\phi,\infty}(\mathscr{F}_{\widetilde{r}})$ and hence $\Gamma(\mathcal{A}) \subset S_{\mathcal{A},\infty}(\mathscr{F}_{\widetilde{r}})$. Also, it is easy to see that $\mathscr{F}_{\widetilde{r}}|_{[0,m]}$ is uniformly integrable in $L^1([0, m], X)$ for each $m \in \mathbb{N} \setminus \{0\}$. Applying Lemma 2.3 (i) gives that $\Gamma(\mathcal{A})|_{[-\tau,m]}(\subset S_{\mathcal{A},m}(\mathscr{F}_{\widetilde{r}}|_{[0,m]}))$ is a relatively compact set in $C([-\tau, m], X)$ for each $m \in \mathbb{N} \setminus \{0\}$. Therefore, by Lemma 1.18 we see that $\Gamma(\mathcal{A})$ is relatively compact in $\widetilde{C}(\widetilde{J}_\tau, X)$, as desired.

What followed is to show that Γ is closed. Let $\{(\phi_n, u_n)\}$ be a sequence in $\text{Gra}(\Gamma)$, which converges to $(\phi, u) \in C([-\tau, 0], X) \times \widetilde{C}(\widetilde{J}_\tau, X)$. It is known that there exists a sequence $\{f_n\} \subset \mathscr{F}_{\widetilde{r}}$ such that $f_n \in Sel_F(u_n)$ and $S_{\phi_n,\infty}(f_n) = u_n$. Then an application of Lemma 2.8 yields that there exists $f \in Sel_F(u)$ and a subsequence of $\{f_n\}$, still denoted by $\{f_n\}$, such that $f_n|_{[0,m]} \rightharpoonup f|_{[0,m]}$ in $L^1([0, m], X)$ for every $m \in \mathbb{N} \setminus \{0\}$. Recalling Lemma 2.1 and the representation of $S_{\phi,m}$, we see that $S_{\phi,m}(f|_{[0,m]}) = u|_{[-\tau,m]}$ for every $m \in \mathbb{N} \setminus \{0\}$, which gives $u = S_{\phi,\infty}(f)$. Thus, it follows that $u \in \Gamma(\phi)$. An application of Lemma 1.9 then completes this proof.

Theorem 2.4 *Suppose that the hypotheses* (H_0)-(H_4) *are satisfied. Then the nonlocal Cauchy problem (2.2) admits at least one* C^0-*solution.*

Proof Since $\overline{D(A)}$ is convex, it follows from Theorem 1.4 that there exists a continuous extension $\widetilde{id}$ of identity mapping $id : \overline{D(A)} \rightarrow \overline{D(A)}$ satisfying $\widetilde{id}(X) \subset \overline{D(A)}$. Let $\rho : X \times C([-\tau, 0], X) \rightarrow \overline{B_r(0, 0)}$ be defined by

$$\rho(x, v) = \begin{cases} (x, v), & \text{if } (x, v) \in \overline{B_r(0, 0)}, \\ r|(x, v)|_\tau^{-1}(x, v), & \text{in rest.} \end{cases}$$

Then we define the multi-value function $F_\rho : \mathbb{R}^+ \times X \times C([-\tau, 0], X) \rightarrow P_{cl,cv}(X)$ by

$$F_\rho(t, x, v) = F(t, \rho(\widetilde{id}(x), \widetilde{id}(v))), \quad (t, x, v) \in \mathbb{R}^+ \times X \times C([-\tau, 0], X),$$

where $\widetilde{id}(v)(s) = \widetilde{id}(v(s))$ for each $s \in [-\tau, 0]$.

Since both ρ and $\widetilde{id}$ are continuous, it follows that F_ρ verifies the condition (H_1). Clearly it satisfies the condition (H_2) (with a modified $L(\cdot)$). Moreover, from (H_3) one has

$$[x, f]_+ \leq 0 \tag{2.20}$$

for each $x \in \overline{D(A)}$ with $|x| \geq r, t \in \mathbb{R}^+, v \in C([-\tau, 0], \overline{D(A)})$ and $f \in F_\rho(t, x, v)$.

In the sequel, let $\Sigma_{\phi,\infty}^{F_\rho}$ be the set of all C^0-solutions to the Cauchy problem of the form

$$\begin{cases} u'(t) \in Au(t) + f(t), & t \in \mathbb{R}^+, \\ f(t) \in F_\rho(t, u(t), u_t), & t \in \mathbb{R}^+, \\ u(t) = \phi(t), & t \in [-\tau, 0]. \end{cases} \tag{2.21}$$

Define the multivalued mapping $\Gamma_\rho : Q_r \rightarrow P(\widetilde{C}(\tilde{J}_\tau, X))$ by

$$\Gamma_\rho(\phi) = \Sigma_{\phi,\infty}^{F_\rho} \quad \text{for each } \phi \in Q_r.$$

Then based on the considerations above with Lemma 2.9 we deduce that Γ_ρ is an R_δ-mapping. Moreover, we claim that $\Gamma_\rho(Q_r) \subset \Omega_r$. In fact, if this is not the case, then we can assume that there exist $\phi \in Q_r, u \in \Gamma_\rho(\phi)$ and $t_0 > 0$ such that $u(t_0) > r$. Therefore, it can find $h \in (0, t_0]$ such that $|u(t)| \geq r$ on $[t_0 - h, t_0]$ and $|u(t_0 - h)| = r$, since u is continuous and $|u(0)| \leq r$. We thus use (2.20) to obtain

$$r < |u(t_0)| = |u(t_0 - h)| + \int_{t_0-h}^{t_0} [u(\sigma), f(\sigma)]_+ d\sigma \leq |u(t_0 - h)| = r,$$

where $f \in Sel_F(u)$ such that $u = S_{\phi,\infty}(f)$, which is a contradiction.

Put

$$Q_r^i = \overline{\text{co}}(g(\Omega_r^i)) \quad \text{and} \quad \Omega_r^{i+1} = \overline{\text{co}}(\Gamma_\rho(Q_r^i)), \quad i = 0, 1,$$

where $\Omega_r^0 := \Omega_r$. Then, by (H_4) (i) we have $Q_r^0 \subset Q_r$, which together with the result $\Gamma_\rho(Q_r) \subset \Omega_r$ implies that $\Omega_r^1 \subset \Omega_r^0$. From this it follows that $\Gamma_\rho(Q_r^1) \subset \Omega_r^2 \subset \Omega_r^1$. Therefore, the following composition is well-defined:

$$\Gamma_\rho \circ g : \quad \Omega_r^1 \xrightarrow{g} Q_r^1 \xrightarrow{\Gamma_\rho} \Omega_r^1.$$

We seek for solutions in Ω_r^1. To do this, we show that the multivalued mapping $\Gamma_\rho \circ g$ has a fixed point in Ω_r^1. Observe that Ω_r^1 and Q_r^1 being respectively convex subset of $\widetilde{C}(\tilde{J}_\tau, X)$ and $C([-\tau, 0], X)$, are AR-spaces. Also, (H_4) (i) implies that g is an R_δ-mapping.

Next, we verify that the set $\Omega_r^1|_{[\delta, \infty)}$ is relatively compact in $\widetilde{C}([\delta, \infty), X)$ for each $\delta > 0$. Assume that $\delta > 0$ and $m \in \mathbb{N}(\delta)$. Let $\mathscr{F}_r$ be defined by (2.19) with r instead of $\tilde{r}$. As

$$\Gamma_\rho(Q_r^0)|_{[0,m]} \subset \{u(\cdot, 0, x, f) \in C([0, m], X) : x \in \overline{D(A)} \text{ with } |x| \leq r, \ f \in \mathscr{F}_r|_{[0,m]}\}$$

and Lemma 2.2, we find that $\Gamma_\rho(Q_r^0)|_{[\delta,m]}$ is relatively compact in $C([\delta, m], X)$. Moreover, using Theorem 1.1 we obtain that $\text{co}(\Gamma_\rho(Q_r^0)|_{[\delta,m]})$ is relatively compact and hence $\overline{\text{co}}(\Gamma_\rho(Q_r^0)|_{[\delta,m]})$ is compact. Now, noticing $\overline{\text{co}}(\Gamma_\rho(Q_r^0)|_{[\delta,m]}) \supset \Omega_r^1|_{[\delta,m]}$ it follows that $\Omega_r^1|_{[\delta,m]}$ is relatively compact, which together with the arbitrariness of m and Lemma 1.18 yields that $\Omega_r^1|_{[\delta,\infty)}$ is relatively compact in $\widetilde{C}([\delta, \infty), X)$. Hence $g(\Omega_r^1)$ is relatively compact in $C([-\tau, 0], X)$ by the arbitrariness of $\delta > 0$ and (H_4) (ii). We thus see, again using Theorem 1.1, that Q_r^1 is compact.

Since Γ_ρ is u.s.c. with compact values, we obtain the compactness of $\Gamma_\rho(Q_r^1)$ due to Lemma 1.11. Therefore, we conclude from the result $\Gamma_\rho(g(\Omega_r^1)) \subset \Gamma_\rho(Q_r^1)$ and Theorem 1.16 that there exists a fixed point u of $\Gamma_\rho \circ g$ in Ω_r^1. Moreover, it is readily checked that $u(t) \in \overline{D(A)}$ and $\max\{|u(t)|, |u_t|_0\} \leq r$ for each $t \in \mathbb{R}^+$. From this we see $F_\rho(t, u(t), u_t) = F(t, u(t), u_t)$ for every $t \in \mathbb{R}^+$, which implies that u is a C^0-solution of the nonlocal Cauchy problem (2.2). The proof is complete.

2.5 Applications

As samples of applications, we consider a system of partial differential inclusions defined on right half-line in this section. The topological characterization of solution set to the system considering a time delay condition is discussed. Then, for the system subject to a nonlocal condition, we establish the existence of C^0-solutions in the absence of nonexpansive condition on nonlocal function. These examples do not aim at generality but indicate how our theorems can be applied to concrete problems.

Our examples are inspired directly from the work of [192, Example 5.1] (see also [189]).

Example 2.1 Let Ω be a nonempty bounded open set in $\mathbb{R}^n$ with C^2-boundary $\partial\Omega$, $p \in [2, \infty)$ and $\lambda > 0$. Consider the system of partial differential inclusions in the form

$$
\begin{cases}
\dfrac{\partial u}{\partial t} - \displaystyle\sum_{i=1}^{n} \dfrac{\partial}{\partial \xi_i}\left(\left|\dfrac{\partial u}{\partial \xi_i}\right|^{p-2}\dfrac{\partial u}{\partial \xi_i}\right) + \lambda|u|^{p-2}u \in F(t, \xi, u(t, \xi), u_t(\xi)), \\
\hspace{6cm} (t, \xi) \in \mathbb{R}^+ \times \Omega, \\
-\displaystyle\sum_{i=1}^{n}\left|\dfrac{\partial u}{\partial \xi_i}\right|^{p-2}\dfrac{\partial u}{\partial \xi_i}\cos(\overrightarrow{n}, \overrightarrow{e_i}) \in \beta(u(t, \xi)), \hspace{0.8cm} (t, \xi) \in \mathbb{R}^+ \times \partial\Omega
\end{cases}
\tag{2.22}
$$

subject to a initial history

$$
u(t, \xi) = \phi(t, \xi), \quad (t, \xi) \in [-\tau, 0] \times \Omega,
\tag{2.23}
$$

where the partial derivatives are taken in the sense of distributions over Ω, $\overrightarrow{n}$ is the outward normal of $\partial\Omega$, $\{\overrightarrow{e_1}, \cdots, \overrightarrow{e_n}\}$ is the canonical base in $\mathbb{R}^n$, $\beta : D(\beta) \subset \mathbb{R} \to 2^{\mathbb{R}}$ is a maximal monotone operator with $0 \in D(\beta)$, $0 \in \beta(0)$, and

$$
F(t, \xi, u, v) = [f_1(t, \xi, u, v) + h(\xi), f_2(t, \xi, u, v) + h(\xi)]
$$

is a closed interval for each $(t, \xi, u, v) \in \mathbb{R}^+ \times \Omega \times \mathbb{R} \times C([-\tau, 0], L^2(\Omega, \mathbb{R}))$, in which $h \in L^2(\Omega, \mathbb{R})$ and $f_i : \mathbb{R}^+ \times \Omega \times \mathbb{R} \times C([-\tau, 0], L^2(\Omega, \mathbb{R})) \to \mathbb{R}$ are given functions such that $f_1(t, \xi, u, v) \leq f_2(t, \xi, u, v)$ for each $(t, \xi, u, v) \in \mathbb{R}^+ \times \Omega \times \mathbb{R} \times C([-\tau, 0], L^2(\Omega, \mathbb{R}))$, f_1 is l.s.c., and f_2 is u.s.c.

Here, our objective is to investigate the topological characterization of solution set to the system (2.22)–(2.23).

Take $X = L^2(\Omega, \mathbb{R})$ and denote its norm by $|\cdot|$ and inner product by $(\cdot, \cdot)$. Assume that f_1, f_2 verify the following hypothesis:

[(A_1)] there exist $L_1, L_2 \in L^\infty(\mathbb{R}^+, \mathbb{R}^+)$ such that

$$
|f_i(t, \xi, u, v)| \leq L_1(t)\,(|u| + |v|_0) + L_2(t), \quad i = 1, 2
$$

for each $(t, \xi, u, v) \in \mathbb{R}^+ \times \Omega \times \mathbb{R} \times C([-\tau, 0], X)$.

Before stating our main results, we first present the following lemma, which can be seen from [192, Lemma 5.1].

Lemma 2.10 *Suppose that (A_1) is satisfied. Define a multivalued function $F : \mathbb{R}^+ \times X \times C([-\tau, 0], X) \to P(X)$ as*

$$
F(t, u, v) = \{x \in X : x(\xi) \in [f_1(t, \xi, u(\xi), v) + h(\xi), f_2(t, \xi, u(\xi), v) + h(\xi)] \text{ a.e.}\}
$$

for each $(t, u, v) \in \mathbb{R}^+ \times X \times C([-\tau, 0], X)$. *Then* F *has nonempty, convex and closed values,* $F(\cdot, u, v)$ *has a strongly measurable selection for every* $(u, v) \in X \times C([-\tau, 0], X)$, *and* $F(t, \cdot, \cdot)$ *is weakly u.s.c. for each* $t \in \mathbb{R}^+$. *Moreover,*

$$|F(t, u, v)| \leq \max\{L_1(t), \operatorname{mes}^{\frac{1}{2}}(\Omega)L_1(t), \operatorname{mes}^{\frac{1}{2}}(\Omega)L_2(t) + |h|\}(1 + |u| + |v|_0)$$

for a.e. $t \in \mathbb{R}^+$, *each* $u \in X$ *and* $v \in C([-\tau, 0], X)$.

Theorem 2.5 *Under the hypothesis* (A_1), *the set of all* C^0-*solutions to the system* (2.22)–(2.23) *is an* R_δ-*set for each* $\phi \in C([-\tau, 0], X)$.

Proof Let $A : D(A) \subset X \to X$ be defined by

$$Au = \sum_{i=1}^{n} \frac{\partial}{\partial \xi_i}\left(\left|\frac{\partial u}{\partial \xi_i}\right|^{p-2}\frac{\partial u}{\partial \xi_i}\right),$$

$$D(A) = \left\{u \in W^{1,p}(\Omega) : \sum_{i=1}^{n} \frac{\partial}{\partial \xi_i}\left(\left|\frac{\partial u}{\partial \xi_i}\right|^{p-2}\frac{\partial u}{\partial \xi_i}\right) \in X, \text{ and}\right.$$

$$\left.-\sum_{i=1}^{n}\left|\frac{\partial u}{\partial \xi_i}\right|^{p-2}\frac{\partial u}{\partial \xi_i}\cos(\overrightarrow{n}, \overrightarrow{e_i}) \in \beta(u(\xi)) \text{ a.e. } \xi \in \partial\Omega\right\}.$$

From [189, Example 1.5.4] and [49, Théorème 1.10, p.43] we see that A is an m-dissipative operator with $0 \in A0$ and $\overline{D(A)} = X$. In addition, as in [189, Example 2.2.4 and Corollary 2.3.2], A generates a compact semigroup of nonexpansive mappings on X, which implies that the hypothesis (H_0) holds. Also, by Lemma 2.10 one finds that F verifies conditions (H_1) and (H_2) with $J = \mathbb{R}^+$ and $L(t) = \max\{L_1(t), \operatorname{mes}^{\frac{1}{2}}(\Omega)L_1(t), \operatorname{mes}^{\frac{1}{2}}(\Omega)L_2(t) + |h|\}$. Therefore, applying Theorem 2.3 gives the result as desired.

Next, we consider the system (2.22) equipped with a nonlocal condition as follows:

$$u(t, \xi) = \int_{\tau}^{\infty} \mathcal{N}(u(t + \theta, \xi))d\mu(\theta), \quad (t, \xi) \in [-\tau, 0] \times \Omega, \tag{2.24}$$

where μ is a σ-finite and complete measure on $[\tau, \infty)$ such that

$$\mu([\tau, \infty)) = 1 \text{ and } \lim_{s \to \tau^+} \mu([\tau, s]) = 0.$$

We assume that $\mathcal{N} : \mathbb{R} \to \mathbb{R}$ is a continuous function satisfying either for some $C_1, C_2 \geq 0$ and $b \in [0, 1)$,

$$|\mathcal{N}(y)| \leq C_1 + C_2|y|^b \text{ for all } y \in \mathbb{R}, \tag{2.25}$$

or

$$|\mathcal{N}(y)| \le |y| \quad \text{for all } y \in \mathbb{R}. \tag{2.26}$$

It can define a Nemytskiĭ operator $\mathcal{N}$ from X into itself by $\mathcal{N}(x)(\xi) = \mathcal{N}(x(\xi))$ for each $x \in X$. Moreover, one finds that $\mathcal{N}$ is continuous on X.

Remark 2.6 If (2.25) is satisfied, then a direct computation upon Hölder's inequality yields that for each $x \in X$,

$$|\mathcal{N}(x)| \le C_1 \mathrm{mes}^{\frac{1}{2}}(\Omega) + C_2 \mathrm{mes}^{\frac{1-b}{2}}(\Omega)|x|^b.$$

Write, for each $l > 0$,

$$\Phi(l) = \max\{C_1 \mathrm{mes}^{\frac{1}{2}}(\Omega) + C_2 \mathrm{mes}^{\frac{1-b}{2}}(\Omega)l^b, l\}.$$

Theorem 2.6 *Let (A_1) and (2.25) or (2.26) hold. Suppose further that the following hypothesis is satisfied.*
(A_2) There exists $c > 0$ such that for every $(t, \xi, u, v) \in \mathbb{R}^+ \times \Omega \times \mathbb{R} \times C([-\tau, 0], X)$,

$$\max\{uf_i(t, \xi, u, v) : i = 1, 2\} \le -cu^2.$$

Then the system (2.22) and (2.24) has at least one C^0-solution.

Proof Let $r > 0$ be such that $r \ge c^{-1}|h|$ and $\Phi(r) \le r$. Take $(t, u, v) \in \mathbb{R}^+ \times X \times C([-\tau, 0], X)$ with $|u| = r$ and $f \in F(t, u, v)$. Noticing (A_2) and using an argument similar to that in [192, Theorem 5.1] we obtain

$$[u, f]_+ \le |u|^{-1} \int_\Omega \left(-c|u(\xi)|^2 + |u(\xi)h(\xi)|\right) d\xi \le -cr + |h| \le 0,$$

which yields that (H_3) remains true.

Next, let us define a mapping $g : \widetilde{C}_b(\tilde{J}_\tau, X) \to C([-\tau, 0], X)$ as

$$g(u)(t) = \int_\tau^\infty \mathcal{N}(u(t + \theta))d\mu(\theta), \quad u \in \widetilde{C}_b(\tilde{J}_\tau, X), \ t \in [-\tau, 0].$$

Taking $u \in \widetilde{C}_b(\tilde{J}_\tau, X)$, we have, in view of $\mu[\tau, \infty) = 1$, that

$$|g(u)(\cdot)|_0 \le \sup_{t \in [-\tau, 0]} \left(\int_\tau^\infty |\mathcal{N}(u(t + \theta))|^2 d\mu(\theta)\right)^{\frac{1}{2}}$$
$$\le \Phi\left(\sup_{t \in \mathbb{R}^+} |u(t)|\right),$$

which implies that $|g(u)|_0 \le r$ for all $u \in \Omega_r$. Also, by means of Lebesgue's dominated convergence theorem it is not difficult to see that $g(u)(\cdot)$ is continuous on $[-\tau, 0]$.

We process to show that g is continuous on Ω_r. Given $\varepsilon > 0$. Let $\{u_n\}$ be a sequence in Ω_r such that u_n converges to $u \in \widetilde{C}(\widetilde{J}_\tau, X)$. Since μ is σ-finite, we can choose $m' \in \mathbb{N} \setminus \{0\}$ such that $\mu[m', \infty) \le \frac{\varepsilon}{4\Phi(r)}$. Therefore, we have that for each $t \in [-\tau, 0]$,

$$|g(u_n)(t) - g(u)(t)| \le \int_\tau^{m'} |\mathcal{N}(u_n(t + \theta)) - \mathcal{N}(u(t + \theta))| d\mu(\theta) + \frac{\varepsilon}{2}. \quad (2.27)$$

On the other hand, noticing that

$$u_n|_{[-\tau, m']} \to u|_{[-\tau, m']} \text{ in } C([-\tau, m'], X) \text{ and } \mathcal{N} \text{ is continuous on } X,$$

we conclude that $\mathcal{N}$ is uniformly continuous on $\overline{\{u_n(t) : n \ge 1, t \in [-\tau, m']\}}$, which implies that

$$\mathcal{N}(u_n|_{[0,m']}) \to \mathcal{N}(u|_{[0,m']}) \text{ in } C([0, m'], X)$$

as $n \to \infty$. So, there exists $N > 0$ such that for all $n \ge N$,

$$\sup_{t \in [-\tau, 0]} \int_\tau^{m'} |\mathcal{N}(u_n(t + \theta)) - \mathcal{N}(u(t + \theta))| d\mu(\theta) \le \frac{\varepsilon}{2}.$$

This together with (2.27) proves the desired result.

Assume that $\mathcal{U} \subset \Omega_r$ and $\mathcal{U}|_{[\delta,\infty)}$ is relatively compact in $\widetilde{C}([\delta, \infty), X)$ for each $\delta > 0$. To prove that $g(\mathcal{U})$ is relatively compact in $C([-\tau, 0], X)$, it suffices to show that $g(\mathcal{U})$ is totally bounded. Given $\varepsilon > 0$, it follows that there exists $\delta_0 > 0$ such that $\mu([\tau, \tau + \delta_0]) \le \frac{\varepsilon}{2\Phi(r)}$.

Next, to construct a finite ε-net of $g(\mathcal{U})$, we need to define an operator

$$g_{\delta_0} : \widetilde{C}_b([\delta_0, \infty), X) \to C([-\tau, 0], X)$$

as

$$g_{\delta_0}(u)(t) = \int_{\tau+\delta_0}^\infty \mathcal{N}(u(t + \theta)) d\mu(\theta), \quad u \in \widetilde{C}_b([\delta, \infty), X), \ t \in [-\tau, 0].$$

The same idea as above can be used to prove that g_{δ_0} is continuous on the set

$$\{u \in \widetilde{C}_b([\delta_0, \infty), X) : |u(t)| \le r \text{ for all } t \in [\delta_0, \infty)\}.$$

Since $\mathcal{U}|_{[\delta_0,\infty)}$ is relatively compact, we obtain that $g_{\delta_0}(\mathcal{U}|_{[\delta_0,\infty)})$ is relatively compact in $C([-\tau, 0], X)$ and thus it admits a finite $\frac{\varepsilon}{2}$-net, denoted by $\mathcal{V}_\varepsilon = \{v_1, \cdots, v_k\}$.

We claim that $\mathcal{V}_\varepsilon$ is a finite ε-net of $g(\mathcal{U})$. Indeed, given $v \in g(\mathcal{U})$, we have that there exists $u \in \mathcal{U}$ such that $v = g(u)$. We choose $v_i \in \mathcal{V}_\varepsilon$ such that

$$|v_i - g_{\delta_0}(u|_{[\delta_0,\infty)})|_0 \leq \frac{\varepsilon}{2}. \tag{2.28}$$

Here we are using the fact $g_{\delta_0}(u|_{[\delta_0,\infty)}) \in g_{\delta_0}(\mathscr{U}|_{[\delta_0,\infty)})$. Also, a direct computation gives

$$|g(u) - g_{\delta_0}(u|_{[\delta_0,\infty)})|_0 \leq \sup_{t \in [-\tau,0]} \int_\tau^{\tau+\delta_0} |\mathscr{N}(u(t+\theta))| d\mu(\theta)$$

$$\leq \frac{\varepsilon}{2},$$

which, together with (2.28), implies that $|v - v_i|_0 \leq \varepsilon$, as desired. Therefore, the desired result follows from Theorem 2.4.

At the end of this chapter, we leave two problems for further research.

(1) Is Theorem 2.4 true for the case when either A is a linear operator generating a C_0-semigroup or A is replaced with a family of linear operators generating an evolution system? More specially, is it true for a linear operator A whose resolvent satisfies the estimate of growth $-\gamma$ $(-1 < \gamma < 0)$ in a sector of the complex plane? Let us note that such operator, generating a semigroup of growth $1 + \gamma$, is called an almost sectorial operator (see e.g., Wang et al. [195]) for more details.
(2) Is Theorem 2.3 true under the weaker condition that the semigroup generated by A is only equicontinuous?

We believe it is possible to find some interesting positive answers.

Remark 2.7 It is noted that if A is a linear operator generating a C_0-semigroup, A is replaced with a family of linear operators generating an evolution system, or A is an almost sectorial operator, then in treating the nonlocal Cauchy problem (2.2), it is inappropriate to impose the invariance condition (H_3) on F.

Chapter 3
Evolution Inclusions with Hille–Yosida Operator

Abstract This chapter deals with a parabolic differential inclusion of evolution type involving a nondensely defined closed linear operator satisfying the Hille–Yosida condition and source term of multivalued type in Banach space. The topological structure of the solution set is investigated in the cases that the semigroup is non-compact. It is shown that the solution set is nonempty, compact and an R_δ-set. It is proved on compact intervals and then, using the inverse limit method, obtained on noncompact intervals. Secondly, the existing solvability and the existence of a compact global attractor for the m-semiflow generated by the system are studied by using measures of noncompactness . As samples of applications, we apply the abstract results to some classes of partial differential inclusions.

3.1 Introduction

Consider the following parabolic evolution inclusion on compact interval

$$\begin{cases} u'(t) \in Au(t) + F(t, u(t)), & t \in [0, b], \\ u(0) = u_0, \end{cases} \tag{3.1}$$

and the corresponding inclusion on noncompact interval

$$\begin{cases} u'(t) \in Au(t) + F(t, u(t)), & t \in \mathbb{R}^+, \\ u(0) = u_0, \end{cases} \tag{3.2}$$

where the state $u(\cdot)$ takes values in a Banach space X with the norm $|\cdot|$, $\mathcal{L}(X)$ stands for the space of all linear bounded operators on Banach space X, with the norm $\|\cdot\|_{\mathcal{L}(X)}$. F is a multivalued function defined on a subset of $\mathbb{R}^+ \times X$, $A : D(A) \subset X \to X$ is a nondensely defined closed linear operator satisfying the Hille–Yosida condition, i.e., there exist two constant $\omega \in \mathbb{R}$ and $M > 0$ such that $(\omega, +\infty) \subset \rho(A)$ and

$$\|(\lambda I - A)^{-k}\|_{\mathcal{L}(X)} \le \frac{M}{(\lambda - \omega)^k}, \quad \lambda > \omega, \ k \ge 1.$$

© Springer Nature Singapore Pte Ltd. 2017
Y. Zhou et al., *Topological Structure of the Solution Set for Evolution Inclusions*,
Developments in Mathematics 51, https://doi.org/10.1007/978-981-10-6656-6_3

The study of (3.2) is justified by a partial differential inclusion of parabolic type

$$
\begin{cases}
\dfrac{\partial}{\partial t} u(t, \xi) \in \Delta u(t, \xi) + F(t, \xi, u(t, \xi)), & t \in \mathbb{R}^+,\ \xi \in \Omega, \\
u(t, \xi) = 0, & t \in \mathbb{R}^+,\ \xi \in \partial\Omega, \\
u(0, \xi) = u_0(\xi), & \xi \in \Omega,
\end{cases}
$$

where Ω is a bounded open set in $\mathbb{R}^n$ ($n \geq 1$) with regular boundary $\partial\Omega$, $u_0 \in L^2(\Omega, \mathbb{R})$ and $F : \mathbb{R}^+ \times \Omega \times C(\mathbb{R}^+ \times \Omega, L^2(\Omega, \mathbb{R})) \to P(\mathbb{R})$ is upper semicontinuous with closed convex values.

As we known, in various problems formed by semilinear differential equations, the nonlinear perturbations may take values lying outside $D(A)$ and one has to consider these problems in a larger space. This leads to models with A being non-densely defined. Under the assumption that A just satisfies the Hille–Yosida condition, there has been many works devoted to solvability and stability for problems like (3.1) and (3.2).

In Sect. 3.3, we consider the following problem

$$
\begin{cases}
u'(t) \in Au(t) + F(u(t), u_t), & t \in \mathbb{R}^+, \\
u(s) = \varphi(s), & s \in [-h, 0],
\end{cases}
\tag{3.3}
$$

where u_t stands for the history of the state function up to time t, i.e., $u_t(s) = u(t + s)$ for $s \in [-h, 0]$, F is a multivalued map defined on a subset of $X \times C([-h, 0], X)$. In this model, $A : D(A) \subset X \to X$ is a linear operator satisfying the Hille–Yosida condition.

It is known that, in the theory of infinite dimensional dynamical systems, the concept of global attractors has been proved to be a powerful tool to study the asymptotics of solutions to various differential systems. We introduce the monographs [73, 183] for a complete reference to this subject.

In dealing with behavior of multivalued dynamical systems associated to differential equations without uniqueness or differential inclusions, some famous theories such as generalized semiflows introduced by Ball [26, 27], multivalued semiflows defined by Melnik and Valero [145] have been used. A comparison of these two theories was given in [109]. Thanks to the framework of Melnik and Valero, there are many works devoted to the investigation of asymptotics for various classes of partial differential equations without uniqueness (see e.g., [15, 187]. We also refer to the theory of trajectory attractors developed by Chepyzhov and Vishik [72] which is an effective tool to study the long-time behavior of solutions of partial differential equations for which the uniqueness is unavailable. In all frameworks, an essential step in formulating global attractors is to verify the asymptotical compactness of the corresponding semiflows. Usually, this property takes place if the semigroup governed by principal parts is compact. However, for partial differential equations in unbounded domains the latter requirement is unrealistic. To overcome this difficulty, we make use of an MNC approach. Let $G(t, \cdot)$ be the m-semiflow generated by (3.3), i.e.,

$$G(t, \varphi) = \{u_t : u(\cdot, \varphi) \text{ is an integral solution of (3.3), }\}$$

and $G_T = G(T, \cdot)$ for $T > h$, the translation operator. We show that G_T is condensing. This property enables us to prove the asymptotical compactness of m-semiflow G.

This chapter is organized as follows. Section 3.2 is devoted to proving that the solution set for inclusion (3.1) on compact interval is a nonempty compact R_δ-set, then proceed to study the R_δ-structure for that of inclusion (3.2) on noncompact interval by the inverse limit method. Section 3.3 provides the existence of integral solutions for system (3.3), and proves that the m-semiflow generated by (3.3) admits a compact global attractor.

The results of Sect. 3.2 are new and due to Zhou and Peng. The contents in Sect. 3.3 are taken from Ke and Lan [132].

3.2 Topological Structure of Solution Set

In this section, we consider the solution set of inclusion (3.1). In the following study, we introduce the following hypothesis:

(H_A) the linear operator $A : D(A) \subset X \to X$ satisfies the Hille–Yosida condition, i.e., there exist two constant $\omega \in \mathbb{R}$ and $M > 0$ such that $(\omega, +\infty) \subset \rho(A)$ and

$$\|(\lambda I - A)^{-k}\|_{\mathcal{L}(X)} \leq \frac{M}{(\lambda - \omega)^k}$$

for all $\lambda > \omega, k \geq 1$.

It is known that (see [134]) if $\{T(t)\}_{t \geq 0}$ is an integrated semigroup generated by a Hille–Yosida operator A, then $t \to T(t)u$ is differentiable for each $u \in \overline{D(A)}$ and $\{T'(t)\}_{t \geq 0}$ is a C_0-semigroup on $\overline{D(A)}$ generated by the part A_0 of A, which is defined by

$$A_0 u = Au \text{ on } D(A_0) = \{u \in D(A) : Au \in X_0\}.$$

(H_T) The C_0-semigroup $\{T'(t)\}_{t \geq 0}$ is norm-continuous, i.e., $t \to T'(t)$ is continuous for $t > 0$.

Let $X_0 = \overline{D(A)}$. Consider Cauchy problem

$$\begin{cases} u'(t) = Au(t) + f(t), & t \in J = (0, b], \\ u(0) = u_0, \end{cases} \tag{3.4}$$

where $f \in C([0, b], X)$ and $u_0 \in X_0$ are given.

Theorem 3.1 ([134]) *Let $f \in C([0, b], X)$ for $u_0 \in X_0$, there exists a unique continuous function $u : [0, b] \to X$ of Cauchy problem (3.4) such that*

(i) $\displaystyle\int_0^t u(s)ds \in D(A)$ *for* $t \in [0, b]$;

(ii) $u(t) = u_0 + A \displaystyle\int_0^t u(s)ds + \int_0^t f(s)ds.$

Moreover, u satisfies the variation of constants formula

$$u(t) = T'(t)u_0 + \frac{d}{dt}\int_0^t T(t-s)f(s)ds \ \ for \ t \in [0, b]. \qquad (3.5)$$

Let $\mathcal{J}_\lambda = \lambda(\lambda I - A)^{-1}$, then for all $u \in X_0$, $\lim_{\lambda\to+\infty}\mathcal{J}_\lambda u = u$ (see [134]). Also from Hille–Yosida condition, it is easy to see that $\lim_{\lambda\to+\infty}|\mathcal{J}_\lambda u| \le M|u|$. Since

$$\|\mathcal{J}_\lambda\|_{\mathcal{L}(X)} = \|\lambda(\lambda I - A)^{-1}\|_{\mathcal{L}(X)} \le \frac{M\lambda}{\lambda - \omega},$$

thus $\lim_{\lambda\to+\infty}\|\mathcal{J}_\lambda\|_{\mathcal{L}(X)} \le M$. Also if u is given by (3.5), then

$$u(t) = T'(t)u_0 + \lim_{\lambda\to+\infty}\int_0^t T'(t-s)\mathcal{J}_\lambda f(s)ds$$

for $t \in [0, b]$.

We assume that the multivalued nonlinearity $F : \mathbb{R}^+ \times X \to P_{cl,cv}(X)$ satisfies:
(H_1) $F(t, \cdot)$ is u.s.c. for a.e. $t \in \mathbb{R}^+$, and the multimap $F(\cdot, x)$ has a strongly measurable selection for every $x \in X$;
(H_2) there exists a function $\alpha \in L^1_{loc}(\mathbb{R}^+, \mathbb{R}^+)$ such that

$$|F(t, x)| \le \alpha(t)(1 + |x|) \ \ \text{for a.e. } t \in \mathbb{R}^+ \text{ and } x \in X.$$

(H_3) there exists a function $k \in L^1_{loc}(\mathbb{R}^+, \mathbb{R}^+)$ such that

$$\beta(F(t, D)) \le k(t)\beta(D)$$

for every bounded set D. Here β denotes Hausdorff MNC.

3.2.1 Existence of Integral Solution

Given $u \in C([0, b], X)$, let us denote

$$Sel_F^b(u) = \{f \in L^1([0, b], X) : f(t) \in F(t, u(t)) \ \text{for a.e. } t \in [0, b]\}.$$

Definition 3.1 A continuous function $u : [0, b] \to X$ is said to be an integral solution of differential inclusion (3.1) if

(i) $\int_0^t u(s)ds \in D(A)$ for $t \in [0, b]$;

(ii) $u(0) = u_0$ and there exists $f(t) \in Sel_F^b(u)(t)$ satisfying the following integral equation

$$u(t) = T'(t)u_0 + \frac{d}{dt}\int_0^t T(t-s)f(s)ds. \qquad (3.6)$$

We notice also that, if u satisfies (3.6), then

$$u(t) = T'(t)u_0 + \lim_{\lambda \to +\infty}\int_0^t T'(t-s)\mathcal{J}_\lambda f(s)ds$$

for $t \in [0, b]$.

Remark 3.1 For any $u \in C([0, b], X_0)$, now define a solution multioperator $\mathscr{F}^b$: $C([0, b], X_0) \to P(C([0, b], X_0))$ as follows:

$$\mathscr{F}^b(u)(t) = \{T'(t)u_0 + \Gamma(f)(t) : f \in Sel_F^b(u)\},$$

where

$$\Gamma(f)(t) = \lim_{\lambda \to +\infty}\int_0^t T'(t-s)\mathcal{J}_\lambda f(s)ds.$$

It is easy to verify that the fixed points of the multioperator $\mathscr{F}^b$ are integral solutions of inclusion (3.1).

Lemma 3.1 *The operator Γ have the following properties:*

(i) *there exists a constant $c_0 > 0$, such that*

$$|\Gamma(f)(t) - \Gamma(g)(t)| \le c_0 \int_0^t |f(s) - g(s)|ds, \quad t \in [0, b]$$

for every $f, g \in L^1([0, b], X)$;

(ii) *for each compact set $K \subset X$ and sequence $\{f_n\} \subset L^1([0, b], X)$ such that $\{f_n(t)\} \subset K$ for a.e. $t \in [0, b]$, the weak convergence $f_n \rightharpoonup f_0$ implies the convergence $\Gamma(f_n) \to \Gamma(f_0)$.*

Proof (i) By calculation, we have

$$|\Gamma(f)(t) - \Gamma(g)(t)| \le \left| \lim_{\lambda \to +\infty}\int_0^t T'(t-s)\mathcal{J}_\lambda\big(f(s) - g(s)\big)ds \right|$$

$$\le MM_1 \int_0^t |f(s) - g(s)|ds$$

$$\le c_0 \int_0^t |f(s) - g(s)|ds,$$

where $c_0 = MM_1$.

(ii) Notice that $\mathcal{J}_\lambda$ is a bounded linear operator and every compact set $K \subset X$. Therefore the set $Q \subset X$ defined by

$$Q = \bigcup_{s \in [0,t]} T'(t-s)\mathcal{J}_\lambda K$$

is relatively compact. For every sequence $\{f_n\} \subset L^1([0,b], X)$ such that $\{f_n(t)\} \subset K$ for a.e. $t \in [0, b]$. We have

$$\{\Gamma(f_n)(t)\}_{n=1}^\infty \subset \lim_{\lambda \to +\infty} t Q$$

and hence, the sequence $\{\Gamma(f_n)(t)\} \subset X$ is relatively compact for every $t \in [0, b]$.

On the other hand, we have

$$|\Gamma(f_n)(t_2) - \Gamma(f_n)(t_1)| \le \left| \lim_{\lambda \to +\infty} \int_{t_1}^{t_2} T'(t-s)\mathcal{J}_\lambda f_n(s)ds \right|$$
$$+ \left| \lim_{\lambda \to +\infty} \int_0^{t_1} \left(T'(t_2 - s) - T'(t_1 - s) \right) \mathcal{J}_\lambda f_n(s)ds \right|.$$

Since $T'(t)$ is strongly continuous and $\{f_n(t)\} \subset K$ for a.e. $t \in [0, b]$, the right-hand side of this inequality tends to zero as $t_2 \to t_1$ uniformly with respect to n. Hence $\{\Gamma(f_n)\}$ is an equicontinuous set. Thus from Lemma 1.3, we obtain that the sequence $\{\Gamma(f_n)\} \subset C([0, b], X_0)$ is relatively compact.

Property (i) ensures that $\Gamma : L^1([0, b], X) \to C([0, b], X_0)$ is a bounded linear operator. Then it is continuous with respect to the topology of weak sequential convergence, that is the weak convergence $f_n \rightharpoonup f_0$ ensuring $\Gamma(f_n) \rightharpoonup \Gamma(f_0)$. Taking into account that $\{\Gamma(f_n)\}$ is relatively compact, we arrive at the conclusion that $\Gamma(f_n) \to \Gamma(f_0)$ in $C([0, b], X_0)$.

Similar to the proof of Theorem 4.2.2 in [130], we have the following result.

Lemma 3.2 *Let* $\{f_n\}$ *be integrably bounded, and*

$$\beta(f_n(t)) \le q(t)$$

for a.e. $t \in [0, b]$, *where* $q \in L^1([0, b], \mathbb{R}^+)$. *Then we have*

$$\beta(\{\Gamma(f_n)(t)\}_{n=1}^\infty) \le 2c_0 \int_0^t q(s)ds$$

for all $t \in [0, b]$, *where* $c_0 > 0$ *is the constant in Lemma 3.1 (i).*

Theorem 3.2 *Let conditions* (H_A), (H_T), (H_1), (H_2) *and* (H_3) *be satisfied. Then the solution set of inclusion (3.1) with initial value* $u_0 \in X_0$ *is a nonempty compact set in* $C([0, b], X_0)$.

Proof Set

$$\mathcal{M}_0 = \{u \in C([0, b], X_0) : |u(t)| \le \psi(t), \ t \in [0, b]\},$$

where $\psi(t)$ is the solution of the integral equation

$$\psi'(t) = MM_1\alpha(t)(1 + \psi(t)) \ a.e. \ on \ [0, b], \ \psi(0) = |u_0|.$$

It is clear that $\mathcal{M}_0$ is a closed and convex subset of $C([0, b], X_0)$. We first show that $\mathscr{F}^b(\mathcal{M}_0) \subset \mathcal{M}_0$. Indeed, taking $u \in \mathcal{M}_0$ and $v \in \mathscr{F}^b(u)$, we have

$$
\begin{aligned}
|v(t)| &\le |T'(t)u_0| + \left| \lim_{\lambda \to +\infty} \int_0^t T'(t-s)\mathcal{J}_\lambda f(s)ds \right| \\
&\le M_1|u_0| + MM_1 \int_0^t \alpha(s)(1 + |u(s)|)ds \\
&\le M_1|u_0| + MM_1 \int_0^t \alpha(s)(1 + |u(s)|)ds \\
&\le \psi(t).
\end{aligned}
$$

Thus $v \in \mathcal{M}_0$. Set $\widetilde{M} = \overline{co}\mathscr{F}^b(\mathcal{M}_0)$, it is clear that $\widetilde{M}$ is a closed, bounded and convex set. Moreover, $\mathscr{F}^b(\widetilde{M}) \subset \widetilde{M}$.

Claim 1 The multioperator $\mathscr{F}^b$ has a closed graph with compact values. Let $u_n \subset \mathcal{M}_0$ with $u_n \to u$ and $v_n \in \mathscr{F}^b(u_n)$ with $v_n \to v$. We shall prove that $v \in \mathscr{F}^b(u)$. By the definition of $\mathscr{F}^b$, there exist $f_n \in Sel_F^b(u_n)$ such that

$$v_n(t) = T'(t)u_0 + \Gamma(f_n)(t).$$

We need to prove that there exists $f \in Sel_F^b(u)$ such that for a.e. $t \in [0, b]$,

$$v(t) = T'(t)u_0 + \Gamma(f)(t).$$

We see that $\{f_n\}$ is integrably bounded by (H_2), and the following inequality holds by (H_3),

$$\beta(\{f_n(t)\}) \le k(t)\beta(\{u_n(t)\}).$$

For the sequence $\{u_n\}$ converges in $C([0, b], X_0)$, thus $\beta(\{f_n(t)\}) = 0$ for a.e. $t \in [0, b]$, then $\{f_n\}$ is a semicompact sequence.

In view of Lemma 1.25, one obtains that $\{f_n\}$ is weakly compact in $L^1([0, b], X)$, so we may assume, without loss of generality, $f_n \rightharpoonup f$ in $L^1([0, b], X)$. By Lemma 3.1 (ii), we have that

$$v_n(t) = T'(t)u_0 + \Gamma(f_n)(t) \to T'(t)u_0 + \Gamma(f)(t) = v(t) \ \text{for a.e. } t \in [0, b].$$

It remains to show that, for $u \in \mathcal{M}_0$ and $\{f_n\}$ chosen in $Sel_F^b(u)$, the sequence $\{\Gamma(f_n)\}$ is relatively compact in $C([0, b], X_0)$. Conditions (H_2) and (H_3) imply that $\{f_n\}$ is semicompact. Using Lemma 3.1 (ii), we obtain that $\{\Gamma(f_n)\}$ is relatively compact in $C([0, b], X_0)$. Thus $\mathscr{F}^b(u)$ is relatively compact in $C([0, b], X_0)$, together with the closeness of $\mathscr{F}^b$, then $\mathscr{F}^b(u)$ has compact values.

Claim 2 The multioperator $\mathscr{F}^b$ is u.s.c. In view of Lemma 1.9, it suffices to check that $\mathscr{F}^b$ is a quasicompact multimap. Let Q be a compact set. We prove that $\mathscr{F}^b(Q)$ is a relatively compact subset of $C([0, b], X_0)$. Assume that $\{v_n\} \subset \mathscr{F}^b(Q)$. Then

$$v_n(t) = T'(t)u_0 + \Gamma(f_n)(t),$$

where $\{f_n\} \in Sel_F^b(u_n)$, for a certain sequence $\{u_n\} \subset Q$. Conditions (H_2) and (H_3) yield the fact that $\{f_n\}$ is semicompact and then it is a weakly compact sequence in $L^1([0, b], X)$. Similar arguments as in the previous proof of the closeness imply that $\{v_n\}$ is relatively compact in $C([0, b], X_0)$. Thus, $\{v_n\}$ converges in $C([0, b], X_0)$, so the multioperator $\mathscr{F}^b$ is u.s.c.

Claim 3 The multioperator $\mathscr{F}^b$ is a condensing multioperator. We first need an MNC constructed suitably for our problem. For a bounded subset $\Omega \subset \mathcal{M}_0$, let $\mathrm{mod}_C(\Omega)$ be the modulus of equicontinuity of the set of functions Ω given by

$$\mathrm{mod}_C(\Omega) = \lim_{\delta \to 0} \sup_{u \in \Omega} \max_{|t_2 - t_1| < \delta} |u(t_2) - u(t_1)|.$$

Given Hausdorff MNC β, let χ be the real MNC defined on bounded D be a subset of $C([0, b], X_0)$ by

$$\chi(D) = \sup_{t \in [0,b]} e^{-Lt} \beta(D(t)).$$

Here, the constant L is chosen such that

$$l := \sup_{t \in [0,b]} \left(2c_0 \int_0^t e^{-L(t-s)} k(s) ds \right) < 1,$$

where $k(t)$ is the function from the condition (H_3).

Consider the function $\nu(\Omega) = \max_{D \in \Delta(\Omega)} \left(\chi(D), \mathrm{mod}_C(D) \right)$ in the space of $C([0, b], X_0)$, where $\Delta(\Omega)$ is the collection of all countable subsets of Ω.

To show that $\mathscr{F}^b$ is ν-condensing, let $\Omega \subset \mathcal{M}_0$ be a bounded set in $\mathcal{M}_0$ such that

$$\nu(\Omega) \le \nu(\mathscr{F}^b(\Omega)). \tag{3.7}$$

We show that Ω is relatively compact. Let $\nu(\mathscr{F}^b(\Omega))$ be achieved on a sequence $\{v_n\} \subset \mathscr{F}^b(\Omega)$, i.e.,

$$\nu(\{v_n\}) = \left(\chi(\{v_n\}), \mathrm{mod}_C(\{v_n\}) \right).$$

Then

$$v_n(t) = T'(t)u_0 + \Gamma(f_n)(t), \quad f_n \in Sel_F^b(u_n),$$

where $\{u_n\} \subset \Omega$. Now inequality (3.7) implies

$$\chi(\{v_n\}) \geq \chi(\{u_n\}). \tag{3.8}$$

It follows from (H_3) that $\beta(\{f_n(t)\}) \leq k(t)\beta(u_n(t))$ for $t \in [0, b]$. Then

$$\beta(\{f_n(t)\}) \leq e^{Lt}k(t)\left(\sup_{s\in[0,t]} e^{-Ls}\beta(u_n(t)) \right) \leq e^{Lt}k(t)\chi(\{u_n\}).$$

Now the application of Lemma 3.2 for Γ yields

$$e^{-Lt}\beta(\{\Gamma(f_n)(t)\}) \leq 2c_0 e^{-Lt}\int_0^t e^{Ls}k(s)ds \cdot \chi(\{u_n\})$$

$$\leq 2c_0\int_0^t e^{-L(t-s)}k(s)ds \cdot \chi(\{u_n\})$$

for any $t \in [0, b]$. Putting this relation together with (3.8), we obtain

$$\chi(\{u_n\}) \leq \chi(\{v_n\}) = \sup_{t\in[0,b]} e^{-Lt}\beta(v_n(t)) \leq l\chi(\{u_n\}).$$

Therefore $\chi(\{u_n\}) = 0$. This implies $\beta(u_n(t)) = 0$ for $t \in [0, b]$.

Using (H_2) and (H_3) again, one gets that $\{f_n\}$ is a semicompact sequence. Then, Lemma 3.1 (ii) ensures that $\{\Gamma(f_n)\}$ is relatively compact in $C([0, b], X_0)$. This yields that $\{v_n\}$ is relatively compact in $C([0, b], X_0)$. Hence $\text{mod}_C(\{v_n\}) = 0$. Finally, $\nu(\{v_n\}) = 0$, and so the map $\mathscr{F}^b$ is ν-condensing.

From Theorem 1.12, we deduce that the fixed point set of Fix $\mathscr{F}^b$ is a nonempty compact set.

3.2.2 Structure of Solution Set

Lemma 3.3 *Under the assumptions in Theorem 3.2, there exists a nonempty compact convex subset $\mathcal{M} \subseteq C([0, b], X_0)$ such that*

(i) $u(0) = u_0$ for all $u \in \mathcal{M}$;

(ii) $T'(t)u_0 + \lim\limits_{\lambda\to+\infty} \int_0^t T'(t-s)\mathcal{J}_\lambda \overline{co} F(s, \mathcal{M}(s))ds \subset \mathcal{M}(t)$ for $t \in [0, b]$,
where $\mathcal{M}(t) = \{u(t) : u \in \mathcal{M}\}$.

Proof Let us construct the decreasing sequence of closed convex sets $\{\mathcal{M}_n\} \subset C([0, b], X_0)$ by the following inductive process. Let

$$\mathcal{M}_0 = \{u \in C([0, b], X_0) : u(0) = u_0, \; |u(t)| \leq N, \; t \in [0, b]\},$$

where $N = \left(M_1|u_0| + M M_1 \|\alpha\|_{L([0,b],\mathbb{R}^+)}\right) \exp(M M_1 \|\alpha\|_{L([0,b],\mathbb{R}^+)})$.
Then $\mathcal{M}_n = \overline{\mathcal{N}_n}$, $n \geq 1$, where $\mathcal{N}_n \subset C([0, b], X_0)$ and

$$\mathcal{N}_n = \left\{v \in C([0, b], X_0) : v(t) = T'(t)u_0 + \lim_{\lambda \to +\infty} \int_0^t T'(t - s)\mathcal{J}_\lambda f(s)ds, \right.$$

$$\left. f \in Sel^b_{\overline{co}F}(\cdot, \mathcal{M}_{n-1}(\cdot))\right\}.$$

First of all, let us note that all $\mathcal{M}_n$, $n \geq 1$ are nonempty since $\Theta^b(u_0) \subset \mathcal{M}_n$ for all $n \geq 0$.

We proceed to verify that the set $\mathcal{M}_n$ is equicontinuous on $[0, b]$. Taking $0 < t_1 < t_2 \leq b$ and $\delta > 0$ small enough, for any $v(t) \in \mathcal{N}_n$, we obtain

$$|v(t_2) - v(t_1)| \leq \|T'(t_2) - T'(t_1)\|_{\mathcal{L}(X)}|u_0| + \left|\lim_{\lambda \to +\infty} \int_{t_1}^{t_2} T'(t - s)\mathcal{J}_\lambda f(s)ds\right|$$

$$+ \left|\lim_{\lambda \to +\infty} \int_0^{t_1} \left(T'(t_2 - s) - T'(t_1 - s)\right)\mathcal{J}_\lambda f(s)ds\right|.$$

The right-hand side tends to zero as $t_2 - t_1 \to 0$, since $T'(t)$ is strongly continuous, and the compactness of $T'(t)$ ($t > 0$), implies the continuity in the uniform operator topology. Thus all sets $\mathcal{M}_n$, $n \geq 1$ are equicontinuous.

Using condition (H_3) we have the following estimation:

$$\beta\left(\overline{co}F(t, \mathcal{M}_{n-1}(t))\right) \leq k(t)e^{Lt}\left(\sup_{s \in [0,t]} e^{-Ls}\beta(\mathcal{M}_{n-1}(s))\right) \leq k(t)e^{Lt}\chi(\mathcal{M}_{n-1}).$$

We have for any $t \in [0, b]$,

$$e^{-Lt}\beta(\mathcal{N}_n(t)) = e^{-Lt}\beta\left(\lim_{\lambda \to +\infty} \int_0^t T'(t - s)\mathcal{J}_\lambda \overline{co}F(s, \mathcal{M}_{n-1}(s))ds\right)$$

$$\leq 2c_0 e^{-Lt}\left(\int_0^t e^{Ls}k(s)ds\right)\chi(\mathcal{M}_{n-1})$$

$$\leq 2c_0\left(\int_0^t e^{-L(t-s)}k(s)ds\right)\chi(\mathcal{M}_{n-1}).$$

Therefore, $\chi(\mathcal{N}_n) \leq l\chi(\mathcal{M}_{n-1})$.

Finally, we have $\chi(\mathcal{M}_n) \leq l\chi(\mathcal{M}_{n-1})$ and therefore $\chi(\mathcal{M}_n) \to 0$ $(n \to \infty)$. We obtain a compact set $\mathcal{M} = \bigcap_{n=0}^{\infty} \mathcal{M}_n$, which has the desired properties.

In the following, let us note that we may assume, without loss of generality, that F satisfies the following estimation:

$(H_2)'$ $|F(t, x)| \leq \eta(t)$ for every $x \in X$ and a.e. $t \in [0, b]$, where $\eta \in L^1([0, b], \mathbb{R}^+)$.

In fact, let $\|\Theta^b(u_0)\| \leq N$, Q_n be a closed ball in the space X and $\rho : X \to Q_n$ be a radial retraction. Then it is easy to see that the multimap $\widetilde{F} : [0, b] \times X \to P_{cp,cv}(X)$, defined by $\widetilde{F}(t, u) = F(t, \rho u)$ satisfies conditions (H_1) and (H_3) (note that ρ is a Lipschitz map), the condition $(H_2)'$ with $\eta(t) = \alpha(t)(1 + N)$. The set $\Theta^b(u_0)$ coincides with the set of all integral solutions of the problem

$$\begin{cases} u'(t) \in Au(t) + \widetilde{F}(t, u(t)), & t \in [0, b], \\ u(0) = u_0. \end{cases}$$

Therefore in what follows we suppose that the multimap $F : [0, b] \times X \to P_{cl,cv}(X)$ satisfies conditions (H_1), (H_3) and $(H_2)'$ instead of (H_2).

Now consider a metric projection $P : [0, b] \times X \to P_{cl,cv}(X)$,

$$P(t, u) = \{v \in \mathcal{M}(t) : \|u - v\| = \text{dist}(u, \mathcal{M}(t))\},$$

and a multimap $\widehat{F} : [0, b] \times X \to P_{cl,cv}(X)$, defined by

$$\widehat{F}(t, u) = \overline{co} F(t, P(t, u)).$$

From Lemma 5.3.2 of [130], we know the multimap P is closed and u.s.c.

Lemma 3.4 ([130]) *The multimap $\widehat{F}$ satisfies conditions (H_1), $(H_2)'$ and (H_3).*

The above result implies that the set $\widehat{\Theta}^b(u_0)$ of all integral solutions of the problem

$$\begin{cases} u'(t) \in Au(t) + \widehat{F}(t, u), & t \in [0, b], \\ u(0) = u_0 \end{cases}$$

is nonempty. Moreover, the following statement is valid.

Lemma 3.5 $\widehat{\Theta}^b(u_0) = \Theta^b(u_0)$.

Proof In fact, let $u \in \widehat{\Theta}^b(u_0)$. Then

$$u(t) \in T'(t)u_0 + \lim_{\lambda \to +\infty} \int_0^t T'(t - s)\mathcal{J}_\lambda \widehat{F}(s, u(s))ds$$

$$= T'(t)u_0 + \lim_{\lambda \to +\infty} \int_0^t T'(t - s)\mathcal{J}_\lambda \overline{co} F(s, P(t, u(s)))ds$$

$$\subset T'(t)u_0 + \lim_{\lambda \to +\infty} \int_0^t T'(t - s)\mathcal{J}_\lambda \overline{co} F(s, \mathcal{M}(s)))ds \subset \mathcal{M}(t),$$

hence $P(t, u(t)) = \{u(t)\}$. Then

$$u(t) = T'(t)u_0 + \lim_{\lambda \to +\infty} \int_0^t T'(t - s)\mathcal{J}_\lambda f(s)ds,$$

where $f \in Sel_{\widehat{F}}^b(u) = Sel_F^b(u)$, and so $u \in \Theta^b(u_0)$.

The inclusion $\Theta^b(u_0) \subset \widehat{\Theta}^b(u_0)$ easily follows from the observation that $\Theta^b(u_0) \subset \mathcal{M}$.

Lemma 3.6 ([130], Lemma 5.3.4) *Let the hypotheses* (H_1) *and* (H_2) *be satisfied. Then there exists a sequence* $\{\widehat{F}_n\}$ *with* $\widehat{F}_n : [0, b] \times X \to P_{cl,cv}(X)$ *such that*

(i) $\widehat{F}(t, x) \subset \cdots \subset \widehat{F}_{n+1}(t, x) \subset \widehat{F}_n(t, x) \subset \cdots \subset \overline{co}(F(t, \mathcal{M}(t)))$, $n \geq 1$ *for each* $t \in [0, b]$ *and* $x \in X$;
(ii) $|\widehat{F}_n(t, x)| \leq \eta(t)$, $n \geq 1$ *for a.e.* $t \in [0, b]$ *and each* $x \in X$;
(iii)

$$F(t, x) = \bigcap_{n=1}^{\infty} F_n(t, x);$$

(iv) $\widehat{F}_n(t, \cdot) : X \to P_{cl,cv}(X)$ *is continuous for a.e.* $t \in [0, b]$ *with respect to Hausdorff metric for each* $n \geq 1$;
(v) *for each* $n \geq 1$, *there exists a selection* $g_n : [0, b] \times X \to X$ *of* $\widehat{F}_n$ *such that* $g_n(\cdot, x)$ *is measurable and* $g_n(t, \cdot)$ *is locally Lipschitz.*

Theorem 3.3 *Let the conditions in Theorem 3.2 be satisfied. Then the solution set of inclusion (3.1) is a compact* R_δ*-set in* $C([0, b], X_0)$.

Proof Now we consider the differential inclusion:

$$\begin{cases} u'(t) \in Au(t) + \widehat{F}_n(t, u), & t \in [0, b], \\ u(0) = u_0. \end{cases} \tag{3.9}$$

Let $\widehat{\Theta}_n^b(u_0)$ denote the solution set of inclusion (3.9). From Lemma 3.4, it follows that each $\widehat{F}_n$ satisfies conditions (H_1), $(H_2)'$ and (H_3), hence each set $\widehat{\Theta}_n^b(u_0)$ is nonempty and compact.

We prove that

$$\widehat{\Theta}^b(u_0) = \bigcap_{n \geq 1} \widehat{\Theta}_n^b(u_0).$$

It is clear that $\widehat{\Theta}^b(u_0) \subset \widehat{\Theta}_n^b(u_0)$ and $\widehat{\Theta}^b(u_0) \subset \bigcap_{n \geq 1} \widehat{\Theta}_n^b(u_0)$.

Let $u \in \bigcap_{n \geq 1} \widehat{\Theta}_n^b(u_0)$, then for each $n \geq 1$, we have

$$u(t) = T'(t)u_0 + \lim_{\lambda \to +\infty} \int_0^t T'(t - s)\mathcal{J}_\lambda g_n(s)ds \text{ for } t \in [0, b],$$

where $g_n \in Sel_{\widehat{F}_n}^b(u)$. From Lemma 3.6 (ii), it follows that $\{g_n\}$ is semicompact and by Lemma 1.25 we may assume, up to subsequence, that $g_n \rightharpoonup f \in L^1([0, b], X)$.

Lemma 3.6 (iii) implies that $f(t) \in \widehat{F}(t, u(t))$ a.e. on $[0, b]$. Applying Lemma 3.1, we derive that

$$u(t) = T'(t)u_0 + \lim_{\lambda \to +\infty} \int_0^t T'(t - s)\mathcal{J}_\lambda f(s)ds \text{ for } t \in [0, b],$$

which means that $u \in \widehat{\Theta}^b(u_0)$.

We show that the set $\Theta_n^b(u_0)$ is contractible for each $n \geq 1$. In fact, let $u_n \in \Theta_n^b(u_0)$ and for any $\varsigma \in [0, 1]$, the function $v_n^\varsigma(t)$ be a unique solution on $[\varsigma b, b]$ of the integral equation

$$v_n(t) = T'(t - \varsigma b)u_n(\varsigma b) + \lim_{\lambda \to +\infty} \int_{\varsigma b}^t T'(t - s)\mathcal{J}_\lambda g_n(s, v_n(s))ds, \qquad (3.10)$$

where g_n is the select of F_n. The functions

$$z_n^\varsigma(t) = \begin{cases} u_n(t), & t \in [0, \varsigma b], \\ v_n^\varsigma(t), & t \in [\varsigma b, b] \end{cases}$$

belong to $\Theta_n^b(u_0)$. Define the deformation $h : [0, 1] \times \Theta_n^b(u_0) \to \Theta_n^b(u_0)$ by the formula

$$h(\varsigma, u_n) = \begin{cases} z_n^\varsigma(t), & \varsigma \in [0, 1), \\ u_n, & \varsigma = 1. \end{cases}$$

Since the function g_n is locally Lipschitz in Lemma 3.6 (v), the solutions of the equation (3.10) depend continuously on (ς, u_n), therefore the definition h is continuous. But $h(0, \cdot) = v_n^0(t)$ and $h(1, \cdot)$ is the identity, hence $\Theta_n^b(u_0)$ is contractible.

Consequently, Theorem 1.14 follows that the solution set of inclusion (3.1) is a compact R_δ-set, completing this proof. $\quad\blacksquare$

Theorem 3.4 *Let the conditions in Theorem 3.2 be satisfied. Then the solution set of inclusion (3.2) is a compact R_δ-set in $C(\mathbb{R}^+, X_0)$.*

Proof Firstly, we introduce the following two inverse systems and their limits. For more details about the inverse system and its limit, we refer the reader to [71] (see also [14, 104]).

For each $p, m > 0$ with $p \geq m$, let us consider a projection $\pi_m^p : C([0, p], X_0) \to C([0, m], X_0)$, defined by

$$\pi_m^p(u) = u|_{[0,m]}, \quad u \in C([0, p], X_0).$$

Put

$$\mathbb{N}_0 = \{m \in \mathbb{N} \setminus \{0\} : m > 0\}, \quad C_m = \{u \in C([0, m], X_0) : u(0) = u_0\}.$$

Then it is readily checked that $\{C_m, \pi_m^p, \mathbb{N}_0\}$ is an inverse system and its limit is

$$\{u \in \widetilde{C}([0, \infty), X_0) : u(0) = u_0\} =: C.$$

Consider the sequence of multivalued maps

$$\mathscr{F}^m(u)(t) = \{T'(t)u_0 + \Gamma(f)(t), \ t \in [0, m] : f \in L^1([0, m], X), \ f(t) \in F(t, u(t))\}.$$

We have the equalities

$$\mathscr{F}^m \pi_m^{m+1}(u)(t) = \{T'(t)u_0 + \Gamma(f)(t), \ t \in [0, m] : f \in L^1([0, m], X),$$
$$f(t) \in F(t, u(t)) \ \text{for a.e. } t \in [0, m]\},$$

and

$$\pi_m^{m+1} \mathscr{F}^{m+1}(u)(t) = \{T'(t)u_0 + \Gamma(f)(t), \ t \in [0, m + 1] : f \in L^1([0, m + 1], X),$$
$$f(t) \in F(t, u(t)) \ \text{for a.e. } t \in [0, m + 1]\}.$$

Noticing that

$$\{f \in L^1([0, m], X) : f(t) \in F(t, u(t)) \ \text{for a.e. } t \in [0, m]\}$$
$$= \{f|_{[0,m]}, \ f \in L^1([0, m + 1], X) : f(t) \in F(t, u(t)) \ \text{for a.e. } t \in [0, m + 1]\},$$

one can find that $\mathscr{F}^m \pi_m^{m+1} = \pi_m^{m+1} \mathscr{F}^{m+1}$, so the family $\{id, \mathscr{F}^m\}$ is the map from the inverse system $\{C_m, \pi_m^p, \mathbb{N}_0\}$ into itself, which enables us to conclude that family $\{id, \mathscr{F}^m\}$ induces a limit mapping

$$\mathscr{F} : C \to P(C),$$

here, for every $u \in C$,

$$\mathscr{F}(u) = \{T'(t)u_0 + \Gamma(f)(t), t \in \mathbb{R}^+ : f \in L^1(\mathbb{R}^+, X),$$
$$f(t) \in F(t, u(t)) \ \text{for a.e. } t \in \mathbb{R}^+\}.$$

Moreover, it follows readily that $\Theta(u_0) := \text{Fix}(\mathscr{F}) = \lim_{\leftarrow} \Theta^m(u_0)$

For every $m \in \mathbb{N} \setminus \{0\}$, the set of all fixed points of $\mathscr{F}^m$ is denoted by $\text{Fix}(\mathscr{F}^m)$, i.e.,

$$\text{Fix}(\mathscr{F}^m) = \{u \in C_m : u \in \mathscr{F}^m(u)\}.$$

Then we see from Theorem 3.3 that $\text{Fix}(\mathscr{F}^m)(= \Theta^m(u_0))$ are compact R_δ-sets. At the end of this step, applying Theorems 1.18 and 1.19 we obtain that the solution set of inclusion (3.2) is a nonempty compact R_δ-set, as claimed.

3.2.3 Applications

Example 3.1 Consider an optimization problem in case when inclusion (3.1) is generated in a separable Banach space X by a semilinear feedback control system of the following form:

$$\begin{cases} u'(t) = Au(t) + f(t, u(t), z(t)), & t \in [0, b], \\ z(t) \in U(t, u(t)), & t \in [0, b], \\ u(0) = u_0 \in D(A). \end{cases} \tag{3.11}$$

We suppose that the linear operator $A : D(A) \subset X \to X$ satisfies the condition (A).

Let Y be a separable Banach space of controls. We will assume that the nonlinear term $f : [0, b] \times X \times Y \to X$ satisfies the following conditions:

(f_1) the function $f(\cdot, u, z) : [0, b] \to X$ is measurable for each $(u, z) \in X \times Y$;
(f_2) $|f(t, u_2, z) - f(t, u_2, z)| \leq k(t)|u_1 - u_2|$ for each $u_1, u_2 \in X$, $z \in Y$, where $k(\cdot) \in L^1([0, b], \mathbb{R}^+)$;
(f_3) the map $f(t, \cdot, \cdot) : X \times Y \to X$ is continuous for a.e. $t \in [0, b]$.

For the feedback multimap

$$U : [0, b] \times X \to P_{cl}(Y),$$

we suppose that he following hypotheses hold true:

(U_1) the multifunction $U(\cdot, u) : [0, b] \to P_{cl}(X)$ is measurable for each $u \in X$;
(U_2) the multimap $U(t, \cdot) : X \to P_{cl}(X)$ is u.s.c. for a.e. $t \in [0, T]$;
(U_3) the multimap U is superpositionally measurable, i.e., the multifunction $t \to P(U(t, u(t)))$ is measurable for each $u \in C([0, b], X)$;
(U_4) the set

$$F(t, u) = f(t, u, U(t, u)) \tag{3.12}$$

is convex for all $(t, u) \in [0, b] \times X$;
(U_5) the multimap F satisfies the boundedness condition (H_2);
(U_6) for every $(t, u) \in [0, b] \times X$ the set $\Omega \subset X$.

Definition 3.2 A continuous function $(u(\cdot), z(\cdot))$, where $u \in C([0, T], X)$ and $z : [0, b] \to Y$ is a measurable function, is said to be an integral solution of problem (3.11) if u has the form

$$u(t) = T'(t)u_0 + \lim_{\lambda \to +\infty} \int_0^t T'(t - s)\mathcal{J}_\lambda f(s, u(s), z(s))ds,$$

and z satisfies the feedback condition $z(t) \in U(t, u(t))$.

The function u is called the trajectory of the system and the function z is the corresponding control.

Theorem 3.5 *Let* $\mathscr{J} : C([0, b], X) \to \mathbb{R}$ *be a lower semicontinuous functional. Then, under conditions* (A), (f_1)–(f_3), *and* (U_1)–(U_6) *there exists an integral solution* $(u^*(\cdot), z^*(\cdot))$ *of problem* (3.11) *such that*

$$\mathscr{J}(v) = \min_{u \in \Sigma_{u_0}} \mathscr{J}(u),$$

where Σ_{u_0} *is the set of all trajectories of the system starting at* u_0.

Proof It is known (see, e.g., [42, 130]) that conditions (f_1)–(f_3) and (U_1)–(U_6) imply that the multimap $F : [0, b] \times X \to P_{cl,cv}(X)$ defined by relation (3.12) satisfies conditions (H_1)–(H_3), and (F_4). Applying Theorem 3.2, we conclude that the set of all integral solutions of the associated differential inclusion

$$\begin{cases} u'(t) \in Au(t) + F(t, u(t)), & t \in [0, b], \\ u(0) = u_0, \end{cases}$$

is nonempty compact and hence it contains a minimizer u^* of the functional $\mathscr{J}$. The existence of the corresponding control z^* realizing u^* as the trajectory of the system follows from the Filippov implicit function lemma (see, e.g., [42, 130]). ∎

Example 3.2 Consider a controllability problem for semilinear differential inclusions

$$\begin{cases} u'(t) \in Au(t) + Bw(t) + F(t, u(t)), & t \in [0, b], \ 0 < q \leq 1, \\ u(0) = u_0. \end{cases} \qquad (3.13)$$

For system (3.13), we assume

(H_B) the control function $w(\cdot)$ takes its value in $\overline{U}$, a Banach space of admissible control functions, where $\overline{U} = L^2([0, b], U)$ and U is a Banach space. $B : U \to X$ is a bounded linear operator, with

$$\|B\|_{\mathcal{L}(U,X)} \leq M_2.$$

Definition 3.3 A continuous function $u : [0, b] \to X$ is said to be an integral solution of system (3.13) if $u(0) = u_0$ and there exists $f(t) \in Sel_F^b(u)(t)$ for a.e. $t \in [0, b]$ satisfying the following integral equation

$$u(t) = T'(t)u_0 + \lim_{\lambda \to +\infty} \int_0^t T'(t - s)\mathcal{J}_\lambda Bw(s)ds + \lim_{\lambda \to +\infty} \int_0^t T'(t - s)\mathcal{J}_\lambda f(s)ds.$$

We consider the controllability problem for system (3.13), i.e., we study conditions which guarantee the existence of an integral solution to system (3.13) satisfying

$$u(b) = x_1, \qquad (3.14)$$

where $x_1 \in X_0$ is a given point. A pair (u, w) consisting of an integral solution $u(\cdot)$ to (3.13) satisfying (3.14) and of the corresponding control $w(\cdot) \in L^2([0, b], U)$ is called a solution of the controllability problem.

We assume the standard assumption that the corresponding linear problem (i.e., when $F(t, u) \equiv 0$) has a solution. More precisely, we suppose that (H_W) the controllability operator $W : \overline{U} \to X$ given by

$$Ww = \lim_{\lambda \to +\infty} \int_0^b T'(b - s) \mathcal{J}_\lambda Bw(s) ds$$

has a bounded inverse which takes values in $\overline{U}/ker(W)$, and there exists a positive constant $M_3 > 0$ such that

$$\|W^{-1}\|_{\mathcal{L}(X, \overline{U})} \le M_3.$$

Theorem 3.6 *Under the assumptions in Theorem 3.2, furthermore, we assume that (H_B) and (H_W) hold. Then the solution set of problem (3.13) and (3.14) is a nonempty compact R_δ-set.*

Proof We denote by $\Gamma_i : L^1([0, b], X) \to C([0, b], X_0)$ $(i = 1, 2)$ the following integral operators: for $t \in [0, b]$,

$$\Gamma_1 f(t) = \lim_{\lambda \to +\infty} \int_0^t T'(t - s) \mathcal{J}_\lambda f(s) ds,$$

$$\Gamma_2 f(t) = \lim_{\lambda \to +\infty} \int_0^t T'(t - s) \mathcal{J}_\lambda BW^{-1} \left(-\lim_{\lambda \to +\infty} \int_0^b T'(b - s) \mathcal{J}_\lambda f(\eta) d\eta \right)(s) ds.$$

Then we define the solution multioperator $\Gamma_3 : C([0, b], X_0) \to P(C([0, b], X_0))$ as

$$\Gamma_3 u = \left\{ \begin{array}{l} v(t) \in C([0, b], X_0) : v(t) = T'(t) u_0 + \Gamma_1 f(t) + \Gamma_2 f(t) \\ \quad + \lim_{\lambda \to +\infty} \int_0^t T'(t - s) \mathcal{J}_\lambda BW^{-1}(x_1 - T'(b) u_0)(s) ds, \ f \in Sel_F^b(u) \end{array} \right\}.$$

Similar to Theorem 3.3, we can see that the solution set of controllability problem (3.13) and (3.14) is a nonempty compact R_δ-set.

3.3 Global Attractor

In this section we study the global attractor of system (3.3). Here β is Hausdorff MNC on X. We recall the concept of β-norm of a bounded linear operator S $(S \in \mathcal{L}(X))$ as follows

$$\|S\|_\beta = \inf\{\gamma > 0 : \beta(S(B)) \le \gamma\beta(B) \text{ for all bounded set } B \subset X\}. \qquad (3.15)$$

It is noted that the β-norm of S can be formulated by

$$\|S\|_\beta = \beta(S(\mathbf{B_1})) = \beta(S(\mathbf{S_1})),$$

where $\mathbf{B_1}$ and $\mathbf{S_1}$ are a unit ball and a unit sphere in X, respectively. It is known that

$$\|S\|_\beta \le \|S\|_{\mathcal{L}(X)},$$

where the last norm is understood as the operator norm in $\mathcal{L}(X)$. Obviously, S is a compact operator if and only if $\|S\|_\beta = 0$.

Denote

$$\mathcal{C}_h = \{\varphi \in C([-h, 0], X) : \varphi(0) \in \overline{D(A)}\},$$
$$\mathcal{C}_\varphi = \{v : [0, b] \to \overline{D(A)} : v \in C([0, b], X), \ v(0) = \varphi(0)\}.$$

For $v \in \mathcal{C}_\varphi$, we denote the function $v[\varphi] \in C([-h, b], X)$ as follows

$$v[\varphi](t) = \begin{cases} v(t), & \text{if } t \in [0, b], \\ \varphi(t), & \text{if } t \in [-h, 0]. \end{cases}$$

In what follows, we assume that the multivalued function $F : D(A) \times \mathcal{C}_h \to P_{cl}(X)$ satisfies:

(F_1) F is u.s.c. with weakly compact and convex values;
(F_2) $|F(x, y)| := \sup\{|\xi| : \xi \in F(x, y)\} \le a_1|x| + a_2|y|_{\mathcal{C}_h} + a_3$ for all $x \in \overline{D(A)}$, $y \in \mathcal{C}_h$, where a_1, a_2, a_3 are nonnegative constants;
(F_3) if $\{T'(t)\}$ is noncompact, then $\beta(F(B, C)) \le p\beta(B) + q \sup_{t \in [-h,0]} \beta(C(t))$, for all $B \subset \overline{D(A)}$, $C \subset \mathcal{C}_h$, where $p, q \in \mathbb{R}^+$.

3.3.1 Existence of Integral Solution

Putting

$$Sel_F(v) = \{f \in L^1([0, b], X) : f(t) \in F(v(t), v[\varphi]_t) \text{ for a.e. } t \in [0, b]\}.$$

Similar to the proof in [45, Theorem 1], we have the following properties.

Lemma 3.7 *Let (F_1) and (F_2) hold. Then $Sel_F(u) \ne \emptyset$ for each $u \in \mathcal{C}_\varphi$. In addition, $Sel_F : C([0, b], X) \to P(L^1([0, b], X))$ is weakly u.s.c. with weakly compact and convex values.*

Definition 3.4 For a given $\varphi \in \mathcal{C}_h$, a continuous function $u : [-h, b] \to X$ is said to be an integral solution of system (3.3) with each initial datum $\varphi \in \mathcal{C}_h$. if there exists $f(t) \in Sel_F(u)(t)$ such that

$$
u(t) = \begin{cases} T(t)\varphi(0) + \lim\limits_{\lambda \to +\infty} \int_0^t T'(t-s)\mathcal{J}_\lambda f(s)ds, & \text{for } t \in [0, b], \\ \varphi(t), & \text{for } t \in [-h, 0]. \end{cases}
$$

We now define a solution multioperator $\mathscr{F} : \mathcal{C}_\varphi \to P(\mathcal{C}_\varphi)$ as follows:

$$
\mathscr{F}(v)(t) = \{T'(t)\varphi(0) + \Gamma(f)(t) : f \in Sel_F(v)\},
$$

where

$$
\Gamma(f)(t) = \lim\limits_{\lambda \to +\infty} \int_0^t T'(t-s)\mathcal{J}_\lambda f(s)ds, \quad t \geq 0.
$$

It is easy to see that $v \in \mathcal{C}_\varphi$ is a fixed point of F iff $v[\varphi]$ is an integral solution of (3.3). In order to prove the existence result for system (3.3).

Theorem 3.7 *Let hypotheses (H_A) and (F_1)–(F_3) hold. Then system (3.3) has at least one integral solution for each initial datum $\varphi \in \mathcal{C}_h$.*

Proof Firstly, we have $\lim_{\lambda \to \infty} \|\mathcal{J}_\lambda\|_{\mathcal{L}(X)} \leq M$, in view of the fact that $\lim_{\lambda \to \infty} \mathcal{J}_\lambda v = v$ for all $v \in X$. Now, we will show that there exists a closed convex set $\mathcal{M}_0 \subset \mathcal{C}_h$ satisfying that $\mathscr{F}(\mathcal{M}_0) \subset \mathcal{M}_0$. Let $z \in \mathscr{F}(u)$. Then from the definition of the solution operator one has

$$
\begin{aligned}
|z(t)| &\leq \|T'(t)\|_{\mathcal{L}(X)}|\varphi(0)| + \left| \lim_{\lambda \to +\infty} \int_0^t T'(t-s)\mathcal{J}_\lambda f(s)ds \right| \\
&\leq M_1|\varphi(0)| + \lim_{\lambda \to +\infty} \int_0^t \|T'(t-s)\|_{\mathcal{L}(X)}\|\mathcal{J}_\lambda\|_{\mathcal{L}(X)}|f(s)|ds \Big| \\
&\leq M_1|\varphi(0)| + MM_1 \int_0^t \left(a_1|u(s)| + a_2|u_s|_{\mathcal{C}_h} + a_3\right)ds,
\end{aligned}
$$

where $M_1 = \sup_{t \in [0, b]} \|T'(t)\|_{\mathcal{L}(X)}$.

On the other hand, we have

$$
|u_s|_{\mathcal{C}_h} = \sup_{\theta \in [-h, 0]} |u(s+\theta)| \leq |\varphi|_{\mathcal{C}_h} + \sup_{\theta \in [0, s]} |u(\theta)|.
$$

It follows that

$$|z(t)| \leq M_1|\varphi(0)| + MM_1 \int_0^t \left(a_1|u(s)| + a_2|\varphi|_{\mathcal{C}_h} + a_2 \sup_{\theta \in [0,s]} |u(\theta)| + a_3\right)ds$$

$$\leq M_1|\varphi(0)| + MM_1 b(a_2|\varphi|_{\mathcal{C}_h} + a_3) + MM_1 \int_0^t \left(a_1|u(s)| + a_2 \sup_{\theta \in [0,s]} |u(\theta)|\right)ds$$

$$\leq M_2 + MM_1(a_1 + a_2) \int_0^t \sup_{\theta \in [0,s]} |u(\theta)|ds \text{ for } t \in [0, b],$$

where $M_2 = M_1|\varphi(0)| + MM_1 b(a_2|\varphi|_{\mathcal{C}_h} + a_3)$. Since the last term is increasing in t, we have

$$\sup_{\theta \in [0,t]} |z(\theta)| \leq M_2 + MM_1(a_1 + a_2) \int_0^t \sup_{\theta \in [0,s]} |u(\theta)|ds. \tag{3.16}$$

Denote

$$\mathcal{M}_0 = \left\{ u \in \mathcal{C}_h : \sup_{s \in [0,t]} |u(s)| \leq \psi(t), \ t \in [0, b] \right\},$$

where ψ is the unique solution of the integral equation

$$\psi(t) = M_2 + MM_1(a_1 + a_2) \int_0^t \psi(s)ds.$$

It is clear that $\mathcal{M}_0$ is a closed convex subset of $\mathcal{C}_\varphi$ and estimate (3.16) ensures that $\mathscr{F}(\mathcal{M}_0) \subset \mathcal{M}_0$.

Set

$$\mathcal{M}_{k+1} = \overline{co}F(\mathcal{M}_k), \ k = 0, 1, 2, \dots.$$

We see that $\mathcal{M}_k$ is closed, convex and $\mathcal{M}_{k+1} \subset \mathcal{M}_k$ for all $k \in \mathbb{N}$.

Let $\mathcal{M} = \bigcap_{k=0}^\infty \mathcal{M}_k$, then $\mathcal{M}$ is a closed convex subset of $\mathcal{C}_\varphi$ and $\mathscr{F}(\mathcal{M}) \subset \mathcal{M}$. We show that $\mathcal{M}$ is compact by using Arzela–Ascoli's theorem. Indeed, for each $k \geq 0$, $Sel_F(\mathcal{M}_k)$ is integrably bounded thanks to (F_2). Then the proof of Lemma 3.1 ensures that $\mathscr{F}(\mathcal{M}_k)$ is equicontinuous. It follows that $\mathcal{M}_{k+1}$ is equicontinuous for all $k \geq 0$. Thus $\mathcal{M}$ is equicontinuous as well.

In order to apply Arzela–Ascoli's theorem, we have to prove that $\mathcal{M}(t)$ is compact for each $t \geq 0$. This will be done if we show that $\mu_k(t) = \beta(\mathcal{M}_k(t)) \to 0$ as $k \to \infty$, where β is the Hausdorff MNC on X. If $\{T'(t)\}$ is compact then it is easily checked that

$$\mu_{k+1}(t) = \beta(\mathcal{M}_{k+1}(t)) \leq \lim_{\lambda \to +\infty} \beta\left(\int_0^t T'(t-s)\mathcal{J}_\lambda Sel_F(\mathcal{M}_k)(s)ds\right)$$

$$\leq 4 \lim_{\lambda \to +\infty} \int_0^t \beta(T'(t-s)\mathcal{J}_\lambda Sel_F(\mathcal{M}_k)(s))ds$$

$$= 0$$

according to Property 1.5. In the opposite case, we have

$$
\mu_{k+1}(t) \leq 4MM_1 \int_0^t \beta(Sel_F(\mathcal{M}_k)(s))ds
$$

$$
\leq 4MM_1 \int_0^t \left(p\beta(\mathcal{M}_k(s)) + q \sup_{\theta \in [-h,0]} \beta(\mathcal{M}_k[\varphi](s+\theta)) \right)ds
$$

$$
\leq 4MM_1 \int_0^t \left(p\beta(\mathcal{M}_k(s)) + q \sup_{\theta \in [0,s]} \beta(\mathcal{M}_k(\theta)) \right)ds
$$

$$
\leq 4MM_1(p+q) \int_0^t \sup_{\theta \in [0,s]} \mu_k(\theta)ds
$$

according to (F_3). Since the right-hand side is nondecreasing in t, we get

$$
\sup_{\theta \in [0,t]} \mu_{k+1}(\theta) \leq 4MM_1(p+q) \int_0^t \sup_{\theta \in [0,s]} \mu_k(\theta)ds.
$$

Putting $\eta_k(t) = \sup_{\theta \in [0,t]} \mu_{k+1}(\theta)$, we have

$$
\eta_k(t) \leq 4MM_1(p+q) \int_0^t \eta_k(s)ds.
$$

Let $\eta_\infty(t) = \lim_{k\to\infty} \eta_k(t)$, then we have

$$
\eta_\infty(t) \leq 4MM_1(p+q) \int_0^t \eta_\infty(s)ds.
$$

Hence Gronwall–Bellman's inequality ensures that $\eta_\infty(t) = 0$ for all $t \in [0, b]$. Since $0 \leq \mu_k(t) \leq \eta_k(t) \to 0$ as $k \to \infty$, we have $\mu_k(t) \to 0$ as $k \to \infty$ as desired.

We consider $F : \mathcal{M} \to P(\mathcal{M})$. In order to apply Theorem 1.6, it remains to show that $\mathscr{F}$ is u.s.c. with convex closed values. Since Sel_F has convex values, so F does. By Lemma 1.9, it suffices to show that $\mathscr{F}$ has a closed graph. Let $\{u_n\} \subset \mathcal{M}$ with $u_n \to u^*$ and $v_n \in \mathscr{F}(u_n)$ with $v_n \to v^*$. Then we have

$$
v_n(t) = T'(t)\varphi(0) + \Gamma \circ Sel_F(u_n)(t).
$$

Let $f_n \in Sel_F(u_n)$ be such that

$$
v_n(t) = T'(t)\varphi(0) + \Gamma(f_n)(t). \tag{3.17}
$$

Since Sel_F is weakly u.s.c. by Lemma 3.7, we can employ Lemma 1.7 (ii) to state that $f_n \rightharpoonup f^*$ in $L^1(J, X)$ and $f^* \in Sel_F(u^*)$. Moreover, let $K(t) = F(\{u_n(t), u_n[\varphi]_t\})$ then $\{f_n(t)\} \subset K(t)$ for a.e. $t \in [0, b]$ with $K(t)$ being compact in X thanks to the fact that F is u.s.c. and has closed values. Taking into account (F_2), we see that $\{f_n\}$

is integrably bounded. Therefore $\{f_n\}$ is a semicompact sequence. Now applying Lemma 3.1 (ii) gives the compactness of $\{\Gamma(f_n)\}$ in $C([0, b], X)$. Thus one can pass the limit in (3.17) to get $v^*(t) = T'(t)\varphi(0) + \Gamma(f^*)(t)$, where $f^* \in Sel_F(u^*)$. It infers that $v^* \in \mathscr{F}(u^*)$ and the proof is complete.

In the rest of this section, we prove some properties of the solution set, which will be used later.

Let $\pi_b, b > 0$, be the truncation operator to $[0, b]$ acting on $C([0, +\infty), X)$, that is, for $z \in C([0, +\infty), X)$, $\pi_b(z)$ is the restriction of z on interval $[0, b]$. Denote

$$\Sigma(\varphi) = \{u \in C([0, \infty), X) : u[\varphi] \text{ is an integral solution of (3.3)}$$

$$\text{on } [-h, b] \text{ for any } b > 0\}.$$

Lemma 3.8 *Let (H_A) and (F_1)–(F_3) hold and $\varphi_n \subset C_h$ be a convergent sequence. Then $\pi_b \circ \Sigma(\{\varphi_n\})$ is relatively compact in $C([0, b], X)$. In particular, $\pi_b \circ \Sigma(\varphi)$ is a compact set for each $\varphi \in C_h$.*

Proof Let $\varphi_n \in \pi_b \circ \Sigma(\varphi_n)$, $n \in \mathbb{N}$ be a sequence. Then we have

$$v_n(t) \in T'(t)\varphi_n(0) + \Gamma \circ Sel_F(v_n)(t), \quad t \in [0, b].$$

We have to show that $\{v_n\}$ is equicontinuous and $\{v_n(t)\}$ is relatively compact for each $t \in [0, b]$. The same estimate as (3.16) gives

$$|v_n(t)| \leq M_2 + MM_1(a_1 + a_2) \int_0^t \sup_{\theta \in [0,s]} |u(\theta)| ds, \quad \forall\, t \in [0, b], \qquad (3.18)$$

where $M_1 = \sup_{t \in [0,b]} \|T'(t)\|_{\mathcal{L}(X)}$ and

$$M_2 = M_1 \sup_n |\varphi_n(0)| + MM_1 b(a_2 \sup_n |\varphi_n|_{C_h} + a_3).$$

Since the right-hand side of (3.18) is nondecreasing in t, we have

$$|w_n(t)| \leq M_2 + MM_1(a_1 + a_2) \int_0^t w(s)ds, \quad \forall\, t \in [0, b],$$

where $w_n(t) = \sup_{\theta \in [0,t]} |u(\theta)|$. By Gronwall–Bellman's inequality, we get the boundedness of $\{w_n\}$, that implies the boundedness of $\{v_n\}$ in $C([0, b], X)$. Let $f_n \in Sel_F(v_n)$ such that

$$v_n(t) = T'(t)\varphi_n(0) + \Gamma(f_n)(t).$$

Using (F_2), we see that $\{f_n\}$ is integrably bounded due to the boundedness of $\{v_n\}$. If $\{T'(t)\}$ is compact, then $\{v_n\}$ is compact. In the case when $\{T'(t)\}$ is noncompact, we have that $\{\Gamma(f_n)\}$ is an equicontinuous set in view of the proof of Lemma 3.1

(ii). This means that $\{v_n\}$ is equicontinuous as well. In addition, one has

$$
\begin{aligned}
\beta(\{v_n(t)\}) &= \beta\left(\left\{T'(t)\varphi_n(0) + \lim_{\lambda \to +\infty} \int_0^t T'(t-s)\mathcal{J}_\lambda f_n(s)ds\right\}\right) \\
&\leq \lim_{\lambda \to +\infty} \beta\left(\left\{\int_0^t T'(t-s)\mathcal{J}_\lambda f_n(s)ds\right\}\right) \\
&\leq 2 \lim_{\lambda \to +\infty} \int_0^t \|T'(t-s)\mathcal{J}_\lambda\|_{\mathcal{L}(X)}\beta(f_n(s))ds \\
&\leq 2MM_1 \int_0^t \left(p\beta(\{v_n(s)\}) + q \sup_{\theta \in [0,s]} \beta(\{v_n[\varphi_n](\theta)\})\right)ds \\
&\leq 2MM_1(p+q) \int_0^t \sup_{\theta \in [0,s]} \beta(\{v_n(\theta)\})ds.
\end{aligned}
$$

Arguing as in the proof of Theorem 3.4, we obtain that $\{v_n(t)\}$ is relatively compact for each $t \in [0, b]$.

We now prove that $\pi_b \circ \Sigma(\varphi)$ is a compact set for each $\varphi \in \mathcal{C}_h$. It suffices to show that it is closed. Assume that $v_n \in \pi_b \circ \Sigma(\varphi), v_n \to v^*$ in $C([0, b], X)$. By the same arguments as in the proof of Theorem 3.7, we get $v^* \in \pi_b \circ \Sigma(\varphi)$. That is, $\pi_b \circ \Sigma(\varphi)$ is closed. The proof is complete.

The m-semiflow governed by (3.3) is defined as follows

$$
G : \mathbb{R}^+ \times \mathcal{C}_h \to P(\mathcal{C}_h),
$$
$$
G(t, \varphi) = \{u_t : u[\varphi] \text{ is an integral solution of (3.3)}\}.
$$

By the same argument as in [55], we see that

$$
G(t_1 + t_2, \varphi) = G(t_1, G(t_2, \varphi)) \text{ for all } t_1, \ t_2 \in \mathbb{R}^+, \varphi \in \mathcal{C}_h.
$$

We will prove that G is u.s.c. in the following lemma.

Lemma 3.9 *Under assumptions (H_A) and (F_1)–(F_3), $G(t, \cdot)$ is u.s.c. with compact values for each $t > 0$.*

Proof We have that $\pi_t \circ \Sigma(\varphi)$ is compact in $C([0, t], X)$ for any $t > 0$ as proved in Lemma 3.8. This deduces that $G(t, \varphi)$ is a compact set for each $\varphi \in \mathcal{C}_h$, that is $G(t, \cdot)$ has compact values. By Lemma 1.9, it remains to show that $G(t, \varphi)$ is quasicompact and has a closed graph.

We first show that $G(t, \cdot)$ is quasicompact. Assume that $K \subset \mathcal{C}_h$ is a compact set. Let $\{z_n\} \subset G(t, K)$, then one can find a sequence $\{\varphi_n\} \subset K$ such that $z_n \in G(t, \varphi_n)$. One can assume that $\{\varphi_n\}$ converges to φ^* in $\mathcal{C}_h$. Let $u_n \in \Sigma(\varphi_n)$ be such that

$$
z_n(s) = u_n[\varphi_n](t+s), \quad s \in [-h, 0]. \tag{3.19}
$$

By Lemma 3.8, we obtain that $\pi_t \circ \Sigma(\{\varphi_n\})$ is relatively compact in $C([0, t], X)$. Then, there is a subsequence of $\{u_n\}$ (denoted again by $\{u_n\}$) such that

$$\pi_t(u_n) \to u^* \text{ in } C([0, t], X).$$

Therefore, relation (3.19) implies that $\{z_n\}$ converges to $u^*[\varphi^*]_t$.

We now prove that $G(t, \cdot)$ has a closed graph. Let $\{\varphi_n\}$ be a sequence in $\mathcal{C}_h$ converging to φ^* and $\xi_n \in G(t, \varphi_n)$ such that $\xi_n \to \xi^*$. We have to show that $\xi^* \in G(t, \varphi^*)$. Choose $u_n \in \Sigma(\varphi_n)$ such that $\xi_n(s) = u_n[\varphi_n](t+s)$. By the result of Lemma 3.8, $\{u_n\}$ has a convergent subsequence (also denoted by $\{u_n\}$). Assume that $u^* = \lim_{n\to\infty} u_n$, then $u_n[\varphi_n] \to u^*[\varphi^*]$ in $C([-h, t], X)$ and $\xi^*(s) = u^*[\varphi^*](t+s)$, $s \in [-h, 0]$. Accordingly, it suffices to show that $u^* \in \pi_t \circ \Sigma(\varphi^*)$. Let $f_n \in Sel_F(u_n)$ be such that

$$u_n(r) = T'(r)\varphi_n(0) + \Gamma(f_n)(r), \quad r \in [0, t]. \tag{3.20}$$

By (F_2) and the fact that $\{u_n\}$ is bounded, we see that $\{f_n\} \subset L^1([0, t], X)$ is integrably bounded. Furthermore, $K(r) = F(\overline{\{u_n(r), u_n[\varphi_n]_r\}})$, $r \in [0, t]$ is compact and $\{f_n(r)\} \subset K(r)$. Then $\{f_n\}$ is a semicompact sequence. Applying Lemma 3.1 (ii), we have $f_n \rightharpoonup f^*$ and $\Gamma(f_n) \to \Gamma(f^*)$. Thus one can pass into limits equality (3.20) to obtain

$$u^*(r) = T'(t)\varphi^*(0) + \Gamma(f^*)(r), \quad r \in [0, t].$$

Since Sel_F is weakly u.s.c., one has $f^* \in Sel_F(u^*)$. So the last relation tells us that $u^* \in \pi_t \circ \Sigma(\varphi^*)$. The proof is complete.

3.3.2　Existence of Global Attractor

In this subsection, we need an additional assumption as following.
(S) $\{T'(t)\}_{t\geq 0}$ is a contraction semigroup. In addition, it is exponentially stable with exponent α, and is β-decreasing with exponent γ, that is

$$\|T'(t)\|_{\mathcal{L}(X)} \leq e^{-\alpha t}, \quad \|T'(t)\|_\beta \leq Ne^{-\gamma t}, \quad \forall\, t > 0,$$

where $\alpha, \gamma > 0$, $N \geq 1$, $\|\cdot\|_\beta$ is the β-norm defined in (3.15). It should be noted that if $\{T'(t)\}$ is a compact semigroup then $\|T'(t)\|_\beta = 0$ for all $t > 0$. In this case one can take $\gamma = +\infty$.

Denote by β_C the Hausdorff MNC on $\mathcal{C}_h$. We have the following properties of β_C (see [9]):

(1) $\sup_{s\in[-h,0]} \beta(D(s)) \leq \beta_C(D)$, for all $D \in \mathcal{C}_h$;
(2) if D is equicontinuous then $\beta_C(D) = \sup_{s\in[-h,0]} \beta(D(s))$.

For $b > h$, we define the translation operator $G_b = G(b, \cdot)$. We first prove the condensing property of G_b. Then using this fact, we show that the m-semiflow G admits a compact global attractor in $\mathcal{C}_h$.

Lemma 3.10 *Let hypotheses (H_A), (F_1)–(F_3) and (S) hold. If $\gamma - 4N(p + q) > 0$ then there exist $b > h$ and a number $\zeta \in [0, 1)$ such that*

$$\beta_C(G_b(B)) \leq \zeta \beta_C(B) \text{ for all bounded set } B \subset \mathcal{C}_h.$$

Proof Let $B \subset \mathcal{C}_h$ be a bounded set. Putting $D = \Sigma(B)$, we have

$$D(t) \subset \left\{ T'(t)B(0) + \lim_{\lambda \to +\infty} \int_0^t T'(t - s) \mathcal{J}_\lambda Sel_F(D)(s)ds \right\}, \quad t \geq 0. \quad (3.21)$$

It is readily seen that $\pi_t(D)$ is bounded in $C([0, t], X)$ for each $t > 0$. Thus, if $\{T'(t)\}$ is compact then $\beta(D(t)) = 0$. Assume opposite, i.e., $\{T'(t)\}$ is noncompact. Let

$$v(t) = \begin{cases} \beta(D(t)), & t \geq 0, \\ \beta(B(t)), & t \in [-h, 0]. \end{cases} \quad (3.22)$$

It follows from (3.21) that

$$v_n(t) \leq \beta(T'(t)B(0)) + \beta\left(\lim_{\lambda \to +\infty} \int_0^t T'(t - s) \mathcal{J}_\lambda Sel_F(D)(s)ds \right)$$

$$\leq Ne^{-\gamma t}\beta(B(0)) + \lim_{\lambda \to +\infty} \beta\left(\int_0^t T'(t - s) \mathcal{J}_\lambda Sel_F(D)(s)ds \right)$$

$$\leq Ne^{-\gamma t}\beta(B(0)) + 4MN \lim_{\lambda \to +\infty} \int_0^t e^{-\gamma(t-s)}\beta(Sel_F(D)(s))ds$$

according to assumption (S) and Property 1.5. Thus

$$v_n(t) \leq Ne^{-\gamma t}\beta(B(0)) + \lim_{\lambda \to +\infty} \beta\left(\int_0^t T'(t - s) \mathcal{J}_\lambda Sel_F(D)(s)ds \right)$$

$$\leq e^{-\gamma t}\left[N\beta(B(0)) + 4MN \int_0^t e^{\gamma s}\left(p\beta(D(s)) + q \sup_{\theta \in [s-h,s]} \beta(D[B](\theta)) \right)ds \right],$$

where

$$D[B](t) = \begin{cases} D(t), & t \geq 0, \\ B(t), & t \in [-h, 0]. \end{cases}$$

Denoting by $z(t)$ the right-hand side of the last inequality and setting $z(s) = Nv(s)$ for $s \in [-h, 0]$, we have $v(t) \leq z(t)$ for all $t \geq -h$, and

$$z'(t) = - \gamma z(t) + 4N \left(p v(t) + q \sup_{s \in [t-h,t]} v(s) \right)$$

$$\leq - (\gamma - 4Np)z(t) + 4Nq \sup_{s \in [t-h,t]} z(s), \quad t \geq 0.$$

Applying Halanay's inequality for z, we have

$$z(t) \leq \sup_{s \in [-h,0]} z(s)e^{-lt} = N \sup_{s \in [-h,0]} v(s)e^{-lt}, \quad t \geq 0,$$

where l is the solution of the equation $\gamma - 4Np = l + 4Nqe^{lh}$. Therefore

$$v(t) \leq z(t) \leq N \sup_{s \in [-h,0]} \beta(B(s))e^{-lt} \leq N\beta_C(B)e^{-lt}, \quad t \geq 0,$$

according to the definition of v in (3.22). This implies

$$\sup_{\theta \in [-h,0]} v(b + \theta) \leq N\beta_C(B)e^{-l(b-h)}.$$

Taking into account (3.21), one has

$$D_b(\theta) \subset T'(b + \theta)B(0) + \lim_{\lambda \to +\infty} \int_0^{b+\theta} T'(b + \theta - s) \mathcal{J}_\lambda Sel_F(D)(s)ds$$

$$\subset T'(b + \theta)B(0) + \Gamma \circ Sel_F(D)(b + \theta), \quad \theta \in [-h, 0].$$

Since $b > h$ and $T(\cdot)$ is norm continuous, the set $T(b + \cdot)B(0)$ is equicontinuous in $\mathcal{C}_h$. By the proof of Lemma 3.1 (ii), the set $\Gamma \circ Sel_F(D)(b + \cdot)$ is also equicontinuous in $\mathcal{C}_h$. So is D_b. This fact leads to

$$\beta_C(D_b) = \sup_{\theta \in [-h,0]} \beta(D(b + \theta)) = \sup_{\theta \in [-h,0]} v(b + \theta) \leq \zeta \beta_C(B),$$

where $\zeta = Ne^{-l(b-h)}$. Finally, choosing $b > h + \frac{1}{l} \ln N$ and taking into account that

$$G_b(B) = \Sigma(B)_b = D_b,$$

we have the conclusion of the lemma as desired.

Lemma 3.11 *Assume* (H_A), (F_1)–(F_3) *and* (S) *hold. Then* G *has an absorbing set provided that* $\alpha > a_1 + a_2$.

Proof Let $t > 0$ and $B \subset \mathcal{C}_h$ be a bounded set. Taking $\varphi \in B, |\varphi|_{\mathcal{C}_h} \leq C$, we consider the solution $u[\varphi]$ given by

$$u(t) = T'(t)\varphi(0) + \lim_{\lambda \to +\infty} \int_0^t T'(t - s) \mathcal{J}_\lambda f(s)ds$$

for $f \in Sel_F(u)$. Using (F_2) and (S), we have

$$|u(t)| \le e^{-\alpha t}|\varphi(0)| + \int_0^t e^{-\alpha(t-s)}\big(a_1|u(s)| + a_2|u_s|_{C_h} + a_3\big)ds. \qquad (3.23)$$

Since $\alpha - a_1 > a_2$, one can choose $R > 0$ such that $a_2 + \frac{a_3}{R} = d < \alpha - a_1$. We first prove that the solution $u[\varphi]$ has the following property: $\forall\, t_0 > 0$ such that $|u_{t_0}|_{C_h} \le R$. Assume the opposite: for all $t > 0$, $|u_t|_{C_h} > R$ then

$$a_2|u_s|_{C_h} + a_3 = |u_s|_{C_h}\left(a_2 + \frac{a_3}{|u_s|_{C_h}}\right) \le |u_s|_{C_h}\left(a_2 + \frac{a_3}{R}\right) = d|u_s|_{C_h}, \quad \forall\, s \ge 0.$$

Thus, (3.23) implies

$$|u(t)| \le e^{-\alpha t}|\varphi(0)| + \int_0^t e^{-\alpha(t-s)}\big(a_1|u(s)| + d|u_s|_{C_h}\big)ds, \quad t \ge 0.$$

Let

$$v(t) = \begin{cases} e^{-\alpha t}|\varphi(0)| + \int_0^t e^{-\alpha(t-s)}\big(a_1|u(s)| + d|u_s|_{C_h}\big)ds, & t \ge 0, \\ |\varphi(t)|, & t \in [-h, 0]. \end{cases}$$

Then we have $|u(t)| \le v(t)$ for all $t \ge -h$ and the following estimate:

$$v'(t) \le -(\alpha - a_1)v(t) + d \sup_{s \in [t-h,t]} v(s), \quad t \ge 0.$$

Applying Halanay's inequality yields

$$|u(t)| \le |\varphi|_{C_h} e^{-lt} \le Ce^{-lt} \quad \text{for all } t \ge 0,$$

where l is a positive number. Then we have

$$R < |u_t|_{C_h} = \sup_{\theta \in [-h,0]} |u(t+\theta)| \le Ce^{-l(t+\theta)}, \quad \forall\, t \ge h.$$

Hence for t large enough, we get the contradiction.

We just proved that if $\|\varphi\| \le C$ then there exists $t_0 > 0$ such that $|u_{t_0}|_{C_h} \le R$. We assert that $|u_t|_{C_h} \le R$, $\forall\, t \ge t_0$. Assume to the contrary that there exists $t_1 \ge t_0$ such that

$$|u_{t_1}|_{C_h} \le R \text{ but } |u_t|_{C_h} > R \text{ for all } t \in (t_1, t_1 + \theta),$$

where $\theta > 0$. Regarding the solution $u[\varphi]$ on $[t_1, t_1 + \theta)$, we have

$$u(t) = S(t - t_1)u(t_1) + \lim_{\lambda \to +\infty} \int_{t_1}^{t} T'(t - s)\mathcal{J}_\lambda f(s)ds,$$

where $f \in Sel_F(u)$. Then

$$|u(t)| \le e^{-\alpha(t-t_1)}|u(t_1)| + \lim_{\lambda \to +\infty} \int_{t_1}^{t} e^{-\alpha(t-s)}(a_1|u(s)| + d|u_s|_{\mathcal{C}_h})ds, \quad t \in [t_1, t_1 + \theta).$$

Then using the same arguments as above, we see that for all $t \in [t_1, t_1 + \theta)$,

$$|u(t)| \le |u_{t_1}|_{\mathcal{C}_h} e^{-l(t-t_1)} \le |u_{t_1}|_{\mathcal{C}_h} \le R.$$

Hence for $t \in (t_1, t_1 + \theta)$, we have

$$
\begin{aligned}
|u_t|_{\mathcal{C}_h} = \sup_{s \in [-h,0]} |u(t+s)| &= \sup_{r \in [t-h,t]} |u(r)| \\
&\le \sup_{r \in [t_1-h,t]} |u(r)| \\
&= \max \left\{ \sup_{r \in [t_1-h,t_1]} |u(r)|, \; \sup_{r \in [t_1,t]} |u(r)| \right\} \\
&= \max \left\{ |u_{t_1}|_{\mathcal{C}_h}, \; \sup_{r \in [t_1,t]} |u(r)| \right\} \\
&\le R.
\end{aligned}
$$

This is a contradiction. In summary, one can take the ball centered at 0 with radius R as an absorbing set for the m-semiflows G, where R is chosen such that $R > \frac{c}{\alpha - a_1 - a_2}$.

Lemma 3.12 *Let* (H_A), (F_1)–(F_3) *and* (S) *hold. If* $\gamma - 4N(p+q) > 0$, *then* G *is asymptotically upper semicompact.*

Proof Let $B \subset \mathcal{C}_h$ be a bounded set and Ξ_B be the collection of all sequences $\{\xi_k : \xi_k \in G(t_k, B), \ t_k \to \infty\}$. Denote

$$\mu = \sup\{\beta_C(\Omega) : \Omega \in \Xi_B\}.$$

We show that $\mu = 0$. Assume the opposite, then for $\theta \in (0, (1 - \zeta)\mu)$ there exists $\Omega_\theta = \{\xi_k\} \in \Xi_B$ such that

$$\beta_C(\Omega_\theta) > \mu - \theta.$$

Here ζ is given in Lemma 3.10. Fixed $b > h$, for each $t_k \in (b, \infty)$ there is a number $m_k \in \mathbb{N}$ such that $t_k = m_k b + r_k$, $r_k \in [0, b)$. Let $\tau_k = (m_k - 1)b + r_k$, then $\xi_k \in G(t_k, B) = G(b + \tau k, B) = G_b(G(\tau_k, B))$, and one can take $\eta_k \in G(\tau_k, B)$ such that $\xi_k \in G_b(\eta_k)$. It follows that

$$\beta_C(\Omega_\theta) = \beta_C(\xi_k) \le \beta_C(G_b(\{\eta_k\})) \le \zeta\beta_C(\{\eta_k\}) \le \zeta\mu < \mu - \theta.$$

This contradiction completes the proof.

Combining Lemmas 3.9, 3.11 and 3.12, we arrive at the main result.

Theorem 3.8 *Let hypotheses* (H_A), (F_1)–(F_3) *and* (S) *hold. Then the m-semiflow* G *generated by system* (3.3) *admits a compact global attractor provided that*

$$\min\{\alpha - (a_1 + a_2), \gamma - 4N(p + q)\} > 0.$$

3.3.3 Applications

Example 3.3 Let Ω be a bounded open set in $\mathbb{R}^n$ with smooth boundary $\partial\Omega$ and $\mathcal{O} \subset \Omega$ be an open subset. Consider the following problem

$$\frac{\partial u}{\partial t}(t, x) - \Delta_x u(t, x) + \lambda u(t, x) = f(x, u(t, x)) + \sum_{i=1}^{m} b_i(x)v_i(t), \quad x \in \Omega, \ t > 0,$$

(3.24)

$$v_i(t) \in \left[\int_{\mathcal{O}} k_{1,i}(y)u(t - h, y)dy, \int_{\mathcal{O}} k_{2,i}(y)u(t - h, y)dy \right], \quad 1 \le i \le m, \quad (3.25)$$

$$u(t, x) = 0, \quad x \in \partial\Omega, \ t \ge 0, \tag{3.26}$$

$$u(s, x) = \varphi(s, x), \quad x \in \Omega, \ s \in [-h, 0], \tag{3.27}$$

where Δ_x is the Laplace operator with respect to variable x, $\lambda > 0$, $f : \Omega \times \mathbb{R} \to \mathbb{R}$ is a continuous function, $b_i \in C(\overline{\Omega}, \mathbb{R})$, $k_{j,i} \in L^1(\mathcal{O}, \mathbb{R})$ for $i \in 1, \ldots, m$, $j = 1, 2$ and $\varphi \in \mathcal{C}_h = C([-h, 0], C(\overline{\Omega}, \mathbb{R}))$.

The above model can be seen as a control problem with control $v = (v_1, \ldots, v_m)$, which is taken in the form of feedbacks expressed by inclusion (5.2). Here the interval $[z_1, z_2] = \{\tau z_1 + (1 - \tau)z_2 : \tau \in [0, 1]\}$.

In this example, by putting $u(t)(x) = u(t, x)$, we consider the unknown function u as a vector-valued function acting on $\mathbb{R}^+$ and taking values in $C(\overline{\Omega}, \mathbb{R})$. Let

$$X = C(\overline{\Omega}, \mathbb{R}),$$
$$X_0 = C_0(\overline{\Omega}, \mathbb{R}) = \{v \in C(\overline{\Omega}, \mathbb{R}) : v = 0 \text{ on } \partial\Omega\},$$

here X and X_0 are endowed with the sup-norm $|v| = \sup_{x \in \Omega} |v(x)|$. Define

$$A_1 v = \Delta v,$$
$$v \in D(A_1) = \{v \in C_0(\overline{\Omega}, \mathbb{R}) \cap H_0^1(\Omega, \mathbb{R}) : \Delta v \in C_0(\overline{\Omega}, \mathbb{R})\},$$

where Δ is the Laplacian on Ω. Obviously,

$$\overline{D(A_1)} = X_0 \subsetneq X.$$

Adopting the result in [180, Theorem 5] (for the case $m = 1$ and constant coefficients), we can state that if $\partial\Omega$ is locally regular of class $C^{2,\mu}$, $\mu > 0$ (see [180] for the description), then A_1 generates an analytic semigroup $\{e^{tA_1}\}_{t\geq 0}$ on X_0. In addition, we see that the embedding $D(A_1) \subset X_0$ is compact, here $D(A_1)$ is endowed with the graph norm $|v|_{A_1} = |v|_X + |\Delta v|_X$. Then A_1 has compact resolvent (see, e.g., [100, Proposition 4.25]) and hence the semigroup $\{e^{tA_1}\}_{t\geq 0}$ is compact. On the other hand, it was proved in [190, Theorem 4.1.4] that the semigroup $\{e^{tA_1}\}_{t\geq 0}$ is a contraction one, i.e.,

$$\|e^{tA_1}\|_{\mathcal{L}(X)} \leq 1, \quad \forall\, t > 0.$$

Let $A = A_1 - \lambda I$. Then we see that A generates a compact, analytic semigroup $\{e^{tA}\}_{t\geq 0}$ on X_0 such that

$$\|e^{tA}\|_{\mathcal{L}(X)} \leq e^{-\lambda t}, \quad \forall\, t > 0.$$

Now let $F_1 : X_0 \to X$, $F_2 : C([-h, 0], X) \to P(X)$ be such that

$$F_1(v)(x) = f(x, v(x)), \tag{3.28}$$

$$F_2(w)(x) = \sum_{i=1}^{m} b_i(x)\left[\int_{\mathcal{O}} k_{1,i}(y)w(-h, y)dy, \int_{\mathcal{O}} k_{2,i}(y)w(-h, y)dy\right]. \tag{3.29}$$

Then problem (3.24)–(3.27) is exactly a prototype of system (3.3) with $F(v, w) = F_1(v) + F_2(w)$.

We assume, in addition, that there exist functions $a_1, a_2 \in C(\overline{\Omega}, \mathbb{R})$ verifying that

$$|f(x, r)| \leq a_1(x)|r| + a_2(x), \quad \forall\, x \in \Omega, \, r \in R.$$

Then we have

$$|F_1(v)| \leq |a_1||v| + |a_2|, \quad \forall\, v \in C_0(\overline{\Omega}, \mathbb{R}).$$

It is easily seen that F_1 is a continuous (single-valued) map. Regarding F_2, we get

$$|F_2(w)| \leq \sum_{i=1}^{m} |b_i| \max\left\{\int_{\mathcal{O}} |k_{1,i}(y)|dy, \int_{\mathcal{O}} |k_{2,i}(y)|dy\right\} \cdot |w|_{\mathcal{C}_h}.$$

In addition, F_2 is a multimap with closed convex values and the range lying in a finite dimensional space $\text{span}\{b_1, \ldots, b_m\} \subset X$. Then one sees that F_2 maps any bounded set into a relatively compact set. The fact that F_2 has a closed graph can be testified by a simple argument. Thus F_2 is a u.s.c. multimap (thanks to Lemma 1.9 again) with convex closed and compact values, and so is F.

Following Theorem 3.8, the m-semiflow generated by (3.24)–(3.27) has a compact global attractor in $C([-h, 0], C(\overline{\Omega}, \mathbb{R}))$ if

$$|a| + \sum_{i=1}^{m} |b_i| \max \left\{ \int_{\mathcal{O}} |k_{1,i}(y)| dy, \int_{\mathcal{O}} |k_{2,i}(y)| dy \right\} < \lambda.$$

Example 3.4 It is worth noting that if A generates a C_0-semigroup $\{S(t)\}_{t \geq 0}$ on X then it also generates an integrated semigroup $\{T(t)\}_{t \geq 0}$, which is given by

$$T(t)v = \int_0^t S(s)v ds, \quad v \in X.$$

Hence our results, obviously, can be applied to this situation. We consider again problem (3.24)–(3.27) but now with $\Omega = \mathbb{R}^n$ and $\mathcal{O}$ is a bounded domain in $\mathbb{R}^n$. It is rewritten as follows

$$\frac{\partial u}{\partial t}(t, x) - \Delta_x u(t, x) + \lambda u(t, x) = f(x, u(t, x)) + \sum_{i=1}^{m} b_i(x)v_i(t), \quad x \in \mathbb{R}^n, \ t > 0,$$

$$\tag{3.30}$$

$$v_i(t) \in \left[\int_{\mathcal{O}} k_{1,i}(y)u(t - h, y)dy, \int_{\mathcal{O}} k_{2,i}(y)u(t - h, y)dy \right], \quad 1 \leq i \leq m, \tag{3.31}$$

$$u(s, x) = \varphi(s, x), \quad x \in \mathbb{R}^n, \ s \in [-h, 0]. \tag{3.32}$$

In this model, we assume that

(1) $b_i \in L^2(\mathbb{R}^n, \mathbb{R})$, $k_{j,i} \in L^2(\mathcal{O}, \mathbb{R})$, $j = 1, 2$ and $\varphi \in C([-h, 0], L^2(\mathbb{R}^n, \mathbb{R}))$;
(2) $f : \mathbb{R}^n \times \mathbb{R} \to \mathbb{R}$ such that $f(\cdot, z)$ is measurable for each $z \in \mathbb{R}$ and there exists $\kappa \in L^2(\mathbb{R}^n, \mathbb{R})$ verifying

$$|f(x, z_1) - f(x, z_2)| \leq \kappa(x)|z_1 - z_2|, \quad \forall x \in \mathbb{R}^n, \ z_1, z_2 \in \mathbb{R}.$$

Let

$$A_1 v = \Delta v, \quad v \in D(A_1) = H^2(\mathbb{R}^n, \mathbb{R}),$$
$$X = L^2(\mathbb{R}^n, \mathbb{R}).$$

Then it is known that A_1 generates an analytic semigroup $S_1(\cdot)$ on X (see, e.g., [100, Theorem 5.15]). Furthermore, $S_1(\cdot)$ is a contraction semigroup. Therefore, $A = A_1 - \lambda I$ generates the analytic semigroup $S(\cdot)$ given by $S(t) = e^{-\lambda t} S_1(t)$ and we have

$$\|S(t)\|_{\mathcal{L}(X)} \leq e^{-\lambda t}, \quad \forall t > 0.$$

This implies that $S(\cdot)$ is exponentially stable and β-decreasing with exponent λ.

Consider F_1, F_2 given by (3.28) and (3.29). Arguing as in the case of bounded domain, we see that $F = F_1 + F_2$ satisfies (F_1), that is F is a u.s.c. multimap with convex closed and compact values. It follows from (5.10) that

$$|F_1(v_1) - F_1(v_2)| \leq |\kappa||v_1 - v_2|, \quad \forall \, v_1, v_2 \in X.$$

The last inequality ensures that

$$\beta(F_1(B)) \leq |\kappa|\beta(B) \text{ for bounded set } B \subset X.$$

On the other hand, for a bounded set $C \subset C([-h, 0], X)$ we see that $F_2(C)$ is a bounded subset of the finite dimensional space formed by $\{b_i\}_{i=1}^m$. So

$$\beta(F_2(C)) = 0.$$

Let $F(v, w) = F_1(v) + F_2(w)$, then

$$\beta(F(B, C)) \leq \beta(F_1(B)) + \beta(F_2(C)) \leq |\kappa|\beta(B),$$

for all bounded set $B \subset X$, all bounded set $B \subset C([-h, 0], X)$. Thus F satisfies (F_3) with $p = |\kappa|$, $q = 0$.

We now check (F_2). It is easily seen that

$$|F_1(v)| \leq |\kappa||v| + |f(\cdot, 0)|,$$

$$|F2(w)| \leq \sum_{i=1}^m |b_i| \max \left\{ |k_{1,i}|_{L^2(\mathcal{O}, \mathbb{R})}, |k_{2,i}|_{L^2(\mathcal{O}, \mathbb{R})} \right\} \cdot |w|_{C([-h,0], X)}.$$

Then (F_2) takes place with

$$a_1 = |\kappa|, \quad a_2 = \sum_{i=1}^m |b_i| \max \left\{ |k_{1,i}|_{L^2(\mathcal{O}, \mathbb{R})}, |k_{2,i}|_{L^2(\mathcal{O}, \mathbb{R})} \right\}.$$

We have the following result due to Theorem 3.8.

Theorem 3.9 *The m-semiflow generated by* (3.30)–(3.32) *admits a compact global attractor in* $C([-h, 0], L^2(\mathbb{R}^n, \mathbb{R}))$ *provided that*

$$\max \left\{ 4|\kappa|, \sum_{i=1}^m |b_i| \max\{|k_{1,i}|_{L^2(\mathcal{O}, \mathbb{R})}, |k_{2,i}|_{L^2(\mathcal{O}, \mathbb{R})}\} \right\} < \lambda.$$

Chapter 4
Quasi-autonomous Evolution Inclusions

Abstract This chapter deals with a kind of semilinear differential inclusions in general Banach spaces. Firstly, we study different types of generalized solutions including limit and weak solutions. Under appropriate assumptions, we show that the set of the limit solutions is a compact R_δ-set. When the right-hand side satisfies the one-sided Perron condition, a variant of the well-known lemma of Filippov-Pliś, as well as a relaxation theorem, are proved. Secondly, we study a kind of semilinear evolution inclusions. If the nonlinearity is one-sided Perron with sublinear growth, then we establish the relation between the solutions of the considered differential inclusion and the solutions of the relaxed one. A variant of the well known Filippov-Pliś lemma is also proved. Finally, we analyze the existence of pullback attractor for non-autonomous differential inclusions with infinite delays by using measures of noncompactness. As samples of applications, we apply the abstract results to control systems driven by semilinear partial differential equations and multivalued feedbacks.

4.1 Introduction

Consider the following semilinear evolution inclusion

$$\begin{cases} u'(t) \in Au(t) + F(t, u(t)), & t \in I = [t_0, T] \subset \mathbb{R}, \\ u(t_0) = u_0, \end{cases} \qquad (4.1)$$

where the state $u(\cdot)$ takes values in a Banach space X with the norm $|\cdot|$, $A : D(A) \subset X \to X$ is a (unbounded) linear operator that generates a C_0- semigroup $\{S(t)\}_{t \geq 0}$, and $F : I \times X \to P(X)$ is a multifunction with nonempty closed values.

By a (mild) solution of (4.1) on $[t_0, \tau]$, where $\tau \in (t_0, T]$, we mean a continuous function $u : [t_0, \tau] \to X$ for which there exists $f_u \in L^1([t_0, \tau], X)$ such that $f_u(s) \in F(s, u(s))$ for a.a. $s \in [t_0, \tau]$ and

$$u(t) = S(t - t_0)u_0 + \int_{t_0}^t S(t - s) f_u(s) ds \qquad (4.2)$$

© Springer Nature Singapore Pte Ltd. 2017

Y. Zhou et al., *Topological Structure of the Solution Set for Evolution Inclusions*, Developments in Mathematics 51, https://doi.org/10.1007/978-981-10-6656-6_4

for all $t \in [t_0, \tau]$. It will be denoted by $u(\cdot, t_0, u_0)$ or simply by $u(\cdot)$, depending on the context. The function $f_u(\cdot)$ satisfying (4.2) is called the pseudo-derivative of $u(\cdot)$. Notice that the pseudo-derivatives coincide with the derivatives when $A = 0$. In Sect. 4.2 we shall refer to mild solutions as solutions.

Notice that a large class of semilinear parabolic partial differential equations (inclusions) can be rewritten as (4.1). This is the main reason for extensively studying (4.1). See, e.g., [57, 62, 64, 102, 130, 190] and references therein.

To prove the existence of solutions one needs either compactness type assumptions or dissipative type assumptions. We refer the reader to [139], where these conditions are comprehensively studied in the case $A = 0$. In the compactness case one assumes that either $F(\cdot, \cdot)$ satisfies compactness type conditions or the semigroup $S(t)$ is compact. In the last case, further assumptions are needed, i.e., X is reflexive, or the values of F are contained in some weakly compact subset of X. Regarding the dissipative type conditions, one assumes that $F(t, \cdot)$ is locally Lipschitz or locally Perron (see, e.g., [102]).

One approach to getting solutions for differential inclusions is to find a converging sequence of approximate solutions to some continuous function, which, under further assumptions, is a solution. Section 4.2 is devoted to the study of the limits of the approximate solutions, called here limit solutions, which are not necessarily mild solutions. For example, a limit solution may have no pseudo-derivative. This phenomena has been considered in [91], where, for ordinary (functional) differential inclusions with the right-hand side $F(\cdot, \cdot)$ continuous, weak solutions are defined as continuous functions $u(\cdot)$ with $u(t) \in u(s) + \int_s^t F(\tau, u(\tau))d\tau$ for every $t_0 \le s \le t \le T$. Later, in [181], directional solutions are defined as absolutely continuous functions $u(\cdot)$ which satisfy

$$\lim_{h \to 0} \text{dist}\left(\frac{u(t+h) - u(t)}{h}, F(t, u(t))\right) = 0 \tag{4.3}$$

for a.a. $t \in I$ and the equivalence between them and weak solutions is proved, under the hypothesis that $F(\cdot, \cdot)$ is upper semicontinuous.

In Sect. 4.2 we consider different types of (generalized) solutions for the differential inclusion (4.1), which are generalizations of the mild one. We are interested in the topological properties of the set of limit solutions of (4.1) and we obtain that, under appropriate assumptions, this set is compact R_δ. Further, we give an integral representation of the limit solution and define the so-called weak solutions.

Another goal of Sect. 4.2 is to extend the classical relaxation theorem for limit solutions. To this end, we first prove a variant of the Filippov-Pliś lemma when the right-hand side is one-sided Perron.

One of the motivations of Sect. 4.2 is the application to optimal control. Consider the optimal control system

$$\min\left\{g(T, u(T)) + \int_{t_0}^T f(t, u(t))dt\right\} \tag{4.4}$$

subject to the solution $u(\cdot)$ of (4.1), Suppose that A generates a compact semigroup. If $F(\cdot, \cdot)$ does not have compact values, then the optimal control problem (4.4) may have no optimal solution, even if $F(\cdot, \cdot)$ has closed, bounded, convex values and $F(t, \cdot)$ is Lipschitz continuous. Clearly, the optimal limit solution always exists when $g(\cdot, \cdot)$ is lower semicontinuous and $f(\cdot, \cdot)$ is Carathéodory.

Notice that the optimal control problems for ordinary differential inclusions (even in more general form as written above) have been studied by many authors. In the fundamental paper [149] the technique of discrete approximations is used to derive strong approximation of the solutions (with their derivatives)and necessary optimality conditions. We recall [94], where the approximation results from [149] are extended in the case of Hilbert spaces, without Lipschitz assumption. Also, we refer to [151, 152], where the necessary optimal conditions in the case of evolution inclusions are derived using the discrete approximations technique. For a comprehensive discussion on that approach we refer the reader to [150]. Also, among others, notice the recent book [141]. The discrete approximations technique can be also applied here.

In Sect. 4.3, we consider the Cauchy problem

$$\begin{cases} u'(t) \in Au(t) + F(t, u(t)), & t \in [0, T], \\ u(0) = u_0 \in \overline{D(A)}, \end{cases} \tag{4.5}$$

where $A : D(A) \subseteq X \to P(X)$ is an m-dissipative operator, generating the (possibly nonlinear) semigroup $\{S(t) : \overline{D(A)} \to \overline{D(A)}, \ t \geq 0\}$. Let $F : [0, T] \times X \to P(X)$ be a multifunction with nonempty values. Then we proceed to study the relation between the solutions to the problem (4.5) and the solutions to the relaxed (convexified) problem

$$\begin{cases} u'(t) \in Au(t) + \overline{co}F(t, u(t)), \\ u(0) = u_0. \end{cases} \tag{4.6}$$

The relaxation problem consists in proving that the set of solutions to the nonconvex problem is dense in the set of solutions to the convexified problem. To obtain such results it is not enough to assume that F is continuous. In [170], a counterexample which illustrates this fact was provided. However, it is possible to get relation between so-called ε-solutions (i.e., solutions to the problems with the right-hand sides $F(t, u(t) + \varepsilon \mathbb{B})$ and $\overline{co}F(t, u(t) + \varepsilon \mathbb{B})$, respectively, where $\mathbb{B}$ is the closed unit ball in X). See [140], for the semilinear case, and [53], for the nonlinear case.

The existing relaxation results for the nonlinear differential inclusions involve the Lipschitz condition on the multifunction F. Moreover, we should point out that all these results were obtained under additional hypothesis on the involved Banach spaces, that is, reflexivity. We refer, for details, to the following papers: [164], for a differential inclusions involving time-dependent convex subdifferentials in a separable Hilbert space; [166], for parametric nonlinear evolution inclusions defined on an evolution triple of spaces; [147], for the case of a separable Hilbert space and the map $A(\cdot, u)$ being measurable, $A(t, \cdot)$ monotone and hemicontinuous, and $A(\cdot, \cdot)$ sublinear and of coercive type.

We should mention that our results on relaxation are established under an weaker than Lipschitz hypothesis on the multifunction F, that is, one-sided Perron hypothesis. The latter property is much weaker than the former, since almost all (in the Baire sense) Carathéodory multifunctions in a finite dimensional space are locally one-sided Perron, but the set of all locally Lipschitz Carathéodory multifunctions is of first Baire category. This result was proved in [92].

There were few relaxation results under assumptions weaker than Lipschitz one. For instance, in the finite dimensional case with $A = 0$, Pianigiani [169] imposed full Perron assumption, which is weaker than Lipschitz, but still stronger than one-sided Perron. Tateishi [182] extended to the infinite dimensional case the argument used by Pianigiani. Recently, the relaxation result under one-sided Perron assumption in the finite dimensional case with $A = 0$ was established in [95].

We would like to stress that all the above mentioned results were obtained for reflexive Banach spaces. In this section, we extend these results to the infinite dimensional nonlinear case without reflexivity assumptions on the space. Notice that in this case, even when F is almost continuous with compact values, the set of solutions to (4.6) is not necessarily closed. We refer the reader to [45].

Our approach here is based on an existence result from [23]. To this end, we construct a suitable almost lower semicontinuous multifunction G with $G(t, u) \subseteq F(t, u)$ for all $(t, u) \in [0, T] \times X$.

In Sect. 4.4, we consider the following non-autonomous differential inclusions with infinite delays

$$\begin{cases} u'(t) \in Au(t) + F(t, u_t), & t \geq \tau, \\ u_\tau(s) = \varphi^\tau(s), & s \in (-\infty, 0], \end{cases} \tag{4.7}$$

where the state function $u(\cdot)$ takes values in X, A is a closed linear operator generating a C_0-semigroup $\{S(t)\}_{t \geq 0}$ on X, F is a multivalued function defined on $\mathbb{R} \times \mathscr{B}$, u_t is the history of the state function up to the time t, i.e., $u_t(s) = u(t + s)$ for $s \in (-\infty, 0]$. The function $\varphi^\tau \in \mathscr{B}$ takes values in X and plays the role of initial datum. Here $\mathscr{B}$ is a phase space.

Let $(\mathscr{B}, |\cdot|_{\mathscr{B}})$ be a semi-normed linear space of functions mapping from $(-\infty, 0]$ into a Banach space X. The definition of the phase space $\mathscr{B}$, introduced in [120], is based on the following axioms. If $v : (-\infty, \sigma + T] \to X$, where $\sigma \in \mathbb{R}$ and T is a positive number, is a function such that $v|_{[\sigma, \sigma+T]} \in C([\sigma, \sigma + T], X)$ and $v_\sigma \in \mathscr{B}$, then

(B_1) $v_t \in \mathscr{B}$ for $t \in [\sigma, \sigma + T]$;

(B_2) the function $t \mapsto v_t$ is continuous on $[\sigma, \sigma + T]$;

(B_3) $|v_t|_{\mathscr{B}} \leq K(t - \sigma) \sup\{|v(s)| : \sigma \leq s \leq t\} + M(t - \sigma)|v_\sigma|_{\mathscr{B}}$ for each $t \geq \sigma$, where $K, M : [0, \infty) \to [0, \infty)$, K is continuous, M is locally bounded and they are independent of v.

In this work, we need an additional assumption on $\mathscr{B}$:

(B_4) there exists $\ell > 0$ such that $|\phi(0)|_X \leq |\ell(0)|_{\mathscr{B}}$ for all $\phi \in \mathscr{B}$.

A typical example of phase spaces is $\mathcal{C}_\gamma$ defined as follows:

$$\mathcal{C}_\gamma = \{\phi \in C((-\infty, 0], X) : \lim_{\theta \to -\infty} e^{\gamma\theta}\phi(\theta) \text{ exists in } X\}.$$

In this case $K(t) = 1$, $M(t) = e^{-\gamma t}$ for all $t \geq 0$. For more examples of phase spaces, see [121].

One of the most important and interesting problems related to differential inclusions is to study the stability of solutions. Since, the question of uniqueness of solutions to differential inclusions is no longer addressed, the Lyapunov theory for stability is not a suitable choice. Thanks to the theories of attractors for multivalued semiflows/processes, one can find a global attractor for semiflows/processes governed by solutions of differential inclusions, which is a compact set attracting all solutions as the time goes to infinity in some contexts.

In Sect. 4.4, we first prove the global solvability for (4.7) under a general setting. Then by giving a new criterion ensuring the asymptotic compactness for multivalued non-autonomous dynamical systems, we show that the multivalued non-autonomous dynamical systems generated by (4.7) admits a pullback attractor in case the phase space $\mathscr{B} = \mathcal{C}_\gamma$. It should be noted that the appearance of infinite delays causes some technical difficulties in proving the dissipativeness as well as the asymptotic compactness of multivalued non-autonomous dynamical systems. These difficulties are due to the complication of phase spaces.

This section is organized as follows. Sections 4.2.1 and 4.2.2 study different types of (generalized) solutions for semilinear evolution inclusion (4.1), called limit and weak solutions. Under appropriate assumptions, we show that the set of the limit solutions is a compact R_δ-set. In Sect. 4.2.3, when the right-hand side satisfies the one-sided Perron condition, a variant of the well-known lemma of Filippov-Pliś, as well as a relaxation theorem, are proved. Section 4.3.1 is devoted to our main results concerning relaxation. In Sect. 4.3.2, some illustrative examples are presented and discussed. Section 4.4.1 is devoted to proving the solvability on $(-\infty, \tau + T]$, $T > 0$. In Sect. 4.4.2, with some additional assumptions, we prove the existence of a compact invariant pullback attractor in $\mathcal{C}_\gamma$ for the multivalued non-autonomous dynamical systems governed by our system. The Sect. 4.4.3 presents an application of the abstract results to a feedback control problem associated with partial differential equations.

The contents of Sect. 4.2 are adopted from Cârjǎ, Donchev and Lazu [59]. The results in Sect. 4.3 are taken from Cârjǎ, Donchev and Postolache [61]. The results in Sect. 4.4 are taken from Dac and Ke [77].

4.2 Generalized Solutions

In the beginning of this section we introduce the standing hypotheses on the multifunction F:

(F_1) there exists a constant $c > 0$ such that $|F(t, u)| \leq c(1 + |u|)$ for any $t \in I$ and any $u \in X$;

(F_2) for any $u \in X$, $F(\cdot, u)$ has an integrable selection, i.e., there exists $f \in L^1([t_0, T], X)$ such that $f(t) \in F(t, u)$ for a.a. $t \in [t_0, T]$;

(F_3) $F(\cdot, \cdot)$ is jointly measurable.

(F_4) $\overline{co} F$ has almost closed graph.

Let us remark that if the multifunction F satisfies (F_1), then the new hypothesis (F_3) is stronger than (F_2). Clearly, every map with almost closed graph is with measurable graph.

Further, we consider the following hypotheses:

(A_1) the semigroup $\{S(t)\}_{t \geq 0}$ is equicontinuous, i.e., $S(\cdot)$ is continuous in the uniform operator topology;

(A_2) the semigroup $\{S(t)\}_{t \geq 0}$ is compact, i.e., $S(t)$ is a compact operator for any $t > 0$.

Remark 4.1 The the hypothesis (A_2) implies that the space X is separable. See, for instance, Theorem 2.4.1 of [190] and [64, p.19, 289].

Lemma 4.1 ([185]) *Let $\varphi : [0, T] \times \mathbb{R}^k \to \mathbb{R}^k$ be a vector function of the variables (t, r) which has the properties*

(i) *φ is a function of Caratheodory type, integrally bounded on bounded subsets from $[t_0, T] \times \mathbb{R}^k$;*

(ii) *$\varphi(t, r)$ for almost every $[t_0, T]$ satisfies Wazewski's condition on the second argument, i.e., does not decrease on extra-diagonal elements of vector*

$$\varphi_l(t, r) \leq \varphi_l(t, s), \quad r_l = s_l, \quad r_m \leq s_m, \quad m \neq l, \quad m, l = 1, \ldots, k.$$

Suppose that $r(t)$, $r(0) = r_0$, is an upper solution of the equation $r' = \varphi(t, r)$ defined on $[t_0, T]$. If the continuous function $s(t)$ is absolutely upper semicontinuous, $s(0) \leq r(0)$, and

$$\underline{D}^+ s(t) \leq \varphi(t, s(t))$$

almost everywhere on $[0, T]$, where the symbol $\underline{D}^+$ denotes the right hand lower derivative, then $s(t) \leq r(t)$, $t \in [t_0, T]$.

4.2.1 Limit Solutions

In this subsection we define the limit solutions of (4.1) and then we study their main properties. To this aim, for a given $\varepsilon > 0$, we introduce first the notion of an ε-solution. Recall that the open unit ball in X is denoted by $\mathbb{B}$ and $\overline{\mathbb{B}}$ denotes the closed unit ball in X.

Definition 4.1 A function $u : [t_0, \tau] \to X$, $\tau \in (t_0, T]$ is called an ε-solution of (4.1) on $[t_0, \tau]$ if it is a solution on $[t_0, \tau]$ of the differential inclusion

$$\begin{cases} u'(t) \in Au(t) + F(t, u(t) + \varepsilon\overline{\mathbb{B}}), \\ u(t_0) = u_0. \end{cases} \tag{4.8}$$

Recall that for every C_0-semigroup, $\{S(t)\}_{t \geq 0}$, there exist $M \geq 1$ and $\omega \in \mathbb{R}$ such that $\|S(t)\|_{\mathcal{L}(X)} \leq Me^{\omega t}$ for any $t \geq 0$ (see, e.g., [190]).

Lemma 4.2 *Under (F_1)-(F_2), for any $\varepsilon > 0$ there exists an ε-solution of (4.1) defined on $[t_0, T]$.*

Proof Let $f_0 \in L^1([t_0, T], X)$ be such that $f_0(t) \in F(t, u_0)$ for a.a. $t \in [t_0, T]$ and define

$$u(t) = S(t - t_0)u_0 + \int_{t_0}^t S(t - s)f_0(s)ds$$

for any $t \in [t_0, T]$. We have that

$$|u(t) - u_0| \leq |S(t - t_0)u_0 - u_0| + Me^{\omega T}c(1 + |u_0|)(t - t_0)$$

for any $t \in [t_0, T]$. Hence for any $\varepsilon > 0$ there exists $t_1 > t_0$ such that $|u(t) - u_0| < \varepsilon$ for any $t \in [t_0, t_1]$, so $F(t, u_0) \subset F(t, u(t) + \varepsilon\overline{\mathbb{B}})$ for any $t \in [t_0, t_1]$. Therefore, u is an ε-solution of (4.1) on $[t_0, t_1]$. Denote by $\mathcal{Y}$ the set of all ε-solutions u_a of (4.1) defined on $[t_0, a]$, $a \in (t_0, T]$. We have proved that $\mathcal{Y}$ is nonempty. We introduce the partial ordering $u_a \preceq u_b$ if $a \leq b$ and $u_a(t) = u_b(t)$ on $[t_0, a]$. It is obvious that every ordered subset of $\mathcal{Y}$ has an upper bound; hence, by Zorn's lemma, $\mathcal{Y}$ has a maximal element u_τ defined on $[t_0, \tau]$. Moreover, we have that $\tau = T$, since otherwise we can extend u_τ beyond τ, repeating the previous construction replacing t_0 by τ and u_0 by $u_\tau(\tau)$.

Lemma 4.3 *Assume that (F_1), (F_2) and (A_1) hold. Then the solution set of*

$$\begin{cases} u'(t) \in Au(t) + \overline{co}F(t, u(t) + \overline{\mathbb{B}}), \\ u(t_0) = u_0 \end{cases} \tag{4.9}$$

is nonempty, uniformly bounded, and equicontinuous. Moreover, if (A_2) holds, then the solution set of (4.9) is precompact in $C(I, X)$.

Proof Denote by Λ the solution set of (4.9), which is nonempty by Lemma 4.2. Let us remark that, by (F_1), we have

$$|\overline{co}F(t, u(t) + \overline{\mathbb{B}})| \leq c(2 + |u|) \tag{4.10}$$

for any $t \in I$ and any $u \in X$. Hence $|\Lambda| \leq k$ for some positive constant k (see [125, Theorem 3.14]). From [125, Theorem 3.12] we know that Λ is equicontinuous. For the last part of the proof, let $u(\cdot)$ be a solution of (4.9). Then there exists $f_u \in L^1([t_0, T], X)$ such that $f_u(s) \in \overline{co}F(s, u(s) + \overline{\mathbb{B}})$ for a.a. $s \in [t_0, T]$ and

$$u(t) = S(t - t_0)u_0 + \int_{t_0}^{t} S(t - t_0) f_u(s) ds$$

for all $t \in [t_0, T]$. Furthermore, $\overline{co} F(s, u(s) + \overline{\mathbb{B}}) \subseteq c(2 + k)\overline{\mathbb{B}}$ for any $t \in [t_0, T]$. For $t \in (t_0, T]$ denote $K(t) = \{\int_{t_0}^{t} S(t - s) f_u(s) ds : u(\cdot) \in \Lambda\}$. For $0 < \varepsilon < t - t_0$ define the set $K_\varepsilon = S(\varepsilon) K(t - \varepsilon)$. Since $K(t - \varepsilon)$ is bounded in X and the operator $S(\varepsilon)$ is compact we get that the set K_ε is relatively compact. This, together with the inequality

$$\left| \int_{t_0}^{t} S(t - s) f_u(s) ds - \int_{t_0}^{t-\varepsilon} S(t - s) f_u(s) ds \right| \leq M e^{\omega T} c(2 + k)\varepsilon,$$

leads to the fact that the set $K(t)$ is precompact for any $t \geq t_0$. Finally, using Arzela-Ascoli's theorem we get that Λ is precompact in $C(I, X)$.

Remark 4.2 The above proof suggests that we may assume that there exists a positive constant μ such that $|F(t, u(t) + \overline{\mathbb{B}})| \leq \mu$ for all $t \in I$ and $u \in X$.

Remark 4.3 It follows from Lemma 4.3 that the solution set of

$$\begin{cases} u'(t) \in Au(t) + \mu \overline{\mathbb{B}}, \\ u(t_0) = u_0 \end{cases} \tag{4.11}$$

is nonempty, uniformly bounded, and equicontinuous. Moreover, under (A_2), the solution set of (4.11) is precompact in $C(I, X)$.

Lemma 4.4 *Suppose* (F_1), (F_3) *and* (A_2). *Then for every* $\varepsilon > 0$ *and every* $\varepsilon' > \varepsilon$ *the closure of the set of* ε'-solutions *of* (4.1) *contains the solution set of*

$$\begin{cases} u'(t) \in Au(t) + \overline{co} F(t, u(t) + \varepsilon \overline{\mathbb{B}}), \\ u(t_0) = u_0. \end{cases} \tag{4.12}$$

Proof Fix $\varepsilon' > \varepsilon > 0$ and let $y(\cdot)$ be a solution of (4.12), i.e.,

$$y(t) = S(t - t_0)u_0 + \int_{t_0}^{t} S(t - s) f_y(s) ds,$$

where $f_y(s) \in \overline{co} F(s, y(s) + \varepsilon \overline{\mathbb{B}})$. Let $\delta = \varepsilon' - \varepsilon$.

Clearly, every solution of (4.12) is a solution of (4.11). By the equicontinuity of the solution set of (4.11) (see Remark 4.3), there exists a (uniform) subdivision $t_0 < t_1 < \cdots < t_n = T$ such that for every solution $z(\cdot)$ of (4.11) we have that $|z(t) - z(\tau)| < \delta/3$ for any $t, \tau \in [t_k, t_k + 1]$ and $k \in \{0, 1, \ldots, n - 1\}$.

Since F is jointly measurable with closed values and $y(\cdot)$ is continuous, by Lemma 1.6, we have that the multifunction $t \to F(t, y(t) + \varepsilon \overline{\mathbb{B}})$ is measurable. Using Theorem 2.2 and Lemma 2.4 of [102] we get that

$$\overline{\int_\tau^t S(t-s)\overline{co}F(s, y(s) + \varepsilon\overline{\mathbb{B}})ds} = \int_\tau^t S(t-s)F(s, y(s) + \varepsilon\overline{\mathbb{B}})ds$$

for every $t_0 \le \tau < t \le T$. Hence for $k \in \{0, 1, \ldots, n-1\}$,

$$y(t_{k+1}) \in S(t_{k+1} - t_k)y(t_k) + \overline{\int_{t_k}^{t_k+1} S(t_{k+1} - s)F(s, y(s) + \varepsilon\overline{\mathbb{B}})ds},$$

so there exists a measurable selection $f_k(s) \in F(s, y(s) + \varepsilon\overline{\mathbb{B}})$ on $[t_k, t_k + 1]$ such that

$$\left| y(t_{k+1}) - S(t_{k+1} - t_k)y(t_k) - \int_{t_k}^{t_k+1} S(t_{k+1} - s)f_k(s)ds \right| < \frac{\delta}{3\mu},$$

where $\mu = (M^n e^{\omega nT} - 1)/(Me^{\omega T} - 1)$. Hence,

$$\left| \int_{t_k}^{t_k+1} S(t_{k+1} - s)(f_y(s) - f_k(s))ds \right| < \frac{\delta}{3\mu}.$$

Define $f(t) = f_k(t)$ for $t \in [t_k, t_k + 1)$ and take

$$u(t) = S(t - t_0)u_0 + \int_{t_0}^t S(t - s)f(s)ds.$$

Clearly, $u(\cdot)$ is a solution of (4.11). Then for every $t \in [t_k, t_k + 1)$, we have that

$$|u(t) - y(t)| \le |u(t_k + 1) - y(t_k + 1)| + |u(t) - u(t_k + 1)| + |y(t) - y(t_k + 1)| < \delta.$$

Hence $f(t) \in F(t, u(t) + (\varepsilon + \delta)\overline{\mathbb{B}})$, i.e., $u(\cdot)$ is $(\varepsilon + \delta)$-solution of (4.1). The proof is therefore complete.

Next, we introduce the notion of a limit solution for the differential inclusion (4.1).

Definition 4.2 A function $u : I \to X$ is called a limit solution of (4.1) if there exist a sequence of positive numbers $\{\varepsilon_n\}_{n=1}^\infty$ decreasing to zero and ε_n-solutions $u^n(\cdot)$ such that $\lim_{n\to\infty} u^n(t) = u(t)$ uniformly on I.

Along with (4.1) we consider the relaxed system

$$\begin{cases} u'(t) \in Au(t) + G(t, u(t)), & t \in I, \\ u(t_0) = u_0, \end{cases} \tag{4.13}$$

where $G(t, u(t)) = \cap_{\varepsilon>0}\overline{co}F(t, u(t) + \varepsilon\overline{\mathbb{B}})$. The ε-solutions and limit solutions of (4.13) are defined analogously to Definitions 4.1 and 4.2.

The next result concerns the topological structure of the set of limit solutions of (4.13).

Theorem 4.1 *Suppose* (F_1), (F_3) *and* (A_2). *Then the set of limit solutions of* (4.13) *is a compact* R_δ-*set.*

Proof The proof will be given using locally Lipshitz approximations of G. We refer the reader to [80], where ordinary differential inclusions are studied.

Let us denote by Θ the set of limit solutions of (4.13). Let $r_n = 3^{-n}$ and $\{U_\mu\}_{\mu \in \mathcal{M}}$ be a locally finite refinement of the open covering $X = \bigcup_{u \in X}(u + r_n \mathbb{B})$. Let $\{\varphi_\mu\}_{\mu \in \mathcal{M}}$ be a locally Lipschitz partition of unity subordinate to $\{U_\mu\}_{\mu \in \mathcal{M}}$ and take u_μ such that $U_\mu \subset u_\mu + r_n \mathbb{B}$. Consider the approximations

$$G_n(t, u) = \sum_{\mu \in \mathcal{M}} \varphi_\mu(u) C_\mu(t), \text{ where } C_\mu(t) = \overline{\text{co}}G(t, u_\mu + 2r_n \overline{\mathbb{B}}).$$

Then we have that

$$G(t, u) \subset G_{n+1}(t, u) \subset G_n(t, u) \subset \overline{\text{co}}G(t, u + 3r_n \mathbb{B}) \text{ on } I \times X. \tag{4.14}$$

For every n, denote by Θ_n the set of mild solutions of the differential inclusion

$$\begin{cases} u'(t) \in Au(t) + G_n(t, u(t)), \\ u(t_0) = u_0. \end{cases}$$

By (4.14) we get that $\Theta_{n+1} \subset \Theta_n$. Moreover, Θ_n is precompact for every n with $3r_n < 1$.

We shall prove that $\overline{\Theta_n}$ is contractible for every n. To this end, let $g_\mu(\cdot)$ be a measurable selection of $F(\cdot, u_\mu)$ for every $\mu \in \mathcal{M}$ and define

$$f(t, u) = \sum_{\mu \in \mathcal{M}} \varphi_\mu(u) g_\mu(t) \text{ on } I \times X.$$

Since for a.a. $t \in I$, $g_\mu(t) \in F(t, u_\mu) \subset C_\mu(t)$ we have that $f(t, u) \in G_n(t, u)$ for a.a. $t \in I$ and any $u \in X$. Since $\{U_\mu\}_{\mu \in \mathcal{M}}$ is locally finite we have that $f(\cdot, u)$ is measurable and $f(t, \cdot)$ is locally Lipschitz.

Take $\tau \in [0, 1]$ and denote $a_\tau = \tau T$. We define the homotopy $H : [0, 1] \times \overline{\Theta_n} \to \overline{\Theta_n}$ as follows:

$$H(\tau, x)(t) = \begin{cases} u(t), & t \in [t_0, a_\tau], \\ \widetilde{v}(t, a_\tau, u(a_\tau)), & t \in (a_\tau, T]. \end{cases}$$

We denoted by $\widetilde{v}(\cdot, s, u)$ the unique solution, defined on $[s, T]$, of

$$\begin{cases} v'(t) = Av(t) + f(t, v(t)), \\ v(s) = u. \end{cases} \tag{4.15}$$

The existence and uniqueness of the solution of (4.15) follows from the locally Lipschitz property and the growth condition satisfied by f (see, e.g., Theorems 3.2 and 4.4 of [46] or Theorem 2.2 (p. 335) of [144]). The latter implies in particular that $\Theta_n \neq \emptyset$ for every n. Moreover, due to Gronwall's inequality, $\tilde{v}$ depends continuously on the initial condition. Then H is continuous, $H(0, u) = \tilde{v}$, and $H(1, u) = u$. So, we have found a decreasing sequence of compact contractible sets $\overline{\Theta}_n \subset C(I, X)$. To complete the proof, by Definition 1.11, we only have to prove that

$$\Theta = \bigcap_{n=1}^{\infty} \overline{\Theta}_n.$$

Notice that $\Theta \subset \overline{\Theta}_n$ for any natural n. Indeed, let $u(\cdot) \in \Theta$ and fix n. Then there exist $\{\varepsilon_m\}_m \downarrow 0$ and $\{u_m\}_m$ a sequence of ε_m-solutions for (4.13) such that $\lim_{m \to \infty} u_m(t) = u(t)$ uniformly on I. Let m_n be such that $\varepsilon_{m_n} < r_n$. For any $m \geq m_n$ we have that

$$G(t, u_m(t) + \varepsilon m \overline{\mathbb{B}}) \subset G(t, u_m(t) + r_n \overline{\mathbb{B}}) \subset G_n(t, u_m(t)),$$

because, if $\varphi_\mu(u_m(t)) > 0$, then $u_m(t) \in U_\mu \subset u_\mu + r_n \overline{\mathbb{B}}$. Hence $u_m \in \Theta_n$ for any $m \geq m_n$.

Now, let $v \in \bigcap_{n=1}^{\infty} \overline{\Theta}_n$, so $v \in \overline{\Theta}_n$ for any n. Then for any n there exists a sequence $\{z_m^n\}_m \subset \Theta_n$ such that $\lim_{m \to \infty} z_m^n(t) = v(t)$ uniformly on I. Let $v_n = z_n^n$. By (4.14), v_n is a solution of $u'(t) \in Au(t) + \overline{co}G(t, u + 3r_n \overline{\mathbb{B}})$. Applying now Lemma 4.4 with G instead of F, we get easily that v is a limit solution for $u'(t) \in Au(t) + G(t, u(t))$.

The locally Lipschitz approximations have been used in the literature for more general systems than (4.1). Among others, notice [45, 46, 198], where m-dissipative differential inclusions in reflexive spaces are studied.

Theorem 4.2 *Suppose* (F_1), (F_3), $(F4)$ *and* (A_2). *Then the limit solution sets of* (4.1) *and* (4.13) *coincide. Hence the limit solution set of* (4.1) *is a compact* R_δ-*set.*

Proof Under $(F4)$, it is easy to prove that $\overline{co}F(t, u) = G(t, u)$ for every $u \in X$ and for a.a. $t \in I$. Therefore, the conclusion follows directly from Lemma 4.4 and Theorem 4.1.

4.2.2 Weak Solutions

In this subsection we provide an integral representation of the limit solutions. To this aim, following [91], we give the definition of weak solutions in the semilinear case.

Definition 4.3 A continuous function $u : I \to X$ is called a weak solution of (4.1) if $u(t_0) = u_0$ and

$$u(t) \in \overline{S(t - s)u(s) + \int_s^t S(t - \tau)F(\tau, u(\tau))d\tau}$$

for any $t_0 \le s \le t \le T$.

Clearly, every (mild) solution of (4.1) is also a weak solution.

Lemma 4.5 *Assume* (F_1), (F_2) *and* (A_1). *Then the set of weak solutions of (4.1) is equicontinuous.*

Proof Let $u(\cdot)$ be a weak solution of (4.1). Then for $t_0 \le s < t \le T$ we have that

$$u(t) - u(s) \in \overline{(S(t - s) - S(0))u(s) + \int_s^t S(t - \tau)F(\tau, u(\tau))d\tau}.$$

Recall that $|F(t, u(t) + \overline{\mathbb{B}})| \le \mu$ for $t \in I$ (see Remark 4.2). Therefore, for $t_0 \le s < t \le T$,

$$|u(t) - u(s)| \le |(S(t - s) - I)u(s)| + (t - s)\mu M e^{\omega T},$$

and using the equicontinuity of the semigroup we get the conclusion.

Now we will prove that the set of weak solutions coincides with the set of limit solutions, when $F(t, \cdot)$ is upper hemicontinuous for a.a. $t \in I$.

Theorem 4.3 *Assume* (F_1), (F_3) *and* (A_1). *Moreover, assume that* $F(t, \cdot)$ *is upper hemicontinuous for a.a.* $t \in I$. *Then* $u(\cdot)$ *is a weak solution of (4.1) if and only if* $u(\cdot)$ *is a limit solution of (4.1).*

Proof Let $u(\cdot)$ be a limit solution of (4.1). Then $u(t) = \lim_{n\to\infty} u_n(t)$ uniformly on $[t_0, T]$, where

$$u_n(t) = S(t - s)u_n(s) + \int_s^t S(t - \tau)f_n(\tau)d\tau$$

for every $t_0 \le s < t \le T$ with $f_n(\tau) \in F(\tau, u_n(\tau) + \varepsilon_n \overline{\mathbb{B}})$ a.e. on I and $\varepsilon_n \downarrow 0$. Fix $l \in X^*$. Then for any $t_0 \le s < t \le T$,

$$\left\langle l, \int_s^t S(t - \tau)f_n(\tau)d\tau \right\rangle \le \sigma\left(l, \int_s^t S(t - \tau)F(\tau, u_n(\tau) + \varepsilon_n \overline{\mathbb{B}})d\tau \right)$$

and, since $F(t, \cdot)$ is upper hemicontinuous for a.a. $t \in I$, we get that

$$\limsup_{n\to\infty}\left\langle l,\ \int_s^t S(t-\tau)f_n(\tau)d\tau\right\rangle\leq\sigma\left(l,\ \int_s^t S(t-\tau)F(\tau,u(\tau))d\tau\right).$$

On the other hand,

$$\limsup_{n\to\infty}\left\langle l,\ \int_s^t S(t-\tau)f_n(\tau)d\tau\right\rangle=\limsup_{n\to\infty}\langle l,u_n(t)-S(t-s)u_n(s)\rangle$$
$$=\langle l,u(t)-S(t-s)u(s)\rangle.$$

Hence we get that

$$\langle l,u(t)-S(t-s)u(s)\rangle\leq\sigma\left(l,\ \int_s^t S(t-\tau)F(\tau,u(\tau))d\tau\right)$$

for any $l\in X^*$. This implies that

$$u(t)\in S(t-s)u(s)+\overline{\int_s^t S(t-\tau)F(\tau,u(\tau))d\tau}.$$

So, we proved that $u(\cdot)$ is a weak solution of (4.1).

For the converse part, let us first recall that the set of solutions of (4.11) is equicontinuous (see Remark 4.3). Then for any $\varepsilon>0$ there exists $\delta>0$ such that $|u(t)-u(s)|<\varepsilon$ for $|t-s|<\delta$, uniformly with respect to the solution $u(\cdot)$ of (4.11). Let $u(\cdot)$ be a weak solution of (4.1), i.e.,

$$u(t)\in S(t-s)u(s)+\overline{\int_s^t S(t-\tau)F(\tau,u(\tau))d\tau}$$

for any $t_0\leq s<t\leq T$. Then for any $\varepsilon>0$ there exists N such that if we divide $[t_0,T]$ into N subintervals $[t_j,t_{j+1}]$ with equal length, less than δ (given before), we have

$$|u(t)-u(s)|<\frac{\varepsilon}{3}\tag{4.16}$$

for any $t,s\in[t_j,t_{j+1}]$. Then for any $j\in\{0,\ldots,N-1\}$ we have that

$$u(t_{j+1})\in S(t_{j+1}-t_j)u(t_j)+\overline{\int_{t_j}^{t_{j+1}}S(t_{j+1}-\tau)F(\tau,u(\tau))d\tau}.$$

There exists a strongly measurable selection $f_u^j(\tau)\in F(\tau,u(\tau)),\tau\in[t_j,t_{j+1}]$, such that

$$\left|u(t_{j+1})-S(t_{j+1}-t_j)u(t_j)-\int_{t_j}^{t_{j+1}}S(t_{j+1}-\tau)f_u^j(\tau)d\tau\right|<\frac{\varepsilon}{3\rho},\tag{4.17}$$

where $\rho = (M^N e^{\omega NT} - 1)/(M e^{\omega T} - 1)$. Define

$$y(t) = S(t - t_0)u_0 + \int_{t_0}^{t} S(t - s) f_y(s) ds,$$

where $f_y(s) = f_u^j(s)$ for $s \in [t_j, t_{j+1})$. So for $t \in [t_j, t_{j+1}]$,

$$y(t) = S(t - t_j) y(t_j) + \int_{t_j}^{t} S(t - s) f_u^j(s) ds.$$

Finally, let $t \in [t_0, T]$. Then $t \in [t_j, t_{j+1}]$ for some j, using (4.16), (4.17), and the fact that $y(\cdot)$ is a solution of (4.11), we have that

$$|y(t) - u(t)| \le |y(t) - y(t_{j+1})| + |y(t_{j+1}) - u(t_{j+1})| + |u(t_{j+1}) - u(t)| < \varepsilon.$$

Then $f_y(t) \in F(t, u(t)) \subset F(t, y(t) + \varepsilon \overline{\mathbb{B}})$ for a.a. $t \in [t_0, T]$. Therefore, $y(\cdot)$ is an ε-solution of (4.1) with $|y(t) - u(t)| < \varepsilon$ on $[t_0, T]$. Thus $u(\cdot)$ is a limit solution of (4.1).

Remark 4.4 The approach of Tabor [181] can be extended to the semilinear case as follows. The continuous function $u(\cdot)$ is said to be a directional solution of (4.1) if $u(t_0) = u_0$ and

$$\lim_{h \to 0^+} \text{dist}\left(\frac{u(t + h) - u(t)}{h}, \frac{S(h) - S(0)}{h} u(t) + F(t, u(t)) \right) = 0$$

for a.a. $t \in [t_0, T]$, i.e.,

$$\lim_{h \to 0^+} \text{dist}\left(\frac{u(t + h) - S(h)u(t)}{h}, F(t, u(t)) \right) = 0.$$

The latter can be rewritten as

$$|u(t + h) - S(h)u(t) - h F(t, u(t))| = o(h)$$

for a.a. $t \in [t_0, T]$, where $\lim_{h \to 0} \frac{o(h)}{h} = 0$. It is not difficult to prove under the hypotheses of Theorem 4.3, that $u(\cdot)$ is a directional solution of (4.1) if and only if $u(\cdot)$ is a weak solution of (4.1), when F has convex values.

4.2.3 One-Sided Perron Condition

In this subsection we study the differential inclusion (4.1) under hypotheses (F_1), (F_2), (A_2) and a one-sided Perron condition.

Definition 4.4 A Carathéodory function $\omega : [0, T] \times \mathbb{R}_+ \to \mathbb{R}_+$, integrally bounded on bounded sets, is called Perron function if for every $T > 0$ the zero function is the only solution on $[0, T]$ to the problem

$$\begin{cases} r'(t) = \omega(t, r(t)), \\ r(0) = 0. \end{cases}$$

We shall denote by $[x, y]_+$ the right directional derivative of the norm calculated at x in the direction y, i.e.,

$$[x, y]_+ = \lim_{h \downarrow 0} \frac{|x + hy| - |y|}{h}.$$

We refer the reader to [139], where this function is comprehensively studied. Notice only that $[\cdot, \cdot]_+$ is upper semicontinuous as a real valued function and $[x, \cdot]_+$ is Lipschitz of constant 1.

Definition 4.5 The multifunction $G : X \to P(X)$ is said to be one-sided Perron with respect to the Perron function $\omega(\cdot, \cdot)$ if for every $x, y \in X$, every $f^x \in G(x)$ and every $\varepsilon > 0$, there exists $f^y \in G(y)$ such that

$$[x - y, f^x - f^y]_+ < \omega(t, |x - y|) + \varepsilon.$$

We are ready now to give the definition of another type of approximate solutions.

Definition 4.6 The continuous function $u(\cdot)$ is said to be an outer ε-solution of (4.1) if

$$u(t) = S(t - t_0)u_0 + \int_{t_0}^t S(t - \tau) f_u(\tau) d\tau,$$

where $f_u(\cdot)$ is strongly measurable and $\text{dist}(f_u(t), F(t, u(t))) \le h_u(t)$ for a.a. $t \in [t_0, T]$ with $h_u(t) \le 2|F(t, u(t) + \overline{\mathbb{B}})|$ for any $t \in [t_0, T]$, $u \in X$ and

$$\int_{t_0}^T h_u(s) ds < \varepsilon.$$

Remark 4.5 Taking into account Remark 4.2, we may assume that $|h_u(t)| \le 2\mu$ for all $t \in [t_0, T]$.

The following result is then valid.

Lemma 4.6 *Let $F(\cdot, \cdot)$ be almost continuous with convex values. Under (F_1), (F_2), and (A_2), for every $\varepsilon > 0$ there exists $\delta > 0$ such that if $u(\cdot)$ is a δ-solution of (4.1), then it is an outer ε-solution of (4.1).*

Proof Notice that, under these hypotheses, we have that the solution set of (4.1) is $C(I, X)$ precompact by Lemma 4.3. Hence there exists a compact set $K \subset X$ such that $u(t) \in K$ for every $t \in I$ and every solution $u(\cdot)$ of (4.1).

Fix $\varepsilon > 0$ and let $\nu \in (0, \frac{\varepsilon}{4\mu})$. Since $F(\cdot, \cdot)$ is almost continuous, there exists a compact set $I_\nu \subset I$ with $\mathrm{mes}(I \setminus I_\nu) < \nu$ such that the restriction $F|_{I_\nu \times K}$ of $F(\cdot, \cdot)$ to $I_\nu \times K$ is continuous. Clearly, $F|_{I_\nu \times K}$ is uniformly continuous on the compact set $I_\nu \times K$. Then there exists $\delta > 0$ such that $d_H(F(t, u), F(t, u + \delta \overline{\mathbb{B}})) < \frac{\varepsilon}{2T}$ for any $(t, u) \in I_\nu \times K$. This implies that $F(t, u + \delta \overline{\mathbb{B}})) \subset F(t, u) + \frac{\varepsilon}{2T}\overline{\mathbb{B}}$ for any $(t, u) \in I_\nu \times K$. Let $u(\cdot)$ be a δ-solution for (4.1), i.e.,

$$u(t) = S(t - t_0)u_0 + \int_{t_0}^t S(t - t_0) f_u(s) ds,$$

where $f_u(s) \in F(s, u(s) + \delta \overline{\mathbb{B}})$ a.e. on $[t_0, T]$. Then $\mathrm{dist}(f_u(t), F(t, u(t))) < \frac{\varepsilon}{2T}$ for a.a. $t \in I_\nu$ and $\mathrm{dist}(f_u(t), F(t, u(t))) \leq 2\mu$ for $t \in I \setminus I_\nu$. Hence $u(\cdot)$ is an outer ε-solution.

Remark 4.6 Due to Lemma 4.6, the definition of outer ε-solutions is more general than Definition 4.1. Similar to limit solutions (see Definition 4.2), one can define outer limit solutions, taking outer ε-solutions instead of ε-solutions in Definition 4.2. Thus the set of outer limit solutions contains the set of limit solutions.

Now we give a variant of the Filippov-Pliś lemma, which has many applications in optimal control, approximation of differential inclusions, etc. We refer the reader to [33, 93].

Theorem 4.4 *Assume* (F_1), (F_2) *and* (A_2). *Further, suppose that* F *is almost lower semicontinuous and* $F(t, \cdot)$ *is one-sided Perron with respect to the Perron function* $\omega(\cdot, \cdot)$. *Let* $h : I \to \mathbb{R}_+$ *be a Lebesgue integrable function. If* $y(\cdot)$ *is a solution of*

$$\begin{cases} y' \in Ay + F(t, y(t)) + h(t)\overline{\mathbb{B}}, \\ y(t_0) = y_0, \end{cases} \tag{4.18}$$

then for every $\delta > 0$ *there exists a solution* $u(\cdot)$ *of* (4.1) *such that*

$$|u(t) - y(t)| < r(t),$$

where $r(\cdot)$ *is the maximal solution of*

$$\begin{cases} r'(t) = \omega(t, r(t)) + h(t) + \delta, \\ r(t_0) = |u_0 - y_0|. \end{cases} \tag{4.19}$$

Proof Let $y(\cdot)$ be a solution of (4.18). Then

$$y(t) = S(t - t_0)y_0 + \int_{t_0}^{t} S(t - s)(f_y(s) + g_y(s))ds,$$

where $f_y(s) \in F(s, y(s))$ and $g_y(s) \in h(s)\overline{\mathbb{B}}$ for a.a. $s \in [t_0, T]$. Fix $\delta > 0$ and define the multifunction

$$F_\delta(t, u) = \overline{\left\{ v \in F(t, u) : [y(t) - u, f_y(t) - v]_+ < \omega(t, |y(t) - u|) + \delta \right\}}.$$

The multifunction F_δ has nonempty values since $F(t, \cdot)$ is one-sided Perron. We shall prove that F_δ is almost lower semicontinuous. To this aim, it is enough to prove that

$$\widetilde{F}_\delta(t, u) = \{v \in F(t, u) : [y(t) - u, f_y(t) - v]_+ < \omega(t, |y(t) - u|) + \delta\}$$

is almost lower semicontinuous. From the almost lower semicontinuity of F and Lusin's property of $f_y(\cdot)$, we have that for every $\varepsilon > 0$ there exists a compact set $I_\varepsilon \subset [t_0, T]$ with mes$(I \setminus I_\varepsilon) < \varepsilon$ such that $F|_{I_\varepsilon \times X}$ is lower semicontinuous, $\omega|_{I_\varepsilon \times X}$ is continuous, and $f_y|_{I_\varepsilon}$ is continuous. Therefore, it remains to show that $\widetilde{F}_\delta|_{I_\varepsilon \times X}$ is lower semicontinuous. To this end, let $(t, u) \in I_\varepsilon \times X$. Let $l \in \widetilde{F}_\delta(t, u)$ and the sequence $\{(t_n, u_n)\}_n \subset I_\varepsilon \times X$ such that $(t_n, u_n) \to (t, u)$. There exists $\gamma > 0$ such that

$$[y(t) - u, f_y(t) - l]_+ \leq \omega(t, |y(t) - u|) + \delta - \gamma. \tag{4.20}$$

Since $F|_{I_\varepsilon \times X}$ is lower semicontinuous at (t, u) there exists a sequence $\{l_n\}_n$ with $l_n \in F(t_n, u_n)$ such that $l_n \to l$. As $f_y|_{I_\varepsilon}$ is continuous and $[\cdot, \cdot]_+$ is upper semicontinuous we get that

$$[y_n(t) - u_n, f_y(t_n) - l_n]_+ \leq [y(t) - u, f_y(t) - l]_+ + \frac{\gamma}{2} \tag{4.21}$$

for n large enough. Also, since $\omega|_{I_\varepsilon \times X}$ is continuous, we have that

$$\omega(t, |y(t) - u|) \leq \omega(t_n, |y(t_n) - u_n|) + \frac{\gamma}{2} \tag{4.22}$$

for n large enough. From (4.20), (4.21), and (4.22) we obtain that $l_n \in \widetilde{F}_\delta(t_n, u_n)$ for n large enough. Hence $\widetilde{F}_\delta$ is almost lower semicontinuous and so is F_δ.

Applying now the main result of [23] we get that the differential inclusion

$$\begin{cases} u'(t) \in Au(t) + F_\delta(t, u(t)), \\ u(t_0) = u_0 \end{cases}$$

has a solution on $[t_0, T]$ given by

$$u(t) = S(t - t_0)u_0 + \int_{t_0}^{t} S(t - s)f_u(s)ds,$$

where $f_u(s) \in F_\delta(s, u(s))$ for a.a. $s \in [t_0, T]$. Using the properties of $[\cdot, \cdot]_+$ it is easy to show that

$$[y(t) - u(t), f_y(t) + g_y(t) - f_u(t)]_+ \leq \omega(t, |y(t) - u(t)|) + h(t) + \delta$$

for a.a. $t \in [t_0, T]$. Since

$$|y(t) - u(t)| \leq |y_0 - u_0| + \int_{t_0}^{t} [y(s) - u(s), f_y(s) + g_y(s) - f_u(s)]_+ds$$

for every $t \in [t_0, T]$, we get that

$$|y(t) - u(t)| \leq |y_0 - u_0| + \int_{t_0}^{t} \omega(s, |y(s) - u(s)|)ds + \int_{t_0}^{t} h(s)ds + \delta t$$

for every $t \in [t_0, T]$. By Lemma 4.1 we obtain that $|y(t) - u(t)| \leq r(t)$ for any $t \in [t_0, T]$, where $r(\cdot)$ is the maximal solution of (4.19). $\blacksquare$

The following result is a relaxation theorem, which is very important in optimal control theory. We refer the reader to [20, 46, 80].

Theorem 4.5 *Assume* (F_1), (F_2) *and* (A_2). *Moreover, suppose that F is almost continuous and $F(t, \cdot)$ is one-sided Perron with respect to the Perron function $\omega(\cdot, \cdot)$. Then the solution set of* (4.1) *is dense in the outer limit solution set of* (4.1).

The proof follows immediately from Lemmas 4.4 and 4.6 and Theorem 4.4.

Recall that in Remark 4.6 we pointed out that the set of outer limit solutions contains the set of limit solutions (given by Definition 4.2). Now, we prove that if F is one-sided Perron, than these two sets coincide.

Corollary 4.1 *Under the hypotheses of Theorem 4.5, the set of limit solutions and the set of outer limit solutions of* (4.1) *coincide.*

Proof From Theorem 4.5 we know that the set of outer limit solutions is the closure of the mild solutions of (4.1). Clearly, every mild solution is also a limit solution and the set of limit solutions is closed. $\blacksquare$

Remark 4.7 In this section, we introduce the notions of limit and weak solutions. The first one is inspired from the fact that every mild solution of (4.1) is a limit solution of a sequence of approximate solutions, but the converse is not true in general. The second extends the fact that for convex compact valued upper semicontinuous $F(\cdot, \cdot)$, $u(\cdot)$ is a solution of $u' \in F(t, u)$ if and only if $u(t) \in u(s) + \int_s^t F(\tau, u(\tau))d\tau$ for any $t_0 \leq s < t \leq T$. If $F(\cdot, \cdot)$ is not compact valued, then $u(\cdot)$ may be not differentiable (if X does not satisfy the Radon−Nikodim property). The first definition is more flexible, but both are equivalent under very mild assumptions.

4.3 Relaxation

Denote by $J(u) = \{u^* \in X^* : \langle u^*, u \rangle = |u|^2 = |u^*|_*^2\}$ the duality map, where $\langle \cdot, \cdot \rangle$ is the duality pairing. In the following, we denote by $[u, v]_+$ the right directional derivative of the norm calculated at u in the direction v, i.e.,

$$[u, v]_+ = \lim_{h \downarrow 0} \frac{|u + hv| - |v|}{h}.$$

Remark 4.8 We shall recall some important facts about the right directional derivative of the norm, which will be used subsequently.

(i) The right directional derivative of the norm is an upper semicontinuous function from $X \times X$ into $\mathbb{R}$ (see, for example, [32, Proposition 3.7 (i)]).

(ii) The right directional derivative of the norm is continuous with respect to the second variable (see, for example, [189, Lemma 1.4.2 (iii)]).

(iii) For each $u, v \in X$, we have $|u|[u, v]_+ = \sup\{\langle u^*, v \rangle : u^* \in J(u)\}$ (see, for example, [189, Lemma 1.4.2 (i) and Lemma 1.4.3]).

Consider f, a Bochner integrable function, and the associated Cauchy problem

$$\begin{cases} u'(t) \in Au(t) + f(t), \\ u(0) = u_0 \in \overline{D(A)}, \end{cases} \tag{4.23}$$

where $A : D(A) \subseteq X \to P(X)$ is an m-dissipative operator (i.e., $R(I - A) = X$ and for every $u_1, u_2 \in D(A)$ and $v_1 \in Au_1, v_2 \in Au_2$, there is $\omega \in J(u_1 - u_2)$ such that $\langle \omega, v_1 - v_2 \rangle \leq 0$; here I denotes the unity operator in X).

Definition 4.7 The continuous function $u(\cdot)$ is said to be a C^0-solution of (4.23) if for every $x \in D(A)$, $y \in Ax$ for each $t \in [0, T]$ the following inequality holds

$$|u(t) - x| \leq |u_0 - x| + \int_s^t [u(\tau) - x, f(\tau) - y]_+ d\tau.$$

In order to stress the dependence of C^0-solution on u_0 and f, we shall denote it by $u(\cdot, 0, u_0, f)$.

Definition 4.8 The continuous function $u(\cdot)$ is said to be a solution to (4.5) on $[0, T]$ if it is a C^0-solution to (4.23) on $[0, T]$ for some Bochner integrable f with $f(t) \in F(t, u(t))$ for a.e. $t \in [0, T]$.

Regarding the existence and properties of C^0-solutions, we shall use the following theorem. It was given first by Crandall and Pazy [75], in spaces with uniformly convex dual, and, in the context of C^0-solutions, by Bénilan [39].

Theorem 4.6 *Let f be Bochner integrable on $[0, T]$ and $u_0 \in \overline{D(A)}$. The Cauchy problem (4.23) has an unique solution $u(\cdot) = u(\cdot, 0, u_0, f)$. Furthermore, if $v(\cdot) = u(\cdot, 0, u_0, g)$ is a solution to (4.23) on $[0, T]$, then*

$$|u(t) - v(t)| \leq |u_0 - \overline{u_0}| + \int_0^t |f(\tau) - g(\tau)| d\tau,$$

$$|u(t) - v(t)| \leq |u_0 - \overline{u_0}| + \int_0^t [u(\tau) - v(\tau), f(\tau) - g(\tau)]_+ d\tau$$

for every $t \in [0, T]$.

Let us define few classes of multifunctions which will be used in the following.

Definition 4.9 The multifunction $F : [0, T] \times X \to P(X)$ is said to have sublinear growth if there exist two nonnegative valued Lebesgue integrable functions $a(\cdot)$ and $b(\cdot)$ such that

$$|y| \leq a(t) + b(t)|z|,$$

for a.e. $t \in [0, T]$, any $z \in X$ and all $y \in F(t, z)$.

The following lemma is standard and useful later on.

Lemma 4.7 *Let $G : [0, T] \times X \to P(X)$ be an almost lower semicontinuous multifunction. Then there exists a countable sequence of pairwise disjoint compact subsets of $[0, T]$, $\{\Delta_m\}_{m \geq 1}$, with Lebesgue measure of $\bigcup_m \Delta_m$ equal to T such that for every $m \in \mathbb{N}^+$, $G(\cdot, \cdot)$ is lower semicontinuous on $\Delta_m \times X$.*

Proof First, we shall prove that there exists a sequence of pairwise disjoint compact subsets of $[0, T]$, $\{\Delta_i\}_{i \in \mathbb{N}^+}$, such that for any $n \in \mathbb{N}^+$, Lebesgue measure of $[0, T] \setminus \bigcup_{i=1}^n \Delta_i$ is less then $\frac{1}{n}$ and for every $i = 1, \ldots, n$, $G(\cdot, \cdot)$ is lower semicontinuous on $\Delta_i \times X$. Indeed, in accordance with Definition 1.4(iv), there exists a compact interval $\Delta_1 \subseteq [0, T]$ with Lebesgue measure of $[0, T] \setminus \Delta_1$ less then 1 such that $G_{|\Delta_1 \times X}$ is lower semicontinuous. Suppose now that there exists a family of pairwise disjoint compact subsets of $[0, T]$, $\{\Delta_i\}_{i=1}^k$, such that, for $k \in \mathbb{N}^+$, Lebesgue measure of $[0, T] \setminus \bigcup_{i=1}^k \Delta_i$ is less then $\frac{1}{k}$ and for every $i = 1, \ldots, k$, $G(\cdot, \cdot)$ is lower semicontinuous on $\Delta_i \times X$. It is easy to see that the set $[0, T] \setminus \bigcup_{i=1}^k \Delta_i$ is open. Thus there exists a compact set $\Delta_{k+1} \subset [0, T] \setminus \bigcup_{i=1}^k \Delta_i$ such that Lebesgue measure of $[0, T] \setminus \bigcup_{i=1}^{k+1} \Delta_i$ is less then $\frac{1}{k+1}$ and $G(\cdot, \cdot)$ is lower semicontinuous on $\Delta_{k+1} \times X$. So, the induction is complete. Since $\frac{1}{n}$ tends to zero, when n tends to infinity, the assertion holds.

Remark 4.9 Any Carathéodory function $\omega : [0, T] \times \mathbb{R}_+ \to \mathbb{R}_+$ has the Scorza Dragoni property: for every $\varepsilon > 0$, there exists a compact interval $\Delta_\varepsilon \subseteq [0, T]$ with Lebesgue measure of $[0, T] \setminus \Delta_\varepsilon$ less then ε such that $\omega_{|\Delta_\varepsilon \times \mathbb{R}_+}$ is continuous (see, for instance, [185, Theorem 1. 2]).

An important feature of a Perron function is given by the following lemma from [95].

Lemma 4.8 *Suppose that* $\omega : [0, T] \times \mathbb{R}_+ \to \mathbb{R}_+$ *is an integrally bounded Perron function and* $\{a_k\}_{k\geq 1}$, $\{b_k\}_{k\geq 1} \subset \mathbb{R}_+$ *are two sequences converging to 0. Let* $r_k(\cdot)$ *be a solution to the problem*

$$\begin{cases} r'(t) \in \omega(t, r(t)) + a_k, \\ r(0) = b_k. \end{cases}$$

Then $\lim_{k \to \infty} r_k(t) = 0$ *uniformly for* $t \in [0, T]$.

4.3.1 Relaxation Results

We introduce the standing hypotheses of the paper.

(H_1) X is a separable Banach space and $A : D(A) \subseteq X \to P(X)$ is an m-dissipative operator, generating the compact semigroup $\{S(t) : \overline{D(A)} \to \overline{D(A)} : t \geq 0\}$;

(H_2) $F : [0, T] \times X \to P(X)$ is an almost lower semicontinuous multifunction with nonempty closed values and sublinear growth.

Remark 4.10 In a general Banach space, it is possible $\{u_n(\cdot)\}_{n\geq 1}$ to be sequence of (integral) solutions to

$$\begin{cases} u'(t) \in Au(t) + f_n(t), \\ u(0) = u_0 \end{cases}$$

with $u_n(\cdot) \to u(\cdot)$ uniformly on $[0, T]$, $f_n(\cdot) \to f(\cdot)$ weakly in $L^1([0, T], X)$, but $u(\cdot)$ not to be a solution to

$$\begin{cases} u'(t) \in Au(t) + f(t), \\ u(0) = u_0. \end{cases}$$

The corresponding example is given in [45]. Evidently, this situation poses several types of problems in general Banach spaces. In particular, it is difficult to prove the existence of solutions (if it is possible at all) under the assumptions in the paper if we replace lower semicontinuity of the right-hand side with upper semicontinuity, requiring that the values of the multifunction F are convex. For instance, in [65] the authors require that the multifunction F has weakly compact values and in [60] the space is supposed to be with uniformly convex dual. Actually, in the former article existence of viable solutions is considered, i.e., solutions remaining in a prescribed set; and in the latter one-near viability of the specified set (notion first introduced in [63]), that is, existence of solutions laying at a given distance from the set.

We begin with a relaxation result for an arbitrary separable Banach space, imposing a stronger than one-sided Perron assumption on the multifunction F. Namely, (H_3) there exists a Perron function $\omega(\cdot, \cdot)$ such that for every $u, v \in X$, a.e. $t \in [0, T]$ and every $f \in \overline{co}F(t, u)$, there exists $g \in F(t, v)$ such that

$$[u - v, f - g]_+ \leq \omega(t, |u - v|).$$

Theorem 4.7 *Assume that* (H_1), (H_2) *and* (H_3) *hold. Then the set of solutions to* (4.5) *is dense in the set of solutions to* (4.6).

Proof For any $\varepsilon > 0$ and let $z(\cdot)$ be a solution to (4.6). So, in accordance with Definition 4.8, it is the solution to (4.23) with $f(\cdot)$ equal to some Bochner integrable function $f_z(\cdot)$ such that $f_z(t) \in \overline{co}F(t, z(t))$ for a.e. $t \in [0, T]$. Since F is an almost lower semicontinuous multifunction, by applying Luzin's theorem to $f_z(\cdot)$, Remark 4.9 and Lemma 4.7, we claim that there exists a sequence of pairwise disjoint compact subsets of $[0, T]$, $\{\Delta_m\}_{m \geq 1}$, with Lebesgue measure of $\bigcup_m \Delta_m$ equal to T such that for every $m \in \mathbb{N}$, $f_z(\cdot)$ is continuous on Δ_m, $\omega(\cdot, \cdot)$ is continuous on $\Delta_m \times \mathbb{R}_+$ and $F(\cdot, \cdot)$ is lower semicontinuous on $\Delta_m \times X$.

Let $m \in \mathbb{N}$ and the multifunction $G_m : \Delta_m \times X \to P(X)$, defined by

$$G_m(t, v) = \overline{\{y \in F(t, v) : [z(t) - v, f_z(t) - y]_+ < \omega(t, |z(t) - v|) + \varepsilon\}}.$$

The multifunction G_m has closed and nonempty values, since F is one-sided Perron. It also has sublinear growth. We shall prove that G_m is lower semicontinuous. We shall first establish that the multifunction

$$\widetilde{G_m}(t, v) = \{y \in F(t, v) : [z(t) - v, f_z(t) - y]_+ < \omega(t, |z(t) - v|) + \varepsilon\}$$

is lower semicontinuous, consequently, G_m has the same property.

Let any $(t_0, v_0) \in \Delta_m \times X$ and any sequence $\{(t_n, v_n)\}_{n \geq 1} \subset \Delta_m \times X$ such that $t_n \to t_0, v_n \to v_0$. Moreover, let any $y_0 \in \widetilde{G_m}(t_0, v_0)$. Thus, $y_0 \in F(t_0, v_0)$ and there exists $\gamma > 0$ such that

$$[z(t_0) - v_0, f_z(t_0) - y_0]_+ < \omega(t_0, |z(t_0) - v_0|) + \varepsilon - \gamma. \tag{4.24}$$

So, since F is lower semicontinuous, there exists a sequence $\{y_n\}_{n \geq 1}$ with $y_n \in F(t_n, v_n)$ such that $y_n \to y_0$. We have to show that $y_n \in \widetilde{G_m}(t_n, v_n)$ for n sufficiently large. Taking into account Remark 4.8 (i), there exists a neighborhood U of (t_0, v_0, y_0) such that

$$[z(t) - v, f_z(t) - y]_+ < [z(t_0) - v_0, f_z(t_0) - y_0]_+ + \frac{\gamma}{2} \tag{4.25}$$

for all $(t, v, y) \in U \cap (\Delta_m \times X \times X)$. Now, it is clear that there exists $n_1 \in \mathbb{N}$ such that $(t_n, v_n, y_n) \in U$ for all $n > n_1$. Since $\omega(\cdot, \cdot)$ is continuous on $\Delta_m \times \mathbb{R}_+$, there exists $n_2 > n_1$ such that

$$\omega(t_0, |z(t_0) - v_0|) < \omega(t_n, |z(t_n) - v_n|) + \frac{\gamma}{2} \qquad (4.26)$$

for all $n > n_2$. From (4.24), (4.25) and (4.26), we get that $y_n \in \widetilde{G_m}(t_n, v_n)$ for all $n > n_2$, as claimed. We have proved that $\widetilde{G_m}$ is lower semicontinuous, therefore, G_m is also lower semicontinuous.

Let us define a new multifunction $G : [0, T] \times X \to P(X)$ with nonempty closed values and sublinear growth by

$$G(t, u) = \begin{cases} G_m(t, u), & (t, u) \in \Delta_m \times X, \ m \in \mathbb{N}, \\ \text{some } y \in X, & \text{otherwise}, \end{cases} \qquad (4.27)$$

which is almost lower semicontinuous. Moreover, for every $t \in [0, T]$, $G(t, \cdot)$ is lower semicontinuous on X. Hence G is measurable (see [185, Theorem 3.2]). Taking into the account that it also has closed values, we have that G is graph measurable (see [125, Proposition 1.7]). Now, we are able to apply the existence theorem from [23] and obtain that there exists a solution $v(\cdot)$ on $[0, T]$ to the problem

$$\begin{cases} v'(t) \in Av(t) + G(t, v(t)), \\ v(0) = u_0. \end{cases} \qquad (4.28)$$

Thus, in accordance with Definition 4.8, $v(\cdot)$ is the solution to (4.23) on $[0, T]$ with $f(\cdot)$ equal to some Bochner integrable function $g(\cdot)$ such that $g(t) \in G(t, v(t))$ for a.e. $t \in [0, T]$. Since $F(\cdot, \cdot)$ has closed values, $v(\cdot)$ is also a solution to (4.5) on $[0, T]$. Moreover, taking into account Remark 4.8 (ii), we have that

$$[z(t) - v(t), f_z(t) - g(t)]_+ \leq \omega(t, |z(t) - v(t)|) + \varepsilon$$

for a.e. $t \in [0, T]$. By means of Theorem 4.6, we obtain that

$$|z(t) - v(t)| \leq \int_0^t \omega(\tau, |z(\tau) - v(\tau)|)d\tau + \varepsilon t \qquad (4.29)$$

for all $t \in [0, T]$.

Let $r_\varepsilon(\cdot)$ be the maximal solution to the Cauchy problem $r'(t) = \omega(t, r(t)) + \varepsilon$, $r(0) = 0$. Since $z(\cdot)$, $v(\cdot)$ are bounded on $[0, T]$, we can assume (without loss of generality) that $\omega(\cdot, \cdot)$ is globally integrally bounded, so any solution of the equation $r'(t) = \omega(t, r(t)) + \varepsilon$ can be continued up to a global one, i.e., defined on the whole interval. So, by (4.29) and Lemma 4.1, we obtain that $|z(t) - v(t)| \leq r_\varepsilon(t)$ for any

$t \in [0, T]$. By Lemma 4.8, $r_\varepsilon(t)$ converges uniformly to 0 on $[0, T]$, when $\varepsilon \downarrow 0$. This completes the proof.

Now we can similarly prove the following variant of Filippov-Pliś lemma.

Theorem 4.8 *Assume that* (H_1) *and* (H_2) *hold. Further, suppose that* $F : [0, T] \times X \to P(X)$ *is one-sided Perron (with respect to a Perron function* ω*) and* $f : [0, T] \to \mathbb{R}$ *is a nonnegative valued Lebesgue integrable function. Then for any* $u_0 \in \overline{D(A)}$*,* $\varepsilon > 0$ *and any solution* $u(\cdot)$ *on* $[0, T]$ *to the problem*

$$\begin{cases} u'(t) \in Au(t) + F(t, u(t)) + f(t)\mathbb{B}, \\ u(0) = u_0, \end{cases} \tag{4.30}$$

there exists a solution to (4.5) *on* $[0, T]$*,* $v(\cdot)$*, such that*

$$|u(t) - v(t)| \le r_\varepsilon(t)$$

for all $t \in [0, T]$*, where* $r_\varepsilon(\cdot)$ *is the maximal solution to the Cauchy problem*

$$\begin{cases} r'(t) = \omega(t, r(t)) + f(t) + \varepsilon, \\ r(0) = 0. \end{cases}$$

Proof For any $\varepsilon > 0$ and let $u(\cdot)$ be a solution to (4.30). So, in accordance with Definition 4.8, it is the solution to (4.23) with $f(\cdot)$ equal to some Bochner integrable function $f_u(\cdot)$ such that $f_u(t) \in F(t, u(t)) + f(t)\mathbb{B}$ for a.e. $t \in [0, T]$. Let $\{\Delta_m\}_{m \ge 1}$ be a sequence of pairwise disjoint compact subsets of $[0, T]$ with Lebesgue measure of $\bigcup_m \Delta_m$ equal to T such that for every $m \in \mathbb{N}$, $f(\cdot)$ and $f_u(\cdot)$ are continuous on Δ_m, $\omega(\cdot, \cdot)$ is continuous on $\Delta_m \times \mathbb{R}_+$, and $F(\cdot, \cdot)$ is lower semicontinuous on $\Delta_m \times X$.

Let $m \in \mathbb{N}$ and the multifunction $G_m : \Delta_m \times X \to P(X)$, defined by

$$G_m(t, v) = \overline{\{y \in F(t, v) : [u(t) - v, f_u(t) - y]_+ < \omega(t, |u(t) - v|) + f(t) + \varepsilon\}}.$$

Let us define a new multifunction $G : [0, T] \times X \to P(X)$ by (4.27). One can show, as in the proof of Theorem 4.7, that G is almost lower semicontinuous with closed nonempty values and graph measurable. Therefore, there exists a solution $v(\cdot)$ on $[0, T]$ to the problem (4.28). Clearly, $v(\cdot)$ satisfies the conclusion of the theorem.

A similar result in a Banach space with uniformly convex dual was proved in [60, Theorem 15]. However, the one-sided Perron hypothesis on F was given in an equivalent form involving the support function. Moreover, the continuity assumption on the multifunction is of upper semicontinuous type.

It is worth mentioning, in addition, that the one-sided Perron condition (or stronger versions, for instance, one-sided Lipschitz) has been used in several papers of the second author, usually under the name of the one-sided Kamke condition (see [86, 92–95]). In all mentioned papers, the problems are considered in a Banach space with uniformly convex dual. If the dual space is not uniformly convex, then additional difficulties arise. These difficulties are the main obstacle to study inclusions with upper semicontinuous F. This is one of the reasons for considering lower semicontinuous multifunctions here.

In the next result we assume that F is one-sided Perron. However, we have to strengthen assumptions on the space X.

Theorem 4.9 *Let X be a Banach space with the duality map $J(\cdot)$ being single valued. Assume that (H_1) and (H_2) hold. Further, suppose that $F : [0, T] \times X \to P(X)$ is one-sided Perron (with the additional assumption that for a.e. $t \in [0, T]$, $\omega(t, \cdot)$ is nondecreasing). Then the set of solutions to (4.5) is dense in the set of solutions to (4.6).*

Proof Let us fix $\varepsilon > 0$. Let $z(\cdot)$, $f_z(\cdot)$, $\{\Delta_m\}_{m \geq 1}$ be as in the proof of Theorem 4.7. In addition, we assume (without loss of generality) that $\omega(\cdot, \cdot)$ is globally integrally bounded. By Lemma 4.8, there exists $\delta > 0$ such that the maximal solution $r_\delta(\cdot)$ to the Cauchy problem

$$\begin{cases} r'(t) = \omega(t, r(t)) + \delta, \\ r(0) = \delta \end{cases}$$

satisfies

$$r_\delta(t) \leq \varepsilon \qquad\qquad (4.31)$$

for all $t \in [0, T]$.

Let $m \in \mathbb{N}$ and the multifunction $G_m : \Delta_m \times X \to P(X)$, defined by

$$G_m(t, v) = \begin{cases} \overline{\{y \in F(t, v) : [z(t) - v, f_z(t) - y]_+ < \omega(t, |z(t) - v|) + \delta\}}, \\ \qquad\qquad\qquad\qquad\qquad\qquad\qquad \text{if } |z(t) - v| \geq \delta, \\ F(t, v), \qquad\qquad\qquad\qquad\qquad\quad \text{if } |z(t) - v| < \delta. \end{cases}$$

The multifunction $G_m(\cdot, \cdot)$ obviously has closed values and sublinear growth. Let us show that $G_m(\cdot, \cdot)$ has nonempty values. Let any $(t_0, v_0) \in \Delta_m \times X$. If $|z(t_0) - v_0| < \delta$, then the conclusion is obvious, since $F(t_0, v_0)$ is nonempty. Suppose that $|z(t_0) - v_0| \geq \delta$. Since $J(\cdot)$ is single valued, by applying Remark 4.8 (iii), we have that

$$[z(t_0) - v_0, y]_+ = \frac{1}{|z(t_0) - v_0|} \langle J(z(t_0) - v_0), y \rangle \qquad\qquad (4.32)$$

for any $y \in X$. So,

$$[z(t_0) - v_0, f_z(t_0)]_+ \leq \frac{1}{|z(t_0) - v_0|} \sup_{y \in \overline{co} F(t_0, z(t_0))} \langle J(z(t_0) - v_0), y \rangle$$

$$\leq \frac{1}{|z(t_0) - v_0|} \sup_{\omega \in F(t_0, z(t_0))} \langle J(z(t_0) - v_0), \omega \rangle.$$

For any $0 < \gamma < \delta$, there exists $\omega \in F(t_0, z(t_0))$ such that

$$\sup_{y \in F(t_0, z(t_0))} \langle J(z(t_0) - v_0), y \rangle < \langle J(z(t_0) - v_0), \omega \rangle + \gamma |z(t_0) - v_0|.$$

Thus,

$$[z(t_0) - v_0, f_z(t_0)]_+ < [z(t_0) - v_0, \omega]_+ + \gamma.$$

Since $F(\cdot, \cdot)$ is one-sided Perron, there exists $y \in F(t_0, v_0)$ such that

$$[z(t_0) - v_0, \omega - y]_+ \leq \omega(t_0, |z(t_0) - v_0|).$$

Therefore,

$$[z(t_0) - v_0, f_z(t_0)]_+ - [z(t_0) - v_0, y]_+ < \omega(t_0, |z(t_0) - v_0|) + \gamma.$$

Since (4.32) holds, we obtain that

$$[z(t_0) - v_0, f_z(t_0) - y]_+ < \omega(t_0, |z(t_0) - v_0|) + \gamma.$$

So, $G_m(\cdot, \cdot)$ has nonempty values.

In the following, we shall prove that G_m is lower semicontinuous. We shall first establish that the multifunction

$$\widetilde{G_m}(t, v) = \begin{cases} \{y \in F(t, v) : [z(t) - v, f_z(t) - y]_+ < \omega(t, |z(t) - v|) + \delta\}, \\ \qquad\qquad\qquad\qquad\qquad\qquad \text{if } |z(t) - v| \geq \delta, \\ F(t, v), \qquad\qquad\qquad\qquad\quad \text{if } |z(t) - v| < \delta \end{cases}$$

is lower semicontinuous, consequently, G_m has the same property. Let any $(t_0, v_0) \in \Delta_m \times X$ and any sequence $\{(t_n, v_n)\}_{n \geq 1} \subset \Delta_m \times X$ such that $t_n \to t_0$, $v_n \to v_0$. Moreover, let any $y_0 \in \widetilde{G_m}(t_0, v_0)$. Thus $y_0 \in F(t_0, v_0)$, and since F is lower semicontinuous, there exists a sequence $\{y_n\}_{n \geq 1}$ with $y_n \in F(t_n, v_n)$ such that $y_n \to y_0$. We have to show that $y_n \in \widetilde{G_m}(t_n, v_n)$ for n sufficiently large.

If $|z(t_0) - v_0| > \delta$, then

$$[z(t_0) - v_0, f_z(t_0) - y_0]_+ < \omega(t_0, |z(t_0) - v_0|) + \delta.$$

Further, as in the proof of Theorem 4.7, we can show that there exists $n_1 \in \mathbb{N}$ such that $|z(t_n) - v_n| > \delta$ and

$$[z(t_n) - v_n, f_z(t_n) - y_n]_+ < \omega(t_n, |z(t_n) - v_n|) + \delta$$

for all $n > n_1$. So, $y_n \in \widetilde{G_m}(t_n, v_n)$ for n sufficiently large.

If $|z(t_0) - v_0| < \delta$, then there exists $n_2 \in \mathbb{N}$ such that $|z(t_n) - v_n| < \delta$ and $y_n \in \widetilde{G_m}(t_n, v_n)$ (since $y_n \in F(t_n, v_n)$) for all $n > n_2$. So, $y_n \in \widetilde{G_m}(t_n, v_n)$ for n sufficiently large.

If $|z(t_0) - v_0| = \delta$, then either there are infinitely many $(t_i, v_i) \in \{(t_n, v_n)\}_{n \geq 1}$ such that $|z(t_i) - v_i| \geq \delta$, or there is a finite number of such (t_i, v_i). In the former case, there exists $n_1 \in \mathbb{N}$ such that

$$[z(t_n) - v_n, f_z(t_n) - y_n]_+ < \omega(t_n, |z(t_n) - v_n|) + \delta$$

for all $n > n_1$. So, as above, $y_n \in \widetilde{G_m}(t_n, v_n)$ for n sufficiently large. In the latter case, there obviously exists $n_2 \in \mathbb{N}$ such that $|z(t_n) - v_n| < \delta$ and $y_n \in \widetilde{G_m}(t_n, v_n)$ (since $y_n \in F(t_n, v_n)$) for all $n > n_2$. So, $y_n \in \widetilde{G_m}(t_n, v_n)$ for n sufficiently large. We have proved that $\widetilde{G_m}$ is lower semicontinuous, therefore, G_m is also lower semicontinuous.

Let us define a new multifunction $G : [0, T] \times X \to P(X)$ with closed nonempty values and sublinear growth by (4.27) which is almost lower semicontinuous (the proof is the same as in Theorem 4.7). Moreover, for every $t \in [0, T]$, $G(t, \cdot)$ is lower semicontinuous on X. We also have that G is graph measurable. Now, we are able to apply the existence theorem from [23] and obtain that there exists a solution $v(\cdot)$ on $[0, T]$ to the problem (4.28). Thus, in accordance with Definition 4.8, $v(\cdot)$ is the solution to (4.23) on $[0, T]$ with f equal to some Bochner integrable function $g(\cdot)$ such that $g(t) \in G(t, v(t))$ for a.e. $t \in [0, T]$. Since $F(\cdot, \cdot)$ has closed values, $v(\cdot)$ is also a solution to (4.5) on $[0, T]$.

Define the set

$$A = \{t \in [0, T] : |z(t) - v(t)| > \delta\}.$$

Since $z(\cdot)$ and $v(\cdot)$ are continuous, A is an open set and, hence it is an union of a countable system of pairwise disjoint open intervals, i.e., $A = \bigcup_{k=1}^{\infty} (a_k, b_k)$.

For any $k \in \mathbb{N}$, it is easy to see that $|z(a_k) - v(a_k)| = |z(b_k) - v(b_k)| = \delta$. Moreover,

$$[z(t) - v(t), f_z(t) - g(t)]_+ \leq \omega(t, |z(t) - v(t)|) + \delta$$

for a.e. $t \in [a_k, b_k]$. Since $z(\cdot), v(\cdot)$ are bounded on $[a_k, b_k]$, we can assume (without loss of generality) that $\omega(\cdot, \cdot)$ is globally integrally bounded, so any solution to the equation $r'(t) = \omega(t, r(t)) + \delta$ can be continued up to a one defined on the whole interval $[a_k, b_k]$. So, by Theorem 4.6 and Lemma 4.1, we have that $|z(t) - v(t)| \leq r_k(t)$ for all $t \in [a_k, b_k]$, where $r_k(\cdot)$ is the maximal solution on $[a_k, b_k]$ of the problem $r'(t) = \omega(t, r(t)) + \delta, r(a_k) = \delta$. Since $\omega(t, \cdot)$ is nondecreasing for a.e. $t \in [0, T]$, we get that $r_k(\cdot)$ is less or equal on $[a_k, b_k]$ to the maximal solution r_δ to the problem $r'(t) = \omega(t, r(t)) + \delta, r(0) = \delta$. Taking into account (4.31), we obtain the conclusion.

Remark 4.11 Regarding Theorem 4.9, when the duality map is required to be single valued, we have to mention that the interesting case is that of the dual norm being not uniformly convex (when X^* is uniformly convex, J is single valued and uniformly continuous on bounded sets; see, for example, [32]). This is possible, since every infinite dimensional separable nonreflexive Banach space admits a Gâteaux differentiable norm the dual of which is not strictly convex (this result was established in [118]). The duality map in this case is single valued, since the norm is Gâteaux differentiable; see, for instance, [205].

4.3.2 Examples

Example 4.1 Let Ω be a bounded domain in $\mathbb{R}^n$ with smooth boundary Γ. Let $r > (n - 2)/n$ and consider the following multivalued, nonlinear, parabolic partial differential equation on $[0, T] \times \Omega$:

$$
\begin{cases}
\dfrac{\partial u(t, z)}{\partial t} - \Delta u(t, z)|u(t, z)|^{r-1} \in F(t, z, u(t, z)), & \\
u(t, z) = 0 & \text{on } [0, T] \times \Gamma, \\
u(0, z) = u_0(z) & \text{in } \Omega.
\end{cases}
\tag{4.33}
$$

Here $F : [0, T] \times \Omega \times \mathbb{R} \to P(\mathbb{R})$ is a multifunction with closed values, which is lower semicontinuous in the third variable. Moreover, we assume that the multifunction $G : [0, T] \times L^1(\Omega, \mathbb{R}) \to P(L^1(\Omega, \mathbb{R}))$, defined by

$$
G(t, v) = \{f \in L^1(\Omega, \mathbb{R}) : f(z) \in F(t, z, v(z)) \text{ a.e. } z \in \Omega\}
$$

is measurable on $[0, T] \times L^1(\Omega, \mathbb{R})$. Also assume that there exist $b(t, \cdot) \in L_+^\infty(\Omega)$ with $t \mapsto |b(t, \cdot)|_{L^\infty(\Omega)}$ belonging to $L^1([0, T], \mathbb{R})$ and $a(\cdot, \cdot) \in L_+^1([0, T] \times \Omega, \mathbb{R})$ such that

$$
|y| \le a(t, z) + b(t, z)|r|
$$

for a.e. $t \in [0, T], z \in \Omega$, all $r \in \mathbb{R}$ and all $y \in F(t, z, r)$.

Consider the nonlinear operator $A : D(A) \subseteq L^1(\Omega, \mathbb{R}) \to P(L^1(\Omega, \mathbb{R}))$ defined by $Av = \Delta v|v|^{r-1}$ with

$$
D(A) = \{v \in L^1(\Omega, \mathbb{R}^n) : v, v^{r-1} \in W_0^{1,1}(\Omega, \mathbb{R}^n), \ \Delta v|v|^{r-1} \in L^1(\Omega)\}.
$$

This operator is m-dissipative, generating a compact nonlinear semigroup (see [23, p. 663]).

As stated in [23, p. 662], the multifunction $G(\cdot, \cdot)$ has closed nonempty values. Moreover, $G(t, \cdot)$ is lower semicontinuous for all $t \in [0, T]$. By means of [125, Proposition 7.17], since $G(\cdot, \cdot)$ is supposed to be measurable, we get that $G(\cdot, \cdot)$ is almost lower semicontinuous. It is also clear that $G(\cdot, \cdot)$ has sublinear growth.

Let us denote $v_0 = u_0(\cdot) \in \overline{D(A)}$.

Now we can rewrite the initial-boundary value problem (4.33) as the equivalent abstract multivalued evolution equation of the form (4.28) with the initial data equal to v_0. Suppose, in addition, that there exists a Perron function $\omega(\cdot, \cdot)$ such that for every $\varphi, \psi \in L^1(\Omega)$, a.e. $t \in [0, T]$ and every $f \in \overline{\text{co}}G(t, \varphi)$, there exists $g \in G(t, \psi)$ such that

$$\int_{\Omega_{\varphi-\psi}^+} (f(z) - g(z))dz - \int_{\Omega_{\varphi-\psi}^-} (f(z) - g(z))dz \pm \int_{\Omega_{\varphi-\psi}^0} (f(z) - g(z))dz$$

$$\leq \omega\left(t, \int_\Omega |\varphi(z) - \psi(z)|\right),$$

where $\Omega_{\varphi-\psi}^{+(-,0)} = \{z \in \Omega : \varphi(z) - \psi(z) > (<, =)0\}$. This assumption ensures that G satisfies (H_3). This fact is due to a characterization of the right directional derivative of the norm in $L^1(\Omega)$ from [189, Example 1.4.3]. The above assumptions ensure that all hypotheses of Theorem 4.7 are satisfied.

Let $u(\cdot, \cdot) \in C([0, T], L^1(\Omega, \mathbb{R}))$ be a generalized solution on $[0, T] \times \Omega$ to the relaxed problem

$$\begin{cases} \dfrac{\partial u(t, z)}{\partial t} - \Delta u(t, z)|u(t, z)|^{r-1} \in \overline{\text{co}}F(t, z, u(t, z)), \\ u(t, z) = 0 & \text{on } [0, T] \times \Gamma, \\ u(0, z) = u_0(z) & \text{in } \Omega. \end{cases}$$

By applying Theorem 4.7 to the abstract problem, for any $\varepsilon > 0$, we can find $u_\varepsilon(\cdot, \cdot)$, a generalized solution to (4.33), such that

$$\sup_{t \in [0,T]} \int_\Omega |u(t, z) - u_\varepsilon(t, z)|dz < \varepsilon.$$

Remark 4.12 The above example is a modification of the example presented in [23, p. 662].

Example 4.2 Let X be a separable Banach space such that the duality map $J(\cdot)$ is single valued, and $A : D(A) \subseteq X \to P(X)$ is an m-dissipative operator, generating a compact semigroup. Consider also two functions $f_1, f_2 : [0, T] \times X \to X$ with the following properties:

(i) for every $\varepsilon > 0$, there exists a compact interval $\Delta_\varepsilon \subseteq [0, T]$ with Lebesgue measure of $[0, T] \setminus \Delta_\varepsilon$ less then ε such that restrictions of f_1, f_2 to $\Delta_\varepsilon \times X$ are continuous;

(ii) there exist nonnegative valued Lebesgue integrable functions $a_1(\cdot), a_2(\cdot)$ and $b_1(\cdot), b_2(\cdot)$ such that

$$|f_1(t, z)| \leq a_1(t) + b_1(t)|z|, \quad |f_2(t, z)| \leq a_2(t) + b_2(t)|z|$$

for a.e. $t \in [0, T]$ and any $z \in X$;

(iii) there exists a Perron function $\omega(\cdot, \cdot)$ (with additional assumption that for a.e. $t \in [0, T]$, $\omega(t, \cdot)$ is nondecreasing) such that for every $u, v \in X$ and a.e. $t \in [0, T]$,

$$[u - v, f_i(t, u) - f_i(t, v)]_+ \leq \omega(t, |u - v|) \text{ for } i = 1, 2.$$

We can not take different Perron functions for f_1 and f_2, since the maximum of two Perron functions is not necessarily a Perron function (see [21, Proposition 7]).

Applying Theorem 4.9, we can conclude that the set of solutions to the problem

$$\begin{cases} u'(t) \in Au + \{f_1(t, u(t)), f_2(t, u(t))\}, \\ u(0) = u_0 \in \overline{D(A)} \end{cases}$$

is dense in the set of solutions to the relaxed problem

$$\begin{cases} u'(t) \in Au + \overline{co}\{f_1(t, u(t)), f_2(t, u(t))\}, \\ u(0) = u_0. \end{cases}$$

Example 4.3 Let $\Omega \subset \mathbb{R}^n$, $n = 1, 2, \ldots$, be a bounded domain with C^2 boundary and $\beta : \mathbb{R} \to \mathbb{R}$ be a maximal monotone operator with $\beta(0) = 0$. Let us consider the Cauchy problem for the nonlinear diffusion equation:

$$\begin{cases} \dfrac{\partial u(t, z)}{\partial t} \in \Delta u(t, z) + \{f_1(t, u(t, z)), f_2(t, u(t, z))\} & \text{in } (0, T) \times \Omega, \\ -u_{\vec{v}}(t, z) \in \beta(u(t, z)) & \text{on } (0, T) \times \Gamma, \\ u(0, z) = u_0(z) & \text{in } \Omega. \end{cases} \quad (4.34)$$

where $u_{\vec{v}}$ is the external normal to and $f_1, f_2 : [0, T] \times \mathbb{R} \to \mathbb{R}_+$ with $f_1 < f_2$.

Consider the nonlinear operator $A : D(A) \subseteq L^2(\Omega, \mathbb{R}) \to P(L^2(\Omega, \mathbb{R}))$ defined by $Av = \Delta v$ with

$$D(A) = \{v \in H^2(\Omega, \mathbb{R}) : -v_{\vec{v}}(z) \in \beta(v(u)) \text{ for a.e. } z \in \Gamma\}.$$

This operator is m-dissipative, generating compact nonlinear semigroup (see, for example, [32, p. 62]).

Let $G : [0, T] \times L^2(\Omega, \mathbb{R}) \to P(L^2(\Omega, \mathbb{R}))$ defined by

$$G(t, v) = \{\widehat{f_1}(t, v), \widehat{f_2}(t, v)\},$$

where $\widehat{f_i}(t, v) = f_i(t, v(z))$ for a.e. $z \in \Omega$ and $i = 1, 2$.

Let us denote $v_0 = u_0(\cdot) \in \overline{D(A)}$.

Now we can rewrite the initial-boundary value problem (4.34) as the equivalent abstract multivalued evolution equation of the form (4.28) with the initial data equal to v_0.

Suppose that the functions $f_1, f_2 : [0, T] \times \mathbb{R} \to \mathbb{R}_+$ have the following properties:

(i) there exist nonnegative valued Lebesgue integrable functions $a(\cdot)$ and $b(\cdot)$ such that

$$f_2(t, u) \leq a(t) + b(t)|u|$$

for a.e. $t \in [0, T]$ and any $u \in \mathbb{R}$;

(ii) for every $\varepsilon > 0$, there exists a compact interval $\Delta_\varepsilon \subseteq [0, T]$ with Lebesgue measure of $[0, T] \setminus \Delta_\varepsilon$ less then ε such that restrictions of f_1, f_2 to $\Delta_\varepsilon \times \mathbb{R}$ are continuous;

(iii) there exist Lebesgue integrable functions $L(\cdot)$ and $M(\cdot)$ such that

$$(u - v)(f_1(t, u) - f_1(t, v)) \leq L(t)(u - v)^2,$$
$$(u - v)(f_2(t, u) - f_2(t, v)) \leq M(t)(u - v)^2$$

for a.e. $t \in [0, T]$ and any $u, v \in \mathbb{R}$.

The above assumptions ensure that all hypotheses of Theorem 4.9 are satisfied.

Let $u(\cdot, \cdot) \in C([0, T], L^2(\Omega, \mathbb{R}))$ be a generalized solution on $[0, T] \times \Omega$ to the relaxed problem

$$\begin{cases} \dfrac{\partial u(t, z)}{\partial t} \in \Delta u(t, z) + [f_1(t, u(t, z)), f_2(t, u(t, z))] & \text{in } (0, T) \times \Omega, \\ -u_{\overrightarrow{v}}(t, z) \in \beta(u(t, z)) & \text{on } (0, T) \times \Gamma, \\ u(0, z) = u_0(z) & \text{in } \Omega. \end{cases}$$

By applying Theorem 4.9 to the abstract problem, for any $\varepsilon > 0$, we can find $u_\varepsilon(\cdot, \cdot)$, a generalized solution to (4.34), such that

$$\sup_{t \in [0,T]} \int_\Omega |u(t, z) - u_\varepsilon(t, z)|^2 dz < \varepsilon.$$

Example 4.4 Let us consider the following modification of the problem considered in the previous example:

$$\begin{cases} \dfrac{\partial u(t, z)}{\partial t} \in \Delta u(t, z) + [f_1(t, u(t, z)), f_2(t, u(t, z))] \cup [g_1(t, u(t, z)), g_2(t, u(t, z))], \\ -u_{\overrightarrow{v}}(t, z) \in \beta(u(t, z)) & \text{on } (0, T) \times \Gamma, \\ u(0, z) = u_0(z) & \text{in } \Omega, \end{cases}$$

$$(4.35)$$

where $u_{\overrightarrow{\jmath}}$ is the external normal to Γ and $f_1, f_2, g_1, g_2 : [0, T] \times \mathbb{R} \to \mathbb{R}_+$ with $f_1 < f_2 < g_1 < g_2$.

Let $G : [0, T] \times L^2(\Omega, \mathbb{R}) \to P(L^2(\Omega, \mathbb{R}))$ defined by

$$G(t, v) = \left\{ f \in L^2(\Omega, \mathbb{R}) : f_1(t, v(z)) \le f(t, v(z)) \le f_2(t, v(u)) \text{ or} \right.$$

$$\left. g_1(t, v(z)) \le f(t, v(z)) \le g_2(t, v(u)) \text{ a.e.} \right\}.$$

Let us denote $v_0 = u_0(\cdot) \in \overline{D(A)}$.

As before, we can rewrite the initial-boundary value problem (4.35) as the equivalent abstract multivalued evolution equation of the form (4.28) with the initial data v_0.

Suppose that the functions $f_1, f_2, g_1, g_2 : [0, T] \times \mathbb{R} \to \mathbb{R}_+$ have properties similar with those in the previous example, which ensure that all hypotheses of Theorem 4.9 are satisfied.

Let $u(\cdot, \cdot) \in C([0, T], L^2(\Omega, \mathbb{R}))$ be a generalized solution on $[0, T] \times \Omega$ to the relaxed problem

$$\begin{cases} \dfrac{\partial u(t, z)}{\partial t} \in \Delta u(t, z) + [f_1(t, u(t, z)), g_2(t, u(t, z))] & \text{in } (0, T) \times \Omega, \\ - u_{\overrightarrow{\jmath}}(t, z) \in \beta(u(t, z)) & \text{on } (0, T) \times \Gamma, \\ u(0, z) = u_0(z) & \text{in } \Omega. \end{cases}$$

By applying Theorem 4.9 to the abstract problem, for any $\varepsilon > 0$, we can find $u_\varepsilon(\cdot, \cdot)$, a generalized solution to (4.35), such that

$$\sup_{t \in [0,T]} \int_\Omega |u(t, z) - u_\varepsilon(t, z)|^2 dz < \varepsilon.$$

4.4   Pullback Attractor

4.4.1   Solvability

In this section, we will show the global existence of integral solution to problem (4.7) under the following assumptions:

(A) The C_0-semigroup $S(t) = e^{tA}$ is norm continuous, i.e., the operator-valued function $t \mapsto S(t)$ is continuous for $t > 0$.

(B) The phase space $\mathscr{B}$ satisfies (B_1)-(B_4).

(F) $F : J \times \mathscr{B} \to P_{cp,cv}(X)$, where $J = [\tau, \tau + T]$ for some positive number T, is a multifunction satisfying that

(i) for each ϕ in $\mathscr{B}$, the multifunction $F(\cdot, \phi) : J \to P_{cp,cv}(X)$ admits a strongly measurable selection;
(ii) the multimap $F(t, \cdot) : \mathscr{B} \to P_{cp,cv}(X)$ is u.s.c. for a.e. $t \in J$;
(iii) there exist two nonnegative functions $m_1, m_2 \in L^1(J, \mathbb{R}^+)$ such that

$$|F(t, \phi)| \le m_1(t) + m_2(t)|\phi|_{\mathscr{B}};$$

(iv) if the semigroup $S(\cdot)$ is noncompact, then there exists a nonnegative function $k \in L^1(J, \mathbb{R}^+)$ such that

$$\beta(F(t, D)) \le k(t) \sup_{\theta \le 0} \beta(D(\theta))$$

for all bounded sets $D \subset \mathscr{B}$ and a.e. $t \in J$.

For a given $\varphi^\tau \in \mathscr{B}$, put

$$C_{\varphi^\tau} = \{u \in C(J, X) : u(\tau) = \varphi^\tau(0)\},$$

then C_{φ^τ} is a closed subspace of $C(J, X)$ with the sup-norm. For $v \in C_{\varphi^\tau}$, we define the function $v[\varphi^\tau] \in C((-\infty, \tau + T], X)$ as follows:

$$v[\varphi^\tau](t) = \begin{cases} v(t), & \text{if } t \in [\tau, \tau + T], \\ \varphi^\tau(t - \tau), & \text{if } t \in (-\infty, \tau]. \end{cases}$$

For $v \in C_{\varphi^\tau}$, put

$$Sel_F(v) = \{f \in L^1(J, X) : f(t) \in F(t, v[\varphi^\tau]_t) \text{ for a.e } t \in J\}.$$

Then we have the following properties.

Lemma 4.9 Let $(F)(i)$–$(F)(iii)$ hold. Then $Sel_F(u) \ne \emptyset$ for each $u \in C_{\varphi^\tau}$. In addition, $Sel_F : C_{\varphi^\tau} \to P(L^1(J, X))$ is weakly u.s.c. with weakly compact and convex values.

Proof The proof is similar to that in [5, Theorem 1] and we omit it.

Remark 4.13 If X is a finite dimensional space, $(F)(iv)$ can be removed since it follows from $(F)(iii)$. Specifically, in this case $F(t, B)$ is bounded whenever B is bounded and then it is relatively compact in X.

Definition 4.10 A function $u : (-\infty, \tau + T] \to X$ is said to be an integral solution of problem (4.7) iff $u(t) = \varphi^\tau(t - \tau)$ for $t \in (-\infty, \tau]$ and there exists $f \in Sel_F(u|_{[\tau, \tau+T]})$ such that

$$u(t) = S(t - \tau)\varphi^\tau(0) + \int_\tau^t S(t - s)f(s)ds \qquad (4.36)$$

for any $t \in [\tau, \tau + T]$.

The solution operator $\mathscr{F} : C_{\varphi^\tau} \to P(C_{\varphi^\tau})$ is defined as follows:

$$\mathscr{F}(u)(t) = \left\{ S(t - \tau)\varphi^\tau(0) + \int_\tau^t S(t - s)f(s)ds : f \in Sel_F(u) \right\}.$$

Defining

$$\Gamma_\tau(f)(t) = \int_\tau^t S(t - s)f(s)ds \text{ for } f \in L^1(J, X), \tag{4.37}$$

we have

$$\mathscr{F}(u)(t) = S(t - \tau)\varphi^\tau(0) + \Gamma_\tau \circ Sel_F(u)(t).$$

It is obviously that $u \in C_{\varphi^\tau}$ is a fixed point of $\mathscr{F}$ iff $u[\varphi^\tau]$ is an integral solution of 4.7 on $(-\infty, \tau + T]$.

We need the following result.

Lemma 4.10 *Let (A) hold. If $D \subset L^1(J, X)$ is integrably bounded, then $\Gamma_\tau(D)$ is equicontinuous in $C(J, X)$, where the operator $\Gamma_\tau : L^1(J, X) \to C(J, X)$ is given by (4.37). In addition, if $\{f_n\} \subset L^1(J, X)$ is a semicompact sequence, then $\{\Gamma_\tau(f_n)\}$ is relatively compact in $C(J, X)$. Moreover, if $f_n \rightharpoonup f^*$ in $L^1(J, X)$, then $\Gamma_\tau(f_n) \to \Gamma_\tau(f^*)$ in $C(J, X)$.*

Proof The proof of the first claim is standard. The second one was proved in [16, Theorem 5.1.1].

We now give the main result of this section.

Theorem 4.10 *Let assumptions (A), (B) and (F) hold. Then problem (4.7) has at least one integral solution for each initial datum $\varphi^\tau \in \mathscr{B}$.*

Proof In order to apply the fixed point theorem stated by Theorem 1.10, we first prove that the solution operator is closed with convex values. Indeed, $\mathscr{F}$ has convex values due to the assumption (F). Let $\{u_n\} \subset C_{\varphi^\tau}$, $u_n \to u^*$ and $z_n \in \mathscr{F}(u_n)$ with $z_n \to z^*$. Then we have

$$z_n(t) \in S(t - \tau)\varphi^\tau(0) + \Gamma_\tau \circ Sel_F(u_n)(t).$$

Let $f_n \in Sel_F(u_n)$ such that

$$z_n(t) = S(t - \tau)\varphi^\tau(0) + \Gamma_\tau(f_n)(t). \tag{4.38}$$

Since Sel_F has compact, convex values and weakly u.s.c., we can employ Lemma 1.7 (ii) to state that $f_n \rightharpoonup f^*$ in $L^1(J, X)$ and $f^* \in Sel_F(u^*)$. Moreover, let $K(t) = \overline{F(t, u_n[\varphi^\tau]_t)}$ then $\{f_n(t)\} \subset K(t)$ for a.e. $t \in J$ with $K(t)$ being compact in X

thanks to the fact that F is u.s.c. Taking into account (F)(iii), we see that $\{f_n\}$ is integrably bounded. Now applying Lemma 4.10 gives the compactness of $\{\Gamma_\tau(f_n)\}$ in $C(J, X)$. Thus one can pass (4.38) into limits to get that

$$z^*(t) = S(t - \tau)\varphi^\tau(0) + \int_\tau^t S(t - s)f^*(s)ds,$$

where $f^* \in Sel_F(u^*)$. It deduces that $z^* \in \mathscr{F}(u^*)$.

In the rest of the proof, we will determine a compact convex subset $\mathcal{M}$ in C_{φ^τ} such that $\mathscr{F}(\mathcal{M}) \subset \mathcal{M}$. We first look for a closed convex set $\mathcal{M}_0 \subset C_{\varphi^\tau}$ satisfying that $\mathscr{F}(\mathcal{M}_0) \subset \mathcal{M}_0$. Let $z \in \mathscr{F}(u)$. Then from the definition of the solution operator, one has

$$|z(t)| \le N|\varphi^\tau(0)| + N \int_\tau^t [m_1(s) + m_2(s)|u[\varphi^\tau]_s|_\mathscr{B}]ds$$

$$\le N|\varphi^\tau(0)| + N\|m_1\|_{L(J,\mathbb{R})} + N \int_\tau^t m_2(s)|u[\varphi^\tau]_s|_\mathscr{B}ds,$$

where $N = \sup\limits_{t \in [0,T]} \|S(t)\|_{\mathcal{L}(X)}$.

Observing that

$$|u[\varphi^\tau]_s|_\mathscr{B} \le K(s - \tau)\sup\{|u(\theta)|, \tau \le \theta \le s\} + M(s - \tau)|u_\tau|_\mathscr{B}$$
$$\le K_T \sup\limits_{r \in [\tau,s]} |u(r)| + M_T|\varphi^\tau|_\mathscr{B},$$

where

$$K_T = \max\limits_{0 \le \theta \le T} K(\theta), \quad M_T = \sup\limits_{\theta \in [0,T]} M(\theta),$$

we deduce

$$|z(t)| \le C_1 + C_2 \int_\tau^t m_2(s) \sup\limits_{r \in [\tau,s]} |u(r)|ds,$$

where $C_1 = N\left(|\varphi^\tau(0)| + \|m_1\|_{L(J,\mathbb{R})} + M_T|\varphi^\tau|_\mathscr{B}T\|m_2\|_{L(J,\mathbb{R})}\right)$, $C_2 = NK_T$.
Since the last term is increasing in t, we have

$$\sup\limits_{\rho \in [\tau,t]} |z(\rho)| \le C_1 + C_2 \int_\tau^t m_2(s) \sup\limits_{\rho \in [\tau,s]} |u(\rho)|ds. \qquad (4.39)$$

Denote

$$\mathcal{M}_0 = \{u \in C_{\varphi^\tau} : \sup\limits_{s \in [\tau,t]} |u(s)| \le \psi(t), \quad t \in J\},$$

where ψ is the solution of the integral equation

$$\psi(t) = C_1 + C_2 \int_\tau^t m_2(s)\psi(s)ds.$$

It is clear that $\mathcal{M}_0$ is a closed convex subset of C_{φ^τ} and estimate (4.39) ensures that $\mathscr{F}(\mathcal{M}_0) \subset \mathcal{M}_0$. Let

$$\mathcal{M}_{k+1} = \overline{\text{co}}\mathscr{F}(\mathcal{M}_k), \quad k = 0, 1, 2, \ldots .$$

Then one sees that $\mathcal{M}_k$ is closed, convex and $\mathcal{M}_{k+1} \subset \mathcal{M}_k$ for all $k \in \mathbb{N}$. Since $Sel_F(\mathcal{M}_k)$ is integrably bounded thanks to (F)(iii), $\mathscr{F}(\mathcal{M}_k)$ is equicontinuous for all $k \geq 0$ by Lemma 4.10. Hence $\mathcal{M}_{k+1}$ is equicontinuous as well.

Let

$$\mathcal{M} = \bigcap_{k=0}^\infty \mathcal{M}_k,$$

then $\mathcal{M}$ is a closed convex and equicontinuous subset of C_{φ^τ} and $\mathscr{F}(\mathcal{M}) \subset \mathcal{M}$. In order to employ Arzela-Ascoli's theorem, we need to prove the compactness of $\mathcal{M}(t)$ for each $t \geq \tau$. This will be done if it is shown that

$$h_k(t) := \beta(\mathcal{M}_k(t)) \to 0 \text{ as } k \to \infty.$$

Using properties of MNC and Property 1.5, one has

$$\begin{aligned}
h_{k+1}(t) &= \beta(\mathcal{M}_{k+1}(t)) \\
&= \beta(\mathscr{F}(\mathcal{M}_k)(t)) \\
&\leq \alpha\left(\int_\tau^t S(t-s)Sel_F(\mathcal{M}_k)(s)ds \right) \\
&\leq 4 \int_\tau^t \beta(S(t-s)Sel_F(\mathcal{M}_k)(s))ds.
\end{aligned}$$

If the semigroup $S(\cdot)$ is compact, then $\beta(S(t-s)Sel_F(\mathcal{M}_k)(s)) = 0$ for a.e. $s \in [\tau, t]$. In this case $h_{k+1}(t) = 0$. In the opposite case, we deduce from (F)(iv) that

$$\begin{aligned}
h_{k+1}(t) &\leq 4N \int_\tau^t k(s) \sup_{\theta \leq 0} \beta(\mathcal{M}_k[\varphi^\tau](s+\theta))ds \\
&\leq 4N \int_\tau^t k(s) \sup_{r \in [\tau,s]} \beta(\mathcal{M}_k(r))ds \\
&\leq 4N \int_\tau^t k(s) \sup_{r \in [\tau,s]} h_k(r)ds.
\end{aligned}$$

Let $v_k(t) = \sup_{r \in [\tau, t]} h_k(r)$, then we have

$$v_{k+1}(t) \le 4N \int_\tau^t k(s) v_k(s) ds.$$

Since $\{v_k(t)\}$ is decreasing, one can pass to the limit in the last inequality to get

$$v_\infty(t) \le 4N \int_\tau^t k(s) v_\infty(s) ds,$$

where $v_\infty(t) = \lim_{k \to \infty} v_k(t)$. Thanks to Gronwall-Bellman's inequality, this implies $v_\infty(t) = 0$ for all $t \in J$. We have the compactness of $\mathcal{M}(t)$ as desired. The proof is done by considering $\mathscr{F} : \mathcal{M} \to P(\mathcal{M})$ and applying Theorem 1.10.

4.4.2 Existence of Pullback Attractor

In this subsection, we need the following assumptions:
(A^*) the C_0-semigroup $S(t) = e^{tA}$ is compact and exponentially stable, i.e., there exist $\alpha > 0$, $N \ge 1$ such that

$$\|S(t)\|_{\mathcal{L}(X)} \le N e^{\alpha t} \quad \text{for all } t \ge 0;$$

(B^*) the phase space $\mathscr{B} = C_\gamma$, $\gamma > 0$;
(F^*) the nonlinearity F satisfies (F) with $m_2 \in L^\infty(\tau, +\infty)$, and m_1 being nondecreasing.

Let

$$\Sigma(\varphi^\tau) = \{ u \in C([\tau, +\infty), X) : u[\varphi^\tau] \text{ is an integral solution}$$
$$\text{of (4.7) on } (-\infty, T] \text{ for any } T > \tau \}.$$

Obviously,

$$\Sigma(\varphi^\tau)(t) \subset S(t - \tau)\varphi^\tau(0) + \Gamma_\tau \circ Sel_F(\Sigma(\varphi^\tau))(t), \quad \forall t > \tau. \tag{4.40}$$

The MNDS generated by problem (4.7) is defined as follows:

$$\mathcal{U} : \mathbb{R}_d^2 \times C_\gamma \to P(C_\gamma), \tag{4.41}$$

$$\mathcal{U}(t, \tau, \varphi^\tau) = \{ u_t : u \in \Sigma(\varphi^\tau) \}. \tag{4.42}$$

Using the same arguments as in [55], one can show that $\mathcal{U}$ is a strict MNDS. We will prove the upper-semicontinuity of $\mathcal{U}$ in the following lemma.

Lemma 4.11 *Under assumptions (A^*), (B^*) and (F^*), $\mathcal{U}(t, \tau, \cdot)$ is u.s.c. with compact values for each $(t, \tau) \in \mathbb{R}_d^2$.*

Proof Since $\mathrm{Fix}(\mathscr{F})$ is compact in $C([\tau, t], X)$ for each $\varphi^\tau \in \mathcal{C}_\gamma$, $\mathcal{U}(t, \tau, \cdot)$ has compact values. By Lemma 1.9, it remains to show that $\mathcal{U}(t, \tau, \cdot)$ is quasicompact and has a closed graph.

We first show that $\mathcal{U}(t, \tau, \cdot)$ is quasicompact. Assume that $K \subset \mathcal{C}_\gamma$ is a compact set. Let $\{z_n\} \subset \mathcal{U}(t, \tau, K)$, then one can find a sequence $\{\phi_n^\tau\} \subset K$ such that $z_n \in \mathcal{U}(t, \tau, \phi_n^\tau)$ and $\{\phi_n^\tau\}$ converges to ϕ^* in $\mathcal{C}_\gamma$. Let $u_n \in \sum(\phi_n^\tau)$ such that

$$z_n(s) = u_n[\phi_n^\tau](t + s), \quad s \in (-\infty, 0]. \tag{4.43}$$

We claim that $\{u_n\}$ is relatively compact in $C([\tau, t], X)$. Observing that

$$u_n(r) \in S(r - \tau)\phi_n^\tau(0) + \Gamma_\tau \circ Sel_F(u_n)(r), \quad r \in [\tau, t],$$

we have the boundedness of $\{u_n\}$ in $C([\tau, t], X)$ by using standard estimates. Let $f_n \in Sel_F(u_n)$ be such that

$$u_n(r) = S(r - \tau)\phi_n^\tau(0) + \Gamma_\tau(f_n)(r).$$

Since $\{u_n\}$ is bounded, one can see that $\{f_n\}$ is integrably bounded by using $(F)(iii)$. Thanks to the compactness of $S(\cdot)$, $\{\Gamma_\tau(f_n)\}$ is relatively compact which implies that $\{u_n\}$ is relatively compact as well. Then there is a subsequence of $\{u_n\}$ (denoted again by $\{u_n\}$) such that $u_n \to u^*$ in $C([\tau, t], X)$. So relation (4.43) ensures that $\{z_n\}$ converges to $u^*[\phi^*]_t$. Equivalently, $\{z_n\}$ is relatively compact in $\mathcal{C}_\gamma$.

We now prove that $\mathcal{U}(t, \tau, \cdot)$ has a closed graph. Let $\{\varphi_n^\tau\}$ be a sequence in $\mathcal{C}_\gamma$ converging to φ^* and $\xi_n \in \mathcal{U}(t, \tau, \varphi_n^\tau)$ such that $\{\xi_n\}$ converges to ξ^* in $\mathcal{C}_\gamma$. We have to show that $\xi^* \in \mathcal{U}(t, \tau, \varphi^*)$. Choose $u_n \in \sum(\varphi_n^\tau)$ such that $\xi_n(s) = u_n[\varphi_n^\tau](t + s)$. By the above arguments, $\{u_n\}$ has a convergent subsequence (still denoted by $\{u_n\}$) converging to u^*, then $\xi^*(s) = u^*[\varphi^*](t + s)$, $s \in (-\infty, 0]$. It remains to show that $u^* \in \sum(\varphi^*)$.

Let $f_n \in Sel_F(u_n)$ such that

$$u_n(r) = S(r - \tau)\varphi_n^\tau(0) + \Gamma_\tau(f_n)(r), \quad r \in [\tau, t]. \tag{4.44}$$

By $(F)(iii)$ and the fact that $\{u_n\}$ is bounded, we see that $\{f_n\} \subset L^1([\tau, t], X)$ is integrably bounded. Furthermore, since $K(r) = F(r, \overline{\{u_n[\varphi_n^\tau]_r\}})$, $r \in [\tau, t]$, is compact then $\{f_n\}$ is a semicompact sequence. Applying Lemma 4.10, we have $f_n \to f^*$ and $\Gamma_\tau(f_n) \to \Gamma_\tau(f^*)$. Thus one can pass to the limit in equality (4.44) to obtain

$$u^*(r) = S(r - \tau)\varphi^*(0) + \Gamma_\tau(f^*)(r), \quad r \in [\tau, t].$$

Since Sel_F is weakly u.s.c., one has $f^* \in Sel_F(u^*)$. So the last relation tells us that $u^* \in \pi_t \circ \sum(\varphi^*)$. The proof is complete.

For our analysis below, we will consider a universe $\mathscr{D}$ which is a family of multifunctions D taking values in $\mathscr{P}_b(\mathcal{C}_\gamma)$ such that $D(\tau) \subset B_{r(\tau)}$, the closed ball centered at the origin with radius $r(\tau)$ in $\mathcal{C}_\gamma$ satisfying

$$\lim_{\tau \to -\infty} r(\tau) e^{(\min\{\alpha,\gamma\}-N\kappa)\tau} = 0, \quad \kappa := \|m_2\|_\infty. \tag{4.45}$$

Denote by β_γ the Hausdorff MNC in $\mathcal{C}_\gamma$.

Lemma 4.12 *Assume (A^*), (B^*) and (F^*). Then the MNDS $\mathcal{U}$ defined by (4.41)–(4.42) admits a monotone pullback $\mathscr{D}$-absorbing set, provided that $\min\{\alpha, \gamma\} > N\kappa$ and*

$$\lim_{\tau \to -\infty} m_1(\tau) e^{(\min\{\alpha,\gamma\}-N\kappa)\tau} = 0.$$

Proof Let $t > \tau$ be a fixed moment. We look for $R(t) > 0$ satisfying that for every $\varphi^\tau \in D(\tau)$, $D \in \mathscr{D}$, we have

$$|\mathcal{U}(t, \tau, \varphi^\tau)|_\gamma = \sup\{|z|_\gamma : z \in \mathcal{U}(t, \tau, \varphi^\tau)\} \leq R(t)$$

as $\tau \to -\infty$.

Let $z \in \mathcal{U}(t, \tau, \varphi^\tau)$. Then one can take $u \in \Sigma(\varphi^\tau)$ such that $z = u_t$. We have

$$u_t(\theta) = \begin{cases} \varphi^\tau(t + \theta - \tau), & \theta < \tau - t, \\ S(t + \theta - \tau)\varphi^\tau(0) + \displaystyle\int_\tau^{t+\theta} S(t + \theta - s) f(s) ds, & \tau - t \leq \theta \leq 0, \end{cases}$$

here $f \in Sel_F(u)$. Then

$$|u_t|_\gamma = \max\left\{ \sup_{\theta < \tau - t} e^{\gamma\theta} |\varphi^\tau(t + \theta - \tau)|, \sup_{\tau - t \leq \theta \leq 0} e^{\gamma\theta} |u_t(\theta)| \right\}. \tag{4.46}$$

One has

$$\sup_{\theta < \tau - t} e^{\gamma\theta} |\varphi^\tau(t + \theta - \tau)| = \sup_{r < 0} e^{r + \tau - t} |\varphi^\tau(r)| \leq e^{-\gamma(t-\tau)} |\varphi^\tau|_\gamma,$$

and for $\theta \in [\tau - t, 0]$,

$$e^{\gamma\theta} |u_t(\theta)| \leq e^{\gamma\theta} \left(\|S(t + \theta - \tau)\|_{\mathcal{L}(X)} \cdot |\varphi^\tau(0)| + \int_\tau^{t+\theta} \|S(t + \theta - s)\|_{\mathcal{L}(X)} \cdot |f(s)| ds \right)$$

$$\leq N e^{\gamma\theta} \left(e^{-\alpha(t+\theta-\tau)} |\varphi^\tau|_\gamma + \int_\tau^{t+\theta} e^{-\alpha(t+\theta-s)} (m_1(s) + m_2(s)|u_s|_\gamma) ds \right)$$

$$\leq N e^{\gamma\theta} \left(e^{-\alpha(t+\theta-\tau)} |\varphi^\tau|_\gamma \right.$$

$$+ \int_\tau^{t+\theta} e^{-\alpha(t+\theta-s)} m_1(s) ds + \kappa \int_\tau^{t+\theta} e^{-\alpha(t+\theta-s)} |u_s|_\gamma ds \bigg).$$

Thus (4.46) implies

$$|u_t|_\gamma \leq \max \bigg\{ e^{-\gamma(t-\tau)} |\varphi^\tau|_\gamma, \; N \sup_{\tau-t \leq \theta \leq 0} e^{\gamma\theta} \bigg(e^{-\alpha(t+\theta-\tau)} |\varphi^\tau|_\gamma$$

$$+ \int_\tau^{t+\theta} e^{-\alpha(t+\theta-s)} m_1(s) ds + \kappa \int_\tau^{t+\theta} e^{-\alpha(t+\theta-s)} |u_s|_\gamma ds \bigg) \bigg\}.$$

Case 1. If $\alpha < \gamma$, then we obtain

$$|u_t|_\gamma \leq \max \bigg\{ e^{-\alpha(t-\tau)} |\varphi^\tau|_\gamma, \; N \sup_{\tau-t \leq \theta \leq 0} e^{\gamma\theta} \bigg(e^{-\alpha(t+\theta-\tau)} |\varphi^\tau|_\gamma$$

$$+ \int_\tau^{t+\theta} e^{-\alpha(t+\theta-s)} m_1(s) ds + \kappa \int_\tau^{t+\theta} e^{-\alpha(t+\theta-s)} |u_s|_\gamma ds \bigg) \bigg\}$$

$$\leq N e^{-\alpha(t-\tau)} |\varphi^\tau|_\gamma + N \int_\tau^t e^{-\alpha(t-s)} m_1(s) ds + N\kappa \int_\tau^t e^{-\alpha t} e^{\alpha s} |u_s|_\gamma ds.$$

Equivalently, we have

$$e^{\alpha t} |u_t|_\gamma \leq N e^{\alpha\tau} |\varphi^\tau|_\gamma + N \int_\tau^t e^{\alpha s} m_1(s) ds + N\kappa \int_\tau^t e^{\alpha s} |u_s|_\gamma ds.$$

Since m_1 is nondecreasing, we get

$$e^{\alpha t} |u_t|_\gamma \leq N e^{\alpha\tau} |\varphi^\tau|_\gamma + \frac{N m_1(t)}{\alpha} (e^{\alpha t} - e^{\alpha\tau}) + N\kappa \int_\tau^t e^{\alpha s} |u_s|_\gamma ds.$$

Applying Gronwall's inequality, we obtain

$$e^{\alpha t} |u_t|_\gamma \leq N e^{\alpha\tau} |\varphi^\tau|_\gamma + \frac{N m_1(t)}{\alpha} (e^{\alpha t} - e^{\alpha\tau})$$

$$+ N\kappa \int_\tau^t \bigg(N e^{\alpha\tau} |\varphi^\tau|_\gamma + \frac{N m_1(s)}{\alpha} (e^{\alpha s} - e^{\alpha\tau}) \bigg) e^{N\kappa(t-s)} ds$$

$$= N e^{\alpha\tau} |\varphi^\tau|_\gamma + \frac{N m_1(t)}{\alpha} (e^{\alpha t} - e^{\alpha\tau}) + N e^{\alpha\tau} |\varphi^\tau|_\gamma e^{N\kappa t} (e^{-N\kappa\tau} - e^{-N\kappa t})$$

$$+ \frac{N^2 \kappa m_1(t)}{\alpha} e^{N\kappa t} \int_\tau^t (e^{(\alpha-N\kappa)s} - e^{(\alpha\tau-N\kappa s)}) ds$$

$$= N e^{\alpha\tau} |\varphi^\tau|_\gamma + \frac{N m_1(s)}{\alpha} (e^{\alpha t} - e^{\alpha\tau}) + N e^{\alpha\tau} |\varphi^\tau|_\gamma (e^{-N\kappa\tau} e^{N\kappa t} - 1)$$

$$+ \frac{N^2 \kappa m_1(t)}{\alpha} e^{N\kappa t} \bigg(\frac{e^{(\alpha-N\kappa)t} - e^{(\alpha-N\kappa)\tau}}{\alpha - N\kappa} + \frac{e^{\alpha\tau}(e^{-N\kappa t} - e^{-N\kappa\tau})}{N\kappa} \bigg).$$

Therefore

$$
\begin{aligned}
|u_t|_\gamma \leq & Ne^{-\alpha(t-\tau)}|\varphi^\tau|_\gamma + \frac{Nm_1(t)}{\alpha}\left(1 - e^{-\alpha(t-\tau)}\right) \\
& + N|\varphi^\tau|_\gamma\left(e^{-(\alpha-N\kappa)(t-\tau)} - e^{-\alpha(t-\tau)}\right) \\
& + \frac{N^2\kappa m_1(t)}{\alpha}e^{-(\alpha-N\kappa)t}\left(\frac{e^{(\alpha-N\kappa)t} - e^{(\alpha-N\kappa)\tau}}{\alpha - N\kappa} + \frac{e^{\alpha\tau}e^{-N\kappa t} - e^{(\alpha-N\kappa\tau)\tau}}{N\kappa}\right) \\
\leq & Ne^{-\alpha(t-\tau)}|\varphi^\tau|_\gamma + \frac{Nm_1(t)}{\alpha}\left(1 - e^{-\alpha(t-\tau)}\right) \\
& + N|\varphi^\tau|_\gamma\left(e^{-(\alpha-N\kappa)(t-\tau)} - e^{-\alpha(t-\tau)}\right) \\
& + \frac{N^2\kappa m_1(t)}{\alpha}\left(\frac{1 - e^{(\alpha-N\kappa)(t-\tau)}}{\alpha - N\kappa} + \frac{e^{-\alpha(t-\tau)} - e^{-(\alpha-N\kappa)(t-\tau)}}{N\kappa}\right) \\
\leq & \frac{Nm_1(t)}{\alpha} + \frac{N^2\kappa m_1(t)}{\alpha(\alpha - N\kappa)} + N|\varphi^\tau|_\gamma e^{-(\alpha-N\kappa)(t-\tau)} \\
\leq & \frac{Nm_1(t)}{\alpha - N\kappa} + Nr(\tau)e^{(\alpha-N\kappa)\tau}e^{-(\alpha-N\kappa)t} \\
\leq & \frac{Nm_1(t)}{\alpha - N\kappa} + 1 \quad \text{as } \tau \to -\infty
\end{aligned}
$$

due to the assumption that $\alpha > N\kappa$, t is fixed, and

$$
\lim_{\tau \to -\infty} r(\tau)e^{(\min\{\alpha,\gamma\}-N\kappa)\tau} = 0.
$$

Case 2. If $\alpha \geq \gamma$, then we get

$$
\begin{aligned}
|u_t|_\gamma \leq & \max\left\{e^{-\gamma(t-\tau)}|\varphi^\tau|_\gamma,\ N\sup_{\tau-t\leq\theta\leq 0}e^{\gamma\theta}\left(e^{-\gamma(t+\theta-\tau)}|\varphi^\tau|_\gamma\right.\right. \\
& \left.\left. + \int_\tau^{t+\theta}e^{-\gamma(t+\theta-s)}m_1(s)ds + \kappa\int_\tau^{t+\theta}e^{-\gamma(t+\theta-s)}|u_s|_\gamma ds\right)\right\} \\
\leq & Ne^{-\gamma(t-\tau)}|\varphi^\tau|_\gamma + N\int_\tau^t e^{-\gamma(t-s)}m_1(s)ds + N\kappa\int_\tau^t e^{-\gamma t}e^{\gamma s}|u_s|_\gamma ds.
\end{aligned}
$$

Using the same arguments as in the first case, we have

$$
|u_t|_\gamma \leq \frac{Nm_1(t)}{\gamma - N\kappa} + 1
$$

as $\tau \to -\infty$.

So we can choose $B(t) = B_{R(t)}$, where $R(t) = \frac{Nm_1(t)}{\ell - N\kappa} + 1$, $\ell = \min\{\alpha, \gamma\}$, as a monotone pullpack $\mathcal{D}$-absorbing set for $\mathcal{U}$.

Lemma 4.13 *Let hypotheses (A^*), (B^*) and (F^*) hold. Then the MNDS $\mathcal{U}$ is β_γ-contracting.*

Proof We begin by decomposing the MNDS $\mathcal{U}$ into two parts:

$$\mathcal{U}(t, \tau, \cdot) = \mathcal{U}_1(t, \tau, \cdot) + \mathcal{U}_2(t, \tau, \cdot),$$

where

$$\mathcal{U}_1(t, \tau, \varphi^\tau)(\theta) = \begin{cases} S(t + \theta - \tau)\varphi^\tau(0), & \text{if } \tau - t \leq \theta \leq 0, \\ \varphi^\tau(t + \theta - \tau), & \text{if } -\infty < \theta < \tau - t, \end{cases}$$

and

$$\mathcal{U}_2(t, \tau, \varphi^\tau)(\theta) = \begin{cases} \Gamma_\tau \circ Sel_F(\Sigma(\varphi^\tau))(t + \theta), & \text{if } \tau - t \leq \theta \leq 0, \\ 0, & \text{if } \theta < \tau - t. \end{cases}$$

It follows that

$$|\mathcal{U}_1(t, \tau, \varphi^\tau)|_\gamma$$

$$\leq \max \left\{ \sup_{\tau - t \leq \theta \leq 0} e^{\gamma\theta} \|S(t + \theta - \tau)\|_{\mathcal{L}(X)} \cdot |\varphi^\tau(0)|, \sup_{\theta < \tau - t} e^{\gamma\theta} |\varphi^\tau(t + \theta - \tau)| \right\}$$

$$\leq \max \left\{ N \sup_{\tau - t \leq \theta \leq 0} e^{\gamma\theta} e^{-\alpha(t + \theta - \tau)} |\varphi^\tau(0)|, e^{-\gamma(t - \tau)} \sup_{r \leq 0} e^{\gamma r} |\varphi^\tau(r)| \right\}$$

$$\leq N \max\{ e^{-\min\{\alpha, \gamma\}(t - \tau)} |\varphi^\tau(0)|, e^{-\gamma(t - \tau)} |\varphi^\tau|_\gamma \}$$

$$\leq N e^{-\min\{\alpha, \gamma\}(t - \tau)} |\varphi^\tau|_\gamma.$$

This implies that

$$\|\mathcal{U}_1(t, \tau, \cdot)\|_{\mathcal{L}(\mathcal{C}_\gamma)} \leq N e^{-\min\{\alpha, \gamma\}(t - \tau)}.$$

We will prove that $\mathcal{U}_2(t, \tau, \cdot)$ is compact for each $(t, \tau) \in \mathbb{R}_d^2$. Let $B(\tau)$ be a bounded set in $\mathcal{C}_\gamma$ and $D = \Sigma(B(\tau))$, then $\mathcal{U}_2(t, \tau, B(\tau))(\theta) = \{0\}$ for $\theta < \tau - t$, and

$$\mathcal{U}_2(t, \tau, B(\tau))(\theta) = \int_\tau^{t+\theta} S(t + \theta - s) Sel_F(D)(s) ds, \quad \theta \in [\tau - t, 0].$$

Then $\mathcal{U}_2(t, \tau, B(\tau))$ can be seen as a subset of

$$\Xi := \left\{ y \in C([\tau, t], X) : y(\xi) \in \int_\tau^\xi S(\xi - s) Sel_F(D)(s) ds \right\},$$

in the sense that for $z \in \mathcal{U}_2(t, \tau, B(\tau))$, there is $u \in \Xi$ verifying $z(\theta) = u(t + \theta)$, $\theta \in [\tau - t, 0]$.

Since $B(\tau)$ is bounded, one can check that D is bounded in $C([\tau, t], X)$. Then $Sel_F(D)$ is integrably bounded. Using assumption (A^*) that $S(\cdot)$ is compact, it is easily seen that $\varXi$ is a relatively compact set in $C([\tau, t], X)$. Let $\{z_n\} \subset \mathcal{U}_2(t, \tau, B(\tau))$ be a sequence. Then there exists $\{u_n\} \subset \varXi$ such that $z_n(\theta) = u_n(t + \theta)$. Moreover,

$$
\begin{aligned}
|z_n - z_m|_\gamma &= \sup_{\theta \leq 0} e^{\gamma \theta} |z_n(\theta) - z_m(\theta)| \\
&= \sup_{\theta \in [\tau - t, 0]} e^{\gamma \theta} |z_n(\theta) - z_m(\theta)| \\
&= \sup_{\theta \in [\tau - t, 0]} e^{\gamma \theta} |u_n(t + \theta) - u_m(t + \theta)| \\
&\leq \sup_{\theta \in [\tau - t, 0]} |u_n(t + \theta) - u_m(t + \theta)| \\
&= \|u_n - u_m\|_{C([\tau, t], X)}.
\end{aligned}
$$

Hence $\{z_n\}$ is relatively compact in $\mathcal{C}_\gamma$. Consequently, we have $\beta_\gamma(\mathcal{U}_2(t, \tau, B(\tau))) = 0$ and then

$$
\begin{aligned}
\beta_\gamma(\mathcal{U}(t, \tau, B(\tau))) &\leq \beta_\gamma(\mathcal{U}_1(t, \tau, B(\tau))) + \beta_\gamma(\mathcal{U}_2(t, \tau, B(\tau))) \\
&\leq \|\mathcal{U}_1(t, \tau, \cdot)\|_{L(\mathcal{C}_\gamma)} \cdot \beta_\gamma(B(\tau)) \\
&\leq N e^{-\min\{\alpha, \gamma\}(t - \tau)} \cdot \beta_\gamma(B(\tau)).
\end{aligned}
$$

Choosing $k(t, \tau) = N e^{-\min\{\alpha, \gamma\}(t - \tau)}$, we see that $k(t, \tau) \to 0$ as $\tau \to -\infty$ and then $\mathcal{U}(t, \tau, \cdot)$ is β_γ-contracting. The proof is complete.

Combining Lemmas 4.11, and Theorem 1.23 allows us to arrive at the following conclusion.

Theorem 4.11 *Let hypotheses (A^*), (B^*) and (F^*) hold. Then the MNDS $\mathcal{U}$ generated by system (4.7) admits a global pullback $\mathscr{D}$-attractor in $\mathcal{C}_\gamma$, provided that $\min\{\alpha, \gamma\} > N\kappa$ and*

$$
\lim_{\tau \to -\infty} m_1(\tau) e^{(\min\{\alpha, \gamma\} - N\kappa)\tau} = 0.
$$

4.4.3 Applications

Let $\varOmega \subset \mathbb{R}^n$ be a bounded domain with smooth boundary $\partial\varOmega$ and $\mathcal{O} \subset \varOmega$ be an open set. Consider the following control system

$$
\frac{\partial u}{\partial t}(t, x) = \Delta u(t, x) + f_0(t, x, u(t, x)) + b(x)v(t), \quad x \in \varOmega, \ t > \tau, \qquad (4.47)
$$

$$
v(t) \in \int_{-\infty}^{0} \int_{\mathcal{O}} v(\theta, y)[f_1(u(t + \theta, y)), f_2(u(t + \theta, y))]dy d\theta, \qquad (4.48)
$$

$$u(t, x) = 0, \quad x \in \partial\Omega, \ t > \tau, \tag{4.49}$$

$$u(\tau + s, x) = \varphi^\tau(s, x), \quad x \in \Omega, \ s \in (-\infty, 0], \tag{4.50}$$

here $[g_1, g_2] = \{\mu g_1 + (1 - \mu)g_2 : \mu \in [0, 1]\}, \ \forall \ g_1, g_2 \in \mathbb{R}$.

This is a control problem with feedback control v, which is taken in the form of energy of state function, up to time t.

Let $X = L^2(\Omega, \mathbb{R})$, $A = \Delta$ with $D(A) = H_0^1(\Omega, \mathbb{R}) \cap H^2(\Omega, \mathbb{R})$. Then A generates a compact semigroup $\{e^{tA}\}_{t \geq 0}$ on X, which satisfies

$$\|e^{tA}\|_{\mathcal{L}(X)} \leq e^{-\lambda_1 t}, \quad t \geq 0,$$

where $\lambda_1 > 0$ is the first eigenvalue of $-\Delta$:

$$\lambda_1 = \sup\{|\nabla u|^2 : |u| = 1\}.$$

Then assumption (A^*) is satisfied for $N = 1$ and $\alpha = \lambda_1$.

Denote $\mathbb{R}_\tau^+ = [\tau, +\infty)$. We are now in a position to give the description for the nonlinearities:

(A_1) $v : (-\infty, 0] \times \Omega \to \mathbb{R}$ is a continuous function and there exist a positive number β and a nonnegative function $k \in L^2(\mathcal{O}, \mathbb{R}^+)$ such that

$$|v(\theta, x)| \leq k(x)e^{\beta\theta};$$

(A_2) $f_0 : \mathbb{R}_\tau^+ \times \Omega \times \mathbb{R} \to \mathbb{R}$ is a continuous function such that

$$|f_0(t, x, z)| \leq l_1(t) + l_2(t)|z|,$$

where $l_1 \in L_{loc}^1(\mathbb{R}_\tau^+, \mathbb{R}^+)$ and $l_2 \in L^\infty(\mathbb{R}_\tau^+, \mathbb{R}^+)$ are nonnegative functions and l_1 is nondecreasing;

(A_3) $f_1, f_2 : \mathbb{R} \to \mathbb{R}$ are continuous functions. In addition, there is a positive number η such that

$$|f_i(z)| \leq \eta|z|, \quad i = 1, 2, \ \forall z \in \mathbb{R};$$

(A_4) $b \in L^2(\Omega, \mathbb{R})$.

We choose the phase space $\mathscr{B} = C_\gamma$ with $\gamma \in (0, \beta)$.

Let $F_0 : \mathbb{R}_\tau^+ \times C_\gamma \to X$ and $F_1 : C_\gamma \to P(X)$ be such that

$$F_0(t, \phi)(x) = f_0(t, x, \phi(0, x)),$$

$$F_1(\phi)(x) = b(x) \int_{-\infty}^0 \int_{\mathcal{O}} v(\theta, y)[f_1(\phi(\theta, y)), f_2(\phi(\theta, y))] dy d\theta.$$

Then using (A_2), one has

$$
\begin{aligned}
|F_0(t, \phi)| &= \left(\int_{\Omega} |f_0(t, x, \phi(0, x)|^2 dx \right)^{\frac{1}{2}} \\
&\leq (2l_1^2(t)|\Omega| + 2l_2^2(t)|\phi(0, \cdot)|^2)^{\frac{1}{2}} \\
&\leq l_1(t)\sqrt{2|\Omega|} + l_2(t)\sqrt{2}|\phi(0, \cdot)| \\
&\leq l_1(t)\sqrt{2|\Omega|} + l_2(t)\sqrt{2}|\phi|_\gamma,
\end{aligned}
\tag{4.51}
$$

where $|\Omega|$ is the volume of Ω.

Regarding F_1, for $g \in F_1(\phi)$, we have

$$
g(x) = b(x) \int_{-\infty}^{0} \int_{\mathcal{O}} v(\theta, y)[\mu f_1(\phi(\theta, y)) + (1 - \mu)f_2(\phi(\theta, y))]dyd\theta
$$

for some $\mu \in [0, 1]$. Then

$$
\begin{aligned}
|g| &\leq \eta|b| \int -\infty^0 e^{\beta\theta} \int_{\mathcal{O}} k(y)|\phi(\theta, y)|dyd\theta \\
&\leq \eta|b|\|k\|_{L^2(\mathcal{O}, \mathbb{R})} \int_{-\infty}^{0} e^{\beta\theta}|\phi(\theta, \cdot)|d\theta \\
&\leq \frac{\eta}{\beta - \gamma}|b|\|k\|_{L^2(\mathcal{O}, \mathbb{R})}|\phi|_\gamma.
\end{aligned}
\tag{4.52}
$$

Let $F(t, \phi) = F_0(t, \phi) + F_1(\phi)$. Then it follows from (4.51) and (4.52) that

$$
|F(t, \phi)| \leq \sqrt{2|\Omega|}l_1(t) + \left(\sqrt{2}l_2(t) + \frac{\eta}{\beta - \gamma}|b|\|k\|_{L^2(\mathcal{O}, \mathbb{R})} \right)|\phi|_\gamma.
$$

Thus F satisfies (F)(iii) with $m_1(t) = \sqrt{2|\Omega|}l_1(t)$ and

$$
m_2(t) = \sqrt{2}l_2(t) + \frac{\eta}{\beta - \gamma}|b|\|k\|_{L^2(\mathcal{O}, \mathbb{R})}.
$$

Noting that if t and ϕ are given, then $F_0(t, \phi)$ is a singleton and $F_1(\phi)$ is a convex, closed and bounded set in the one-dimensional subspace $\text{span}\{b\} \subset L^2(\Omega, \mathbb{R})$. Therefore $F(t, \phi)$ is a convex and compact set in X. That means F has compact and convex values. For each $\xi \in F_1(\phi)$, $f_\xi(\cdot) = F_0(\cdot, \phi) + \xi$ is a (strongly) measurable selection. So one can assert that F fulfills (F)(i). Moreover, (F)(iv) is satisfied obviously.

Observe that for a compact set K in $\mathcal{C}_\gamma$, $F_0(t, K)$ is compact thanks to the fact that $F_0(t, \cdot)$ is a continuous map. In addition, since $F_1(K)$ is a bounded set in $\text{span}\{b\}$, it is relatively compact. Thus $F(t, K) = F_0(t, K) + F_1(K)$ is a relatively compact set, and then F is quasicompact. To verify (F)(ii), it suffices to show that $F(t, \cdot)$ is closed. Let $\{\phi_n\}$ be a sequence in $\mathcal{C}_\gamma$ converging to ϕ^* and $\xi_n \in F(t, \phi_n)$ be such that $\xi_n \to \xi^*$. One gets

$$\xi_n(x) = f_0(t, x, \phi_n(0, x))$$
$$+ b(x) \int_{-\infty}^0 \int_{\mathcal{O}} v(\theta, y)[\mu_n f_1(\phi_n(\theta, y)) + (1 - \mu_n) f_2(\phi_n(\theta, y))] dy d\theta,$$

where $\{\mu_n\} \subset [0, 1]$. Since f_0, f_1, f_2 are continuous and the integrands are integrably bounded, we see that

$$\xi^*(x) = f_0(t, x, \phi^*(0, x))$$
$$+ b(x) \int_{-\infty}^0 \int_{\mathcal{O}} v(\theta, y)[\mu^* f_1(\phi^*(\theta, y)) + (1 - \mu^*) f_2(\phi^*(\theta, y))] dy d\theta,$$

with $\mu^* = \lim_{n \to \infty} \mu_n$. Hence $F(t, \cdot)$ is a closed multimap. By Lemma 1.9, $F(t, \cdot)$ is u.s.c. It follows that $(F)(ii)$ is satisfied.

Employing Theorems 3.3 and 4.4, we obtain the following results.

Theorem 4.12 *Let assumptions (A_1)–(A_4) hold. Then*

(i) *for given datum $\varphi^\tau \in \mathcal{C}_\gamma$, system (4.47)–(4.50) possesses at least one integral solution on $(-\infty, \tau + T]$ for all $T > 0$;*

(i) *the MNDS generated by (4.47)–(4.50) admits a global pullback $\mathcal{D}$-attractor in $\mathcal{C}_\gamma$ if*

$$\min\{\lambda_1, \gamma\} \geq \sqrt{2} \|l_2\|_\infty + \frac{\eta}{\beta - \gamma} \|b\|_{L^2(\mathcal{O}, \mathbb{R})} \|k\|_{L^2(\mathcal{O}, \mathbb{R})},$$

and

$$\lim_{\tau \to -\infty} l_1(\tau) e^{(\min\{\lambda_1, \gamma\} - \kappa)\tau} = 0.$$

Chapter 5
Non-autonomous Evolution Inclusions and Control System

Abstract In this chapter we consider the topological structure of the solution set of non-autonomous parabolic evolution inclusions with time delay, defined on non-compact intervals. The result restricted to compact intervals is then extended to non-autonomous parabolic control problems with time delay. Moreover, as the applications of the information about the structure, we establish the existence result of global mild solutions for non-autonomous Cauchy problems subject to nonlocal condition, and prove the invariance of a reachability set for non-autonomous control problems under single-valued nonlinear perturbations. Finally, some illustrating examples are supplied.

5.1 Introduction

The purpose of this chapter is to study the R_δ-property of solution sets for the Cauchy problems of semilinear parabolic evolution inclusions with time delay applicable to many rather different situations. The main features of problems under consideration are that their linear part is governed by a family of closed linear operators depending on time (possibly unbounded) and one of them is defined on a noncompact interval. Applications include existence problem of global mild solutions for non-autonomous Cauchy problems under nonlocal perturbations and invariance of a reachability set to the control problems under single-valued nonlinear perturbations.

Throughout this chapter, X is a real Banach space with norm $|\cdot|$. As usual, the topological dual of X is denoted by X^*. Let $\mathcal{L}(X)$ denote the Banach space of all bounded linear operators on X. Denote by $|\cdot|_0$ the sup-norm of $C([-\tau, 0], X)$.

We are looking at is a non-autonomous Cauchy problem of semilinear evolution inclusion of the form

$$\begin{cases} u'(t) - A(t)u(t) \in F(t, u(t), u_t), & t \in \mathbb{R}^+, \\ u(t) = \phi(t), & t \in [-\tau, 0], \end{cases} \tag{5.1}$$

© Springer Nature Singapore Pte Ltd. 2017
Y. Zhou et al., *Topological Structure of the Solution Set for Evolution Inclusions*,
Developments in Mathematics 51, https://doi.org/10.1007/978-981-10-6656-6_5

or

$$
\begin{cases}
u'(t) - A(t)u(t) \in F(t, u(t), u_t), & t \in \mathbb{R}^+, \\
u(t) = H(u)(t), & t \in [-\tau, 0],
\end{cases}
\tag{5.2}
$$

where $(A(t), D(A(t)))$ is a family of linear operators on X (possibly unbounded), F : $\mathbb{R}^+ \times X \times C([-\tau, 0], X) \to P(X)$ is a multivalued function with convex, closed values, $\phi \in C([-\tau, 0], X)$, $u_t \in C([-\tau, 0], X)$ is defined by $u_t(s) = u(t + s)$ ($s \in [-\tau, 0]$) for every $u \in \widetilde{C}([-\tau, \infty), X)$, and $H : \widetilde{C}_b([-\tau, \infty), X) \to C([-\tau, 0], X)$ is a function to be specified later.

As one can easily see, the function H, constituting a nonlocal condition, depends on history states, that is, it takes history values. Moreover, the general condition (5.2) contains as particular case (5.1) and it unifies some important specific cases such as periodicity condition ($\tau = 0$, $H(u) = u(T)$), anti-periodicity condition ($\tau = 0$, $g(u) = -u(T)$) and multi-point discrete mean condition ($H(u)(t) = \sum_{i=1}^n \alpha_i u^{\frac{1}{3}}(t_i + t)$ for each $t \in [-\tau, 0]$, where $\sum_{i=1}^n |\alpha_i| \leq 1$ and $\tau < t_1 < t_2 < \cdots < t_n < \infty$ are constants). For some contributions in these topics we refer the reader to Castaing and Monteiro-Marques [69], Hirano [124], Paicu [161], Vrabie [188] for periodic problems, Aizicovici and Reich [7], Aizicovici et al. [6] for anti-periodic problems, and Byszewski [52], Deng [82] for problems involving multi-point discrete condition. It is mentioned in particular that in Deng [82], a partial differential model involving multi-point discrete condition is used to characterize the diffusion phenomenon of a small amount of gas in a transparent tube.

However, for the topological structure of the solution set of the Cauchy problem (5.1) fulfilling the assumption (H_0), the situation is different from those in many previous papers such as [71] and there have been very few applicable results as far as we know. This in fact is the first motivation of this chapter. To make things more applicable, in this chapter we are interested in studying the Cauchy problem (5.1) in some Fréchet spaces. The topological structure of the solution set and its relevant applications are considered. The point that the Cauchy problem (5.1), compared with the semilinear case of the previous research such as [71], being more general because of the non-autonomy, would allow one to deal with a large class of non-autonomous parabolic evolution equations or inclusions. One of the main technical tool is evolution family allowing the direct application of existing results on autonomous problems.

The line, which we go along is that we first study the R_δ-structure of the solution set to the Cauchy problem (5.1). One of the key tools for us is the inverse limit method, sometimes also called the projective limit (see, e.g., Andres and Pavlačková [14], Chen et al. [71] and Gabor [106] for more details). The R_δ-structure of solution sets on compact intervals is, by means of the inverse limit method, extended to noncompact intervals cases.

One of other achievements of this chapter, motivated by applications of the information about the structure to existence problems while certainly significant for its own sake, is establishing some efficient conditions that ensure the existence of global mild solutions to the nonlocal Cauchy problem (5.2). It is worth mentioning that in [197] we derived an existence result of global mild solutions to the nonlocal Cauchy problem (5.2), which generalized the semilinear case of the previous related ones such as Vrabie [192]. Note that in [197], the nonexpansivity condition and uniform boundedness on nonlocal function are needed, though no invariance condition on the nonlinearity is involved. It is also noted that in dealing with nonlocal Cauchy problems, approaches used respectively in [71, 192] accomplishing essentially the same thing, have a very great difference. We emphasize that in the present results, no invariance condition on the nonlinearity and no nonexpansivity condition and uniform boundedness on nonlocal function are involved, which essentially extend some existing results in this area, such as [71, 192, 197].

As the reader can see, the study of the topological structure of solution sets for the Cauchy problem (5.1) on compact intervals is also extended to the control problem given in the form

$$
\begin{cases}
u'(t) - A(t)u(t) \in G(t, u_t) + Bw(t), & t \in [0, b], \\
u(t) = \phi(t), & t \in [-\tau, 0],
\end{cases}
\tag{5.3}
$$

where $\phi \in C([-\tau, 0], X)$, the control function w takes values in the Banach space V, $B \in \mathcal{L}(V, X)$; here $\mathcal{L}(V, X)$ stands for the Banach space of all bounded linear operators from V to X, and $G : [0, b] \times C([-\tau, 0], X) \to P(X)$ is a multivalued function with convex, closed values for which $G(t, \cdot)$ is weakly upper semicontinuous for a.e. $t \in [0, b]$ and $G(\cdot, v)$ has a strongly measurable selection for each $v \in C([-\tau, 0], X)$. Furthermore, under a class of single-valued nonlinearities G which does not guarantee the uniqueness of a mild solution, we prove the invariance of a reachability set of control problem (5.3) under nonlinear perturbations by making use of the information of the topological structure.

In proving that the reachability set is invariant under nonlinear perturbations, it is generally found in the past papers (see, e.g., [153, 154, 174]) that the solution mapping corresponding to each control function w is a single-valued one. In this chapter, the lack of uniqueness, i.e., the solution mapping corresponding to each control function w is a multivalued one, makes impossible the use of the known tools as in [174] to show the invariance of reachability set to control problem (5.3). This determined us to find a new idea to overcome this difficulty. We will see that the R_δ-structure of a solution set fulfills this role.

The results presented in this chapter are new even in the cases of autonomous problems (i.e., $A(t) \equiv A$) and/or if the nonlinearity is a single-valued function. In particular, Theorem 5.5 below presents a solution to one of the problems posed in [71, Problem (1), Remark 5.1 in Page 2071].

Our special goal is to emphasize that in some situations non-autonomous evolution equations can often be rewritten as an autonomous abstract Cauchy problem by means of an appropriate choice for the state-space. Thus, by making use of so-called evolution families, it is possible to apply existing results for autonomous problems.

Remark 5.1 As the reader can see, the results in [71] can not be applied to the study of our problems (see Theorems 5.1, 5.5 and Remarks 5.9, 5.13 below). We in particular mention that the concept of mild solution used in this chapter is different from those in many previous papers such as [71, 192] (where the C^0-solution is defined by means of a set of mild inequalities). This difference shows that when dealing with our problems, it is inappropriate to impose the invariance condition on the nonlinearity F and the compactness characterizations of a solution set in, e.g., [71, Lemmas 2.4, 2.5] is not useful here (see Lemmas 5.1, 5.2 below).

Remark 5.2 We remark that the topological characterizations of a solution set to the Cauchy problem (5.1) play a key role in showing the existence of global mild solutions to the nonlocal Cauchy problem (5.2), which, with the help of the fixed point arguments designed for R_δ-mappings, enable us to get rid of the nonexpansive condition on the nonlocal function. Furthermore, this approach can be easily extended to other functional integral/differential equations or inclusions involving nonlocal conditions.

This chapter continues with a mild solution of nonhomogeneous Cauchy problem in Sect. 5.2. Then R_δ-properties of the solution set for the Cauchy problem (5.1), presented in Sect. 5.3.1, are used to prove the existence of global mild solutions for the nonlocal Cauchy problem (5.2) in Sect. 5.4.1. By means of R_δ-property of the solution set for the control problem (5.3), presented in Sect. 5.3.2, in Sect. 5.4.2 we study the invariance of its reachability set under nonlinear perturbations. Finally, as an illustration of the developed theory, we apply it to the examples of non-autonomous diffusion inclusions.

This results in this chapter are taken from Wang, Ma and Zhou [196].

5.2 Nonhomogeneous Cauchy Problem

Throughout, we assume that the family of closed linear operators $A(t)$ for $t \in \mathbb{R}^+$ on X with domain $D(A(t))$ (possibly not densely defined) satisfies the known Acquistapace-Terreni conditions, which, introduced by Acquistapace and Terreni [2, 3], are widely used in dealing with non-autonomous evolution problems. Then by an obvious rescaling from [2, Theorem 2.3] and [202, Theorem 2.1], it follows that there exists a unique evolution family U on X, which governs the following linear equation

$$u'(t) - A(t)u(t) = 0.$$

In this situation we say that $A(t)$ generate the evolution family U.

From now on, what will be assumed throughout is that $\|U(t, s)\|_{\mathcal{L}(X)} \leq 1$ for all $t \geq s \geq 0$.

Consider the following linear Cauchy problem on X

$$\begin{cases} u'(t) - A(t)u(t) = f(t), & t \in [0, b], \\ u(0) = u_0 \in X, \end{cases} \tag{5.4}$$

where $f \in L^1([0, b], X)$, $b > 0$.

A function $u : [0, b] \to X$ is called a mild solution of the Cauchy problem (5.4), if $u \in C([0, b], X)$ satisfies the integral equation

$$u(t) = U(t, 0)u_0 + \int_0^t U(t, s)f(s)ds, \quad t \in [0, b].$$

Let $u_0 \in X$, $\zeta \in [0, b])$ and $f \in L^1([0, b], X)$. Denote by $v(\cdot, \zeta, u_0, f)$ the unique mild solution $u : [\zeta, b] \to X$ of the Cauchy problem (5.4) which verifies $u(\zeta) = u_0$. We shall also simply denote by $u = \mathscr{H}(u_0, f)$ the unique mild solution of the Cauchy problem (5.4) corresponding to (u_0, f).

We need the following approximation result, which will essentially be used in the sequel.

Lemma 5.1 ([197, Lemma 2.4]) *If the two sequences $\{f_n\} \subset L^1([0, b], X)$ and $\{u_n\} \subset C([0, b], X)$, where u_n is a mild solution of the problem*

$$\begin{cases} u_n'(t) - A(t)u_n(t) = f_n(t), & t \in [0, b], \\ u_n(0) = u_0, \end{cases}$$

$\lim_{n \to \infty} f_n = f$ *weakly in* $L^1([0, b], X)$ *and* $\lim_{n \to \infty} u_n = u$ *in* $C([0, b], X)$, *then* u *is a mild solution of the limit problem*

$$\begin{cases} u'(t) - A(t)u(t) = f(t), & t \in [0, b], \\ u(0) = u_0. \end{cases}$$

Remark 5.3 Notice that in the case of $A(t) \equiv A$ being an m-dissipative operator (possible multivalued and/or nonlinear), the assertion of the lemma above remains true having some additional assumptions: X^* is uniformly convex and the semigroup generated by A is compact.

In order to gain compactness of the solution set to the Cauchy problem (5.1), we need the following compactness result on the linear Cauchy problem (5.4).

Lemma 5.2 ([197, Lemma 3.1]) *Suppose that the evolution family U is compact. If $\mathcal{K}$ is a uniformly integrable subset in $L^1([0, b], X)$, then*

(i) *for each $\eta \in ([0, b])$ and $\mathcal{M}$ being bounded in X, $\mathscr{H}(\mathcal{M} \times \mathcal{K})$ restricted to $[\eta, b]$ is relatively compact in $C([\eta, b], X)$;*

(ii) *for each $\mathcal{M}$ being relatively compact in X, $\mathscr{H}(\mathcal{M} \times \mathcal{K})$ is relatively compact in $C([0, b], X)$.*

Remark 5.4 It is clear that if $\mathcal{K} \subset L^1([0, b], X)$ is integrably bounded, then $\mathcal{K}$ is uniformly integrable. It thus follows that the compactness result above remains true provided that $\mathcal{K}$ is integrably bounded.

Remark 5.5 For the case of $A(t) \equiv A$ being an m-dissipative operator, the compactness result above is due to Baras [30]. See also Vrabie [189, Theorem 2.3.3].

At the end of this section, let us first define the operator $T : L^2([0, b], X) \to C([0, b], X)$ by

$$(Tf)(t) = \int_0^t U(t, s)f(s)ds, \quad f \in L^2([0, b], X).$$

It is clear that T is a linear bounded operator.

The following hypothesis was introduced by Seidman [174].

(S) for each $f \in L^2([0, b], X)$, there exists $w \in L^2([0, b], V)$ such that $(T(Bw))(b) = (Tf)(b)$.

Remark 5.6 It is obvious that the hypothesis (S) is fulfilled if B is bijective.

Lemma 5.3 ([174, Lemma 2]) *If the hypothesis (S) is fulfilled, then there exists a continuous mapping $\mathscr{L} : L^2([0, b], X) \to L^2([0, b], V)$ such that for any $f \in L^2([0, b], X)$,*

$$(T(B\mathscr{L}f))(b) + (Tf)(b) = 0,$$

and

$$\|\mathscr{L}f\|_{L^2([0,b],V)} \leq M_0 \|f\|_{L^2([0,b],X)},$$

where M_0 is a positive number.

5.3   Topological Structure of Solution Set

In this section, we discuss the topological structure of the solution set to the Cauchy problem (5.1) and the control problem (5.3).

5.3.1   Evolution Inclusions

To avoid problems concerning the continuation of local solutions, we impose the growth condition on F:

(H_0) $F(t, \cdot, \cdot)$ is weakly upper semicontinuous for a.e. $t \in \mathbb{R}^+$ and $F(\cdot, x, v)$ has a strongly measurable selection for each $(x, v) \in X \times C([-\tau, 0], X)$.

(H_1) there exists a function $\mu \in L^1_{loc}(\mathbb{R}^+, \mathbb{R}^+)$ such that

$$|F(t, x, v)| := \sup\{|y| : y \in F(t, x, v)\} \le \mu(t)(1 + |x| + |v|_0)$$

for a.e. $t \in \mathbb{R}^+$ and each $(x, v) \in X \times C([-\tau, 0], X)$.

As a key preparation for the proof of Theorem 5.1 below, let us present the result as follows:

Lemma 5.4 *Let hypotheses (H_0) and (H_1) be satisfied. Then there exists a sequence $F_n : [0, b] \times X \times C([-\tau, 0], X) \to P_{cl,cv}(X)$, $n \ge 1$, such that*

(i) $F(t, x, v) \subset F_{n+1}(t, x, v) \subset F_n(t, x, v) \subset \overline{co}(F(t, B_{3^{1-n}}(x, v))$, $n \ge 1$, *for each $t \in [0, b]$, $(x, v) \in X \times C([-\tau, 0], X)$;*

(ii) $|F_n(t, x, v)| \le L(t)(3 + |x| + |v|_0)$, $n \ge 1$, *for a.e. $t \in [0, b]$ and each $(x, v) \in X \times C([-\tau, 0], X)$;*

(iii) *there exists $E \subset [0, b]$ with $mes(E) = 0$ such that for each $x^* \in X^*$, $\varepsilon > 0$ and $(t, x, v) \in [0, b] \backslash E \times X \times C([-\tau, 0], X)$, there exists $N > 0$ such that for all $n \ge N$,*

$$x^*(F_n(t, x, v)) \subset x^*(F(t, x, v)) + (-\varepsilon, \varepsilon);$$

(iv) $F_n(t, \cdot) : X \times C([-\tau, 0], X) \to P_{cl,cv}(X)$ *is continuous for a.e. $t \in [0, b]$ with respect to Hausdorff metric for each $n \ge 1$;*

(v) *for each $n \ge 1$, there exists a selection $G_n : [0, b] \times X \times C([-\tau, 0], X) \to X$ of F_n such that $G_n(\cdot, x, v)$ is strongly measurable for each $(x, v) \in X \times C([-\tau, 0], X)$ and for any compact subset $\mathscr{D} \subset X \times C([-\tau, 0], X)$ there exist constants $C_V > 0$ and $\delta > 0$ for which the estimate*

$$|G_n(t, x_1, v_1) - G_n(t, x_2, v_2)| \le C_V L(t)(|x_1 - x_2| + |v_1 - v_2|_0)$$

holds for a.e. $t \in [0, b]$ and each $(x_1, v_1), (x_2, v_2) \in V$ with $V := \mathscr{D} + B_\delta(0)$;

(vi) F_n *verifies the condition (H_0) with F_n instead of F for each $n \ge 1$, provided that X is reflexive.*

Proof An argument similar to that in the proof of Lemma 2.6 (with X instead of $\overline{D(A)}$) shows that the assertions of the lemma remain true (see also [106, Theorem 3.5], [80, Lemma 2.2]). Here we omit the details for simplicity.

Remark 5.7 As indicated in [71], the condition that $F(t, \cdot, \cdot)$ is weakly u.s.c. for a.e. $t \in [0, b]$ is more easily verified usually in practical applications.

By a mild solution of the Cauchy problem (5.1), we mean a continuous function $u : [-\tau, \infty) \to X$ which restricted to $[-\tau, d]$ is a mild solution to

$$\begin{cases} u'(t) - A(t)u(t) = f(t), & t \in [0, d], \\ u(t) = \phi(t), & t \in [-\tau, 0] \end{cases}$$

for each $d > 0$, where $f \in L^1_{loc}(\mathbb{R}^+, X)$ and $f(t) \in F(t, u(t), u_t)$ for a.e. $t \in \mathbb{R}^+$.

For every $\phi \in C([-\tau, 0], X)$ we let $\Theta(\phi)$ denote the set of all mild solutions of the Cauchy problem (5.1). Now we are able to prove an Aronszajn-type result.

Theorem 5.1 *Assume that X is reflexive and the evolution family U is compact. Let $F : \mathbb{R}^+ \times X \times C([-\tau, 0], X) \to P_{cl,cv}(X)$ be such that (H_0), (H_1) are satisfied. Then $\Theta(\phi)$ is a compact R_δ-set for every $\phi \in C([-\tau, 0], X)$. In particular, it is connected.*

Proof Let $\phi \in C([-\tau, 0], X)$ be given. The proof is divided into three steps.

Step 1. Fix $b > 0$. We first consider the Cauchy problem

$$
\begin{cases}
u'(t) - A(t)u(t) = f(t), & t \in [0, b], \\
u(t) = \phi(t), & t \in [-\tau, 0]
\end{cases}
\tag{5.5}
$$

with some

$$
f \in Sel^b_F(u) := \{f \in L^1([0, b], X) : f(t) \in F(t, u(t), u_t) \text{ for a.e. } t \in [0, b]\}.
\tag{5.6}
$$

We denote by $\Theta_b(\phi)$ the set of all mild solutions of the Cauchy problem (5.5) and (5.6).

With the assumptions (H_0), (H_1) and the fact that X is reflexive, we conclude, analogously to Lemma 2.4, that $Sel^b_F(u) \neq \emptyset$ for every $u \in C([-\tau, b], X)$, and

$$
Sel^b_F : C([-\tau, b], X) \to P(L^1([0, b], X))
$$

is weakly u.s.c. with convex, weakly compact values.

Denote by $S_b f$ the unique mild solution to the Cauchy problem (5.5) corresponding to $f \in L^1([0, b], X)$. Evidently, u is a mild solution of the Cauchy problem (5.5) and (5.6) iff $u \in C([-\tau, b], X)$ is a fixed point of

$$
G_b := S_b \circ Sel^b_F.
$$

We can immediately find a compact convex subset of $C([-\tau, b], X)$ which is invariant under G_b.

In the sequel, taking $u \in C([-\tau, b], X)$ with $u(t) = \phi(t)$ on $[-\tau, 0]$ and $\widetilde{u} \in G_b(u)$, there exists $f \in Sel^b_F(u)$ such that

$$
|\widetilde{u}(t)| \leq |U(t, 0)\phi(0)| + \int_0^t |U(t, s)f(s)|ds, \quad t \in [0, b],
$$

which together with (H_1) enables us to find that

$$
\sup_{s \in [0,t]} |\widetilde{u}(s)| \leq |\phi|_0 + (|\phi|_0 + 1)\int_0^t \mu(s)ds + 2\int_0^t \mu(s) \sup_{r \in [0,s]} |u(r)|ds, \quad t \in [0, b].
$$

Therefore, put

$$E_0 = \{u \in C([-\tau, b], X) : \sup_{s \in [0,t]} |u(s)| \le \psi(t) \text{ on } [0, b], \text{ and } u(t) = \phi(t) \text{ on } [-\tau, 0]\},$$

where ψ is the solution of

$$\psi'(t) = \mu(t)(2\psi'(t) + |\phi|_0 + 1) \text{ a.e. on } [0, b], \quad \psi(0) = |\phi|_0.$$

Then $E_0 \subset C([-\tau, b], X)$ is closed, bounded and convex with $G_b(E_0) \subset E_0$. Put

$$E = \overline{co}G_b(E_0).$$

Note that $G_b(E) \subset E$. We seek for solutions in E.

Firstly, from assumption (H_1) and Lemma 5.2 it follows that $G_b(E_0)$ is relatively compact in $C([-\tau, b], X)$. Also, if $\{(u_n, v_n)\} \subset \text{Gra}(G_b)$ is a sequence such that $(u_n, v_n) \to (u, v)$, then there exists a sequence $\{f_n\} \subset L^1([0, b], X)$ such that $f_n \in Sel_F^b(u_n)$ and $v_n = G_b(f_n)$, which together with Lemma 1.7 (ii) enables us to conclude that there exists $f \in Sel_F^b(u)$ and a subsequence of $\{f_n\}$, still denoted by $\{f_n\}$, such that $f_n \to f$ weakly in $L^1([0, b], X)$. Therefore, from Lemma 1.9 it follows that $v \in G_b(u)$, which yields that G_b is closed. Hence, using Lemma 1.9 we get that G_b is u.s.c. on E_0, so on E by inference. Moreover, we note that $E \subset C([-\tau, b], X)$ is compact and G_b has compact values.

Therefore, applying Theorem 1.17 yields a fixed point of G_b if we are able to prove that G_b has contractible values. To this aim, let $C = G_b(u)$ for some $u \in E$, take $\widetilde{f} \in Sel_F^b(u)$ and define $h : [0, 1] \times C \to C$ by

$$h(s, v)(t) = \begin{cases} v(t), & \text{if } t \in [-\tau, sb], \\ u(t, sb, v(sb), \widetilde{f}), & \text{if } t \in (sb, b]. \end{cases}$$

It is easy to see that $h(s, v) \in C$ for each $(s, v) \in [0, 1] \times C$. Also, it is clear that $h(0, v) = S_b \widetilde{f}$ and $h(1, v) = v$ on C. Moreover, the continuity of h follows from the continuous dependence of $u(\cdot, \zeta, u_0, f)$ on the two variables (ζ, u_0).

Finally, by Theorem 1.17 we get at least a $u \in E$ such that u is a mild solution of the Cauchy problem (5.5) and (5.6).

Step 2. As proved in the above step, the solution set $\Theta_b(\phi)$ of the Cauchy problem (5.5) and (5.6) is a nonempty subset of $C([-\tau, b], X)$. In fact it is a compact R_δ-set as we are going to show in this step.

Since $G_b(E_0)$ is relatively compact in $C([-\tau, b], X)$ and G_b is closed, one can see that $\Theta_b(\phi)$ is compact in $C([-\tau, b], X)$.

In the sequel, we denote by $\Theta_b^n(\phi)$ the solution set of the approximation problem associated with (5.5) and (5.6):

$$\begin{cases} u'(t) - A(t)u(t) \in F_n(t, u(t), u_t), & t \in [0, b], \\ u(t) = \phi(t), & t \in [-\tau, 0], \; n \geq 1, \end{cases}$$

where $\{F_n\}$ is the approximation sequence described in Lemma 5.4. Then noticing Lemma 5.4 (ii), (vi) and using a similar argument as in Step 1 we obtain that for every $n \geq 1$, $\Theta_b^n(\phi)$ is nonempty and it is compact in $C([-\tau, b], X)$. Also, in view of Lemma 5.4 (i) we infer that

$$\Theta_b(\phi) \subset \cdots \subset \Theta_b^n(\phi) \subset \cdots \subset \Theta_b^2(\phi) \subset \Theta_b^1(\phi).$$

Moreover, noticing Lemma 5.4 (iii), (iv) and (v) we obtain, by the similar arguments with the latter half of Theorem 2.2, that $\Theta_b^n(\phi)$ is contractible for each $n \geq 1$ and

$$\Theta_b(\phi) = \bigcap_{n \geq 1} \Theta_b^n(\phi),$$

which implies that $\Theta_b(\phi)$ is a compact R_δ-set.

Step 3. By a similar argument with that in Lemma 2.7 we obtain that for every $u \in \widetilde{C}([-\tau, \infty), X)$, the set $Sel_F(u)$ defined by

$$Sel_F(u) := \{f \in L^1_{loc}(\mathbb{R}^+, X) : f(t) \in F(t, u(t), u_t) \text{ for a.e. } t \in \mathbb{R}^+\}$$

is nonempty.

Below, we introduce the following two inverse systems and their limits. For more details about the inverse system and its limit, we refer the reader to [71] (see also [14, 104]).

For each $p, m > -\tau$ with $p \geq m$, let us consider a projection $\pi_m^p : C([-\tau, p], X) \to C([-\tau, m], X)$, defined by

$$\pi_m^p(u) = u|_{[-\tau, m]}, \quad u \in C([-\tau, p], X).$$

Put

$$\mathbb{N}_{-\tau} = \{m \in \mathbb{N} \setminus \{0\} : m > -\tau\}, \quad D_m = \{u \in C([-\tau, m], X) : u(t) = \phi(t) \text{ on } [-\tau, 0]\}.$$

Then it is readily checked that $\{D_m, \pi_m^p, \mathbb{N}_{-\tau}\}$ is an inverse system and its limit is

$$D := \{u \in \widetilde{C}([-\tau, \infty), X) : u(t) = \phi(t) \text{ on } [-\tau, 0]\},$$

i.e.,

$$\underleftarrow{\lim}\{D_m, \pi_m^p, \mathbb{N}_{-\tau}\} = D.$$

In a similar manner as above, we also obtain that $\{L^1([0, m], X), \dot{\pi}_m^p, \mathbb{N} \setminus \{0\}\}$, $p \geq m$, is an inverse system, where

$$\dot{\pi}_m^p(f) = f|_{[0,m]}, \quad f \in L^1([0, p], X).$$

Moreover,

$$\lim_{\leftarrow}\{L^1([0, m], X), \dot{\pi}_m^p, \mathbb{N} \setminus \{0\}\} = L_{loc}^1(\mathbb{R}^+, X).$$

In the sequel, noticing $\pi_{-\tau,m}^p(S_p f) = S_m \dot{\pi}_m^p(f)$ for all $m \leq p$ and $f \in L^1$ ([0, p], X)$, one can find that the family $\{id, S_m\}$ is a multivalued mapping from $\{L^1([0, m], X), \dot{\pi}_m^p, \mathbb{N} \setminus \{0\}\}$ to $\{D_m, \pi_m^p, \mathbb{N}_{-\tau}\}$. Moreover, the family $\{id, S_m\}$ induces a limit mapping

$$S : L_{loc}^1(\mathbb{R}^+, X) \to P(D),$$

which satisfies $(Sf)|_{[-\tau,m]} = S_m f|_{[0,m]}$ for each $f \in L_{loc}^1(\mathbb{R}^+, X)$ and $m \in \mathbb{N} \setminus \{0\}$.

For every $m \in \mathbb{N} \setminus \{0\}$, the set of all fixed points of $G_m(= S_m \circ Sel_F^m)$ is denoted by $\mathrm{Fix}(G_m)$, i.e.,

$$\mathrm{Fix}(G_m) = \{u \in D_m : u \in G_m(u)\}.$$

Then we see from Step 2 that $\mathrm{Fix}(G_m)(= \Theta_m(\phi))$ are compact R_δ-sets. Also, it is easy to see that the family $\{id, G_m\}$ is a mapping from system $\{D_m, \pi_m^p, \mathbb{N}_{-\tau}\}$ into itself. Also, it follows that $\pi_m^p(G_p(u)) = G_m(\pi_m^p(u))$ for all $p \geq m$ and $u \in D_m$, which enables us to conclude that family $\{id, G_m\}$ induces a limit mapping

$$G : D \to P(D),$$

where for every $u \in D$, $G(u) = \{w \in D : w|_{[-\tau,m]} = S_m f|_{[0,m]}$ for every $m \in \mathbb{N} \setminus \{0\}$ and $f \in Sel_F(u)\}$. Moreover, it follows readily that $G(u) = S \circ Sel_F(u)$ for every $u \in D$.

At the end of this step, applying Theorem 1.19 we obtain that the solution set $\Theta(\phi)$ is a compact R_δ-set, as claimed.

Remark 5.8 The extra condition: X is reflexive in Theorem 5.1 can be dropped in the case of F being a single-valued function.

Remark 5.9 Let us note that Lemmas 5.1 and 5.2 play a key role in the proof of the above theorem just the same as Lemmas 2.5, 2.6 of [71] in the context.

A direct corollary of Theorem 5.1 is the following results.

Theorem 5.2 *Let the hypotheses in Theorem 5.1 hold except that (H_1) is replaced by the following condition:*
$(H_1)'$there exists a function $\mu \in L^1(\mathbb{R}^+, \mathbb{R}^+)$ such that

$$|F(t, x, v)| \leq \mu(t)$$

for a.e. $t \in \mathbb{R}^+$ and each $(x, v) \in X \times C([-\tau, 0], X)$.
Then the assertions in Theorem 5.1 remain true.

Remark 5.10 We mention that Theorem 5.2 will be an important tool for showing the existence of global mild solutions to the nonlocal Cauchy problem (5.2) (see the proof of Theorem 5.5 below).

Remark 5.11 The results of Theorems 5.1 and 5.2 can be easily extended to fractional Cauchy problems with almost sectorial operators in a straight forward way. Please see Wang et al. [195] for more details on this class of problems.

Additional information is contained in the following

Theorem 5.3 *Assume that X is reflexive and the evolution family U is compact. Suppose in addition that $F : \mathbb{R}^+ \times X \times C([-\tau, 0], X) \rightarrow P_{cl,cv}(X)$ has compact values, $F(t, \cdot, \cdot)$ is u.s.c. for a.e. $t \in \mathbb{R}^+$, $F(\cdot, x, v)$ has a strongly measurable selection for each $(x, v) \in X \times C([-\tau, 0], X)$, and (H_1) is satisfied. Then the assertions in Theorem 5.1 remain true.*

5.3.2   Control Problems

In this subsection, the result on topological structure of solution sets for the Cauchy problem (5.1) on compact intervals (see Step 1 and Step 2 in the proof of Theorem 5.1) will be extended to control problem (5.3).

We first consider the following linear control problem on X

$$\begin{cases} u'(t) - A(t)u(t) = g(t) + Bw(t), & t \in [0, b], \\ u(0) = u_0 \in X, \end{cases} \tag{5.7}$$

where $g \in L^1([0, b], X)$ and $w \in L^2([0, b], V)$. By a mild solution of the Cauchy problem (5.7), one means a function $u \in C([0, b], X)$ satisfying

$$u(t) = U(t, 0)u_0 + \int_0^t U(t, s)\big(g(s) + Bw(s)\big)ds, \quad t \in [0, b].$$

In the sequel, we shall simply denote by $u = \mathscr{F}(g, w)$ the unique mild solution of Cauchy problem (5.7) corresponding to (g, w).

By proceeding in a way similar to that in [197, Lemma 2.4], we obtain the following approximation result. For simplicity here we omit the details.

Lemma 5.5 *Let $w \in L^2([0, b], V)$ be given. If the two sequences $\{g_n\} \subset L^1([0, b], X)$ and $\{u_n\} \subset C([0, b], X)$, where u_n is a mild solution of the problem*

$$\begin{cases} u'_n(t) - A(t)u_n(t) = g_n(t) + Bw(t), & t \in [0, b], \\ u_n(0) = u_0, \end{cases}$$

$\lim_{n\to\infty} g_n = g$ weakly in $L^1([0, b], X)$ and $\lim_{n\to\infty} u_n = u$ in $C([0, b], X)$, then u is a mild solution of the limit problem

$$\begin{cases} u'(t) - A(t)u(t) = g(t) + Bw(t), & t \in [0, b], \\ u(0) = u_0. \end{cases}$$

To prove that the set of all solutions to control problem (5.3) is a compact R_δ-set, the following compactness characterization will be helpful.

Lemma 5.6 *Suppose that the evolution family U is compact. If $\mathcal{K}$ is a uniformly integrable subset in $L^1([0, b], X)$ and $\mathcal{M}'$ is bounded subset in $L^2([0, b], V)$, then $\mathcal{F}(\mathcal{K} \times \mathcal{M}')$ is relatively compact in $C([0, b], X)$.*

Proof Take $0 < t_1 < t_2 \le b$ and let $\delta > 0$ small enough. Then it follows that for each $u \in \mathcal{F}(\mathcal{K} \times \mathcal{M}')$,

$$|u(t_2) - u(t_1)| \le |(U(t_2, 0) - U(t_1, 0))u_0|$$
$$+ \left| \int_0^{t_2} U(t_2, s)(g(s) + Bw(s))ds - \int_0^{t_1} U(t_1, s)(g(s) + Bw(s))ds \right|$$
$$\le |(U(t_2, 0) - U(t_1, 0))u_0| + \int_{t_1}^{t_2} |g(s)|ds + \sqrt{t_2 - t_1} \|Bw\|_{L^2([0,b],X)}$$
$$+ \sup_{s \in [0, t_1 - \delta]} \|U(t_2, s) - U(t_1, s)\|_{\mathcal{L}(X)}$$
$$\times \left(\int_0^{t_1 - \delta} |g(s)|ds + \sqrt{t_1 - \delta} \|Bw\|_{L^2([0,b],X)} \right)$$
$$+ 2 \int_{t_1 - \delta}^{t_1} |g(s)|ds + 2\sqrt{\delta} \|Bw\|_{L^2([0,b],X)}$$
$$:= I_1 + I_2 + I_3 + I_4 + I_5 + I_6.$$

From the strong continuity of evolution family U it follows that I_1 tends to zero as $t_1 \to t_2$. Also, it is clear that I_i ($i = 2, 3, 5, 6$) tend to zero as $t_1 \to t_2$, $\delta \to 0$. Moreover, by the continuity of $U(t, s)$ for $t > s$ in the uniform operator topology we see that I_4 tends to zero as $t_1 \to t_2$. Thus, we have

$$|u(t_2) - u(t_1)| \to 0 \text{ as } t_2 - t_1 \to 0.$$

Since $\mathcal{K}$ is uniformly integrable in $L^1([0, b], X)$ and $B(\mathcal{M}')$ is bounded in $L^2([0, b], X)$, the limit above holds uniformly for $u \in \mathcal{F}(\mathcal{K} \times \mathcal{M}')$.

For the case when $0 = t_1 < t_2 \le b$, note that

$$|u(t_2) - u(t_1)| \le |(U(t_2, 0) - I)u_0| + \left| \int_0^{t_2} U(t_2, s)(g(s) + Bw(s))ds \right|$$
$$\le |(U(t_2, 0) - I)u_0| + \int_0^{t_2} |g(s)|ds + \sqrt{t_2} \|Bw\|_{L^2([0,b],X)}.$$

Again by the strong continuity of evolution family U and the facts that $\mathcal{K}$ is uniformly integrable in $L^1([0, b], X)$ and $B(\mathcal{M}')$ is bounded in $L^2([0, b], X)$, $|u(t_2) - u(t_1)|$ can be made small when t_2 is small uniformly for $u \in \mathcal{F}(\mathcal{K} \times \mathcal{M}')$. This gives that the set $\mathcal{F}(\mathcal{K} \times \mathcal{M}')$ is equicontinuous on $[0, b]$.

Next, let $t \in (0, b]$ be arbitrary and $\delta > 0$ such that $t - \delta \in [0, b]$. Since $\mathcal{K}$ is uniformly integrable in $L^1([0, b], X)$, it follows from Lemma 1.1 that $\mathcal{K}$ is bounded in $L^1([0, b], X)$. Moreover, from the boundedness of B we see that $B(\mathcal{M}')$ is bounded in $L^2([0, b], X)$. Therefore, the set $\{\phi_\delta u(t) : u(t) \in \mathcal{F}(\mathcal{K} \times \mathcal{M}')(t)\}$ is relatively compact in X due to the compactness of $U(t, s)$ for $t > s$, where the operator $\phi_\delta : \mathcal{F}(\mathcal{K} \times \mathcal{M}')(t) \to X$ is defined by

$$\phi_\delta u(t) = U(t, 0)u_0$$
$$+ U(t, t - \delta) \int_0^{t-\delta} U(t - \delta, s)\big(g(s) + Bw(s)\big)ds, \quad (g, w) \in \mathcal{K} \times \mathcal{M}'.$$

This yields that ϕ_δ is a compact operator.

Moreover, it readily follows that

$$|\phi_\delta u(t) - u(t)| \leq \int_{t-\delta}^t |g(s)|ds + \sqrt{\delta}\|Bw\|_{L^2([0,b], X)} \to 0 \text{ as } \delta \to 0$$

for each $u(t) \in \mathcal{F}(\mathcal{K} \times \mathcal{M}')(t)$, which together with the facts that $\mathcal{K}$ is uniformly integrable in $L^1([0, b], X)$ and $B(\mathcal{M}')$ is bounded in $L^2([0, b], X)$, enables us conclude that $\phi_\delta u(t) \to u(t)$ as $\delta \to 0$ uniformly for $t \in (0, b]$ and for $u(t) \in \mathcal{F}(\mathcal{K} \times \mathcal{M}')(t)$.

Consequently, the identity operator $I : \mathcal{F}(\mathcal{K} \times \mathcal{M}')(t) \to \mathcal{F}(\mathcal{K} \times \mathcal{M}')(t)$ is a compact operator and hence the set $\mathcal{F}(\mathcal{K} \times \mathcal{M}')(t) = \{u(t) : u \in \mathcal{F}(\mathcal{K} \times \mathcal{M}')\}$ is relatively compact for each $t \in (0, b]$. Thus, it justifies, according to Lemma 1.3, that the assertion of this lemma holds true, thereby completing the proof.

By $\Lambda(w)$ denote the set of mild solutions to control problem (5.3) for every $w \in L^2([0, b], V)$. Now we can obtain the main result in this subsection.

Theorem 5.4 *Let X be reflexive and the evolution family U be compact. Assume that $G : [0, b] \times C([-\tau, 0], X) \to P(X)$ is a multivalued function with convex, closed values for which $G(t, \cdot)$ is weakly u.s.c. for a.e. $t \in [0, b]$ and $G(\cdot, v)$ has a strongly measurable selection for each $v \in C([-\tau, 0], X)$. Suppose in addition that (H_2) there exists a function $\gamma \in L^2([0, b], \mathbb{R}^+)$ such that*

$$|G(t, v)| \leq \gamma(t)(1 + |v|_0)$$

for a.e. $t \in \mathbb{R}^+$ and each $v \in C([-\tau, 0], X)$.
Then for every $w \in L^2([0, b], V)$, $\Lambda(w)$ is a compact R_δ-set.

Proof Let $w \in L^2([0, b], V)$ be given and let $S_b' f$ the unique mild solution to the linear control problem

$$\begin{cases} u'(t) - A(t)u(t) = f(t) + Bw(t), & t \in [0, b], \\ u(t) = \phi(t), & t \in [-\tau, 0]. \end{cases}$$

For every $u \in C([-\tau, b], X)$, write

$$Sel_G(u) := \{g \in L^1([0, b], X) : g(t) \in G(t, u_t) \text{ for a.e. } t \in [0, b]\}.$$

Then $Sel_G(u) \neq \emptyset$ and $Sel_G : C([-\tau, b], X) \to P(L^1([0, b], X))$ is weakly u.s.c. with convex, weakly compact values (see Step 1 in the proof of Theorem 5.1).

Put $G'_b := S'_b \circ Sel_G$ and let $u \in C([-\tau, b], X)$ with $u(t) = \phi(t)$ on $[-\tau, 0]$ and $\tilde{u} \in G'_b(u)$. Then it follows from assumption (H_2) that for each $t \in [0, b]$,

$$\sup_{s \in [0,t]} |\tilde{u}(s)| \leq |U(t, 0)\phi(0)| + \int_0^t |U(t, s)g(s)|ds + \int_0^t |U(t, s)Bw(s)|ds$$

$$\leq |\phi|_0 + (|\phi|_0 + 1)\int_0^t \gamma(s)ds + \int_0^t \gamma(s) \sup_{r \in [0,s]} |u(r)|ds + \sqrt{b}\|Bw\|_{L^2([0,b],X)},$$

where $g \in Sel_G(u)$, which enables us to obtain that $\sup_{s \in [0,t]} |\tilde{u}(s)| \leq \tilde{\psi}(t)$ for all $t \in [0, b]$, where $\tilde{\psi}$ is the solution of

$$\begin{cases} \tilde{\psi}'(t) = \gamma(t)(\tilde{\psi}'(t) + |\phi|_0 + 1) \text{ a.e. on } [0, b], \\ \tilde{\psi}(0) = |\phi|_0 + \sqrt{b}\|Bw\|_{L^2([0,b],X)}. \end{cases}$$

Let

$$\tilde{E}_0 = \{u \in C([-\tau, b], X) : \sup_{s \in [0,t]} |u(s)| \leq \tilde{\psi}(t) \text{ on } [0, b] \text{ and } u(t) = \phi(t), \, t \in [-\tau, 0]\}.$$

Then it is easy to see that $\tilde{E}_0 \subset C([-\tau, b], X)$ is closed, bounded and convex with $G'_b(\tilde{E}_0) \subset \tilde{E}_0$.

Finally, according to Lemmas 5.5, 5.6 and using an argument similar to that in the proof of Theorem 5.1 we obtain that the compact convex set $\overline{co}G'_b(\tilde{E}_0)$ is invariant under G'_b and moreover, the assertion of theorem holds true.

Remark 5.12 The extra condition: X is reflexive in Theorem 5.4 can be dropped in the case of G being a weakly compact valued function.

5.4 Applications

We present two applications of the results about the topological structure established in the previous section.

5.4.1 An Existence Result

Our major effort, in this subsection, is to provide suitable hypotheses ensuring the existence of global mild solutions to the nonlocal Cauchy problem (5.2).

It is assumed in this subsection that the evolution family U satisfies a more strict condition: there exists $\omega > 0$ such that

$$\|U(t, s)\|_{\mathcal{L}(X)} \leq e^{-\omega(t-s)} \tag{5.8}$$

for each $0 \leq s \leq t < +\infty$ (i.e., it is an exponentially stable).

Write, for $r > 0$,

$$\Omega_r = \{u \in \widetilde{C}_b([-\tau, \infty), X) : |u(t)| \leq r \text{ for all } t \geq -\tau\}.$$

Theorem 5.5 *Let the hypotheses in Theorem 5.2 be satisfied. Suppose in addition that $H : \widetilde{C}_b([-\tau, \infty), X) \to C([-\tau, 0], X)$ satisfies the following conditions:*
(H_a) H is continuous, for some $r > 0$, $|H(u)|_0 \leq r$ for all $u \in \Omega_r$, and
(H_b) for each $\mathcal{Q} \subset \Omega_r$ and $\eta > 0$ which restricted to $[\eta, +\infty)$ is relatively compact in $\widetilde{C}_b([\eta, +\infty), X)$, $H(\mathcal{Q})$ is relatively compact in $C([-\tau, 0], X)$.

Then the nonlocal Cauchy problem (5.2) admits at least one mild solution.

Proof Write, for every $r > 0$,

$$\Psi_r = \{v \in C([-\tau, 0], X) : |v|_0 \leq r\},$$
$$\Phi_r = \{(x, v) \in X \times C([-\tau, 0], X) : \max\{|x|, |v|_0\} \leq r\}.$$

Consider the Cauchy problem of the form

$$\begin{cases} u'(t) - A(t)u(t) \in F_\mu(t, u(t), u_t), & t \in \mathbb{R}^+, \\ u(t) = \phi(t), & t \in [-\tau, 0], \end{cases} \tag{5.9}$$

where $F_\mu : \mathbb{R}^+ \times X \times C([-\tau, 0], X) \to P(X)$ is defined by

$$F_\mu(t, x, v) = F(t, \mu(x, v)), \quad (t, x, v) \in \mathbb{R}^+ \times X \times C([-\tau, 0], X),$$

in which $\mu : X \times C([-\tau, 0], X) \to \Phi_r$ is a function such that

$$\mu(x, v) = \begin{cases} (x, v), & \text{if } (x, v) \in \Phi_r, \\ r \max\{|x|, |v|_0\}^{-1}(x, v) & \text{in rest.} \end{cases}$$

Since μ is continuous, it follows that F_μ satisfies the condition (H_0). Also, it is not difficult to see that F_μ satisfies the condition (H_1)'. Hence, by virtue of Theorem 5.2, the solution set, denoted by $\Theta_\mu(\phi)$ of (5.9) is a compact R_δ-set, which enables us to deduce that the multivalued map $\Theta_\mu : \Psi_r \to \mathscr{P}(\widetilde{C}([-\tau, \infty), X))$ is an R_δ-mapping if one can show that Θ_μ is u.s.c.

In fact, noticing Lemma 5.2 and performing a similar argument as Lemma 2.9, one find that Θ_μ is quasicompact. Therefore, in view of Lemma 1.10 it suffices to show that Θ_μ is closed. To this aim, put, for every $u \in \widetilde{C}([-\tau, \infty), X)$,

$$Sel_{F_\mu}(u) := \{f \in L^1_{loc}(\mathbb{R}^+, X) : f(t) \in F_\mu(t, u(t), u_t) \text{ for a.e. } t \in \mathbb{R}^+\},$$

and let $\{(\phi_n, u_n)\} \subset \mathrm{Gra}(\Theta_\mu)$ be a sequence such that $(\phi_n, u_n) \to (\phi_0, u_0)$. Then there exists a sequence $\{f_n\}$ such that $f_n \in Sel_{F_\mu}(u_n)$ and u_n is the mild solution of

$$\begin{cases} u'_n(t) - A(t)u_n(t) = f_n(t), & t \in \mathbb{R}^+, \\ u_n(t) = \phi_n(t), & t \in [-\tau, 0], \ n \geq 1. \end{cases}$$

At the same time, we can show, similar to Lemma 2.8, that if $\{\widetilde{u}_n\} \subset \widetilde{C}([-\tau, \infty), X)$ with $\widetilde{u}_n \to \widetilde{u}_0$ and $\widetilde{f}_n \in Sel_{F_\mu}(\widetilde{u}_n)$, then there exists $\widetilde{f} \in Sel_{F_\mu}(\widetilde{u}_0)$ and a subsequence $\{\widetilde{f}_{n'}\}$ of $\{\widetilde{f}_n\}$ such that $\widetilde{f}_{n'} \to \widetilde{f}$ weakly in $L^1([0, m], X)$ for each $m \in \mathbb{N} \setminus \{0\}$. Therefore, with the help of this result we have that there exists $f \in Sel_{F_\mu}(u_0)$ and a subsequence of $\{f_n\}$, still denoted by $\{f_n\}$, such that

$$f_n|_{[0,m]} \to f|_{[0,m]} \text{ weakly in } L^1([0, m], X)$$

for every $m \in \mathbb{N} \setminus \{0\}$. From Lemma 5.1 it follows that for every $m \in \mathbb{N} \setminus \{0\}$, $S_m(f|_{[0,m]}) = u|_{[-\tau,m]}$, where S_m is defined as in the Step 1 of the proof of Theorem 5.1. Therefore, we have $u_0 = S(f)$, which implies that $u_0 \in \Theta_\mu(\phi_0)$, as desired.

Next, we let $\phi \in \Psi_r$ and take $u \in \Theta_\mu(\phi)$, $f \in Sel_{F_\mu}(u)$ (given $r > 0$). Then for all $t \in [-\tau, 0]$, we see $|u(t)| = |\phi(t)| \leq r$. Moreover, for the case when $t > 0$, by $(H_1)'$ we have

$$|u(t)| \leq |U(t, 0)\phi(0)| + \int_0^t |U(t, s)f(s)|ds$$
$$\leq e^{-\omega t}r + \int_0^t \mu(s)ds. \tag{5.10}$$

Accordingly, we claim that there exists $r_0 > 0$ such that $\Theta_\mu(\Psi_{r_0}) \subset \Omega_{r_0}$. In fact, if this is not true, then for each $r > 0$, there exist $\widehat{\phi} \in \Psi_r$, $\widehat{u} \in \Theta_\mu(\widehat{\phi})$ and $t_0 > 0$ such that $|\widehat{u}(t_0)| > r$. This together with (5.10) enables us to infer that

$$r < |\widehat{u}(t_0)| \leq e^{-\omega t_0}r + \int_0^{+\infty} \mu(s)ds.$$

Dividing by r on both sides and taking the lower limit as $r \to \infty$, we have a contradiction.

Now, to obtain a mild solution of the Cauchy problem (5.9), we consider the following two mappings:

$$\Theta_\mu : \Psi'_{r_0} \to \Omega'_{r_0}$$

and

$$H : \Omega'_{r_0} \to \Psi'_{r_0},$$

where $\Psi'_{r_0} = \overline{co} H(\Omega'_{r_0})$, $\Omega'_{r_0} = \overline{co} \Theta_\mu(\widetilde{\Psi}_{r_0})$ with $\widetilde{\Psi}_{r_0} = \overline{co} H(\Omega_{r_0})$. Since $\widetilde{\Psi}_{r_0} \supset \Psi'_{r_0}$ due to assumption (H_a) and $\Theta_\mu(\Psi_{r_0}) \subset \Omega_{r_0}$, we see $\Theta_\mu(\Psi'_{r_0}) \subset \Omega'_{r_0}$, which implies that the multivalued mapping

$$\mathscr{S} := \Theta_\mu \circ H : \Omega'_{r_0} \to \Omega'_{r_0}$$

is well defined. Moreover, it follows that

- H is an R_δ-mapping due to the assumption (H_a);
- $\Psi'_{r_0}, \Omega'_{r_0}$ are AR-spaces;
- $H(\Omega'_{r_0})$ is relatively compact in $C([-\tau, 0], X)$ due to assumption (H_b) (see [71, Lines 7-16 from the bottom on Page 2066]) and hence Ψ'_{r_0} is compact, which implies that $\Theta_\mu(\Psi'_{r_0})$ is compact (since Θ_μ is u.s.c. with compact values).

Therefore, noticing $\mathscr{S}(\Omega'_{r_0}) \subset \Theta_\mu(\Psi'_{r_0})$ we obtain, thanks to Theorem 1.16, that $\mathscr{S}$ has at least a fixed point u in Ω'_{r_0} and $\max\{|u(t)|, |u_t|_0\} \le r_0$ for each $t \in \mathbb{R}^+$. Since $F_\mu(t, u(t), u_t) = F(t, u(t), u_t)$ for each $t \in \mathbb{R}^+$, u is also a mild solution of the Cauchy problem (5.2), which completes the proof.

Remark 5.13 (1) Theorem 5.5 covers recent results in [197] via a different method. Notice in particular that unlike in the papers [192, 197], no nonexpansive condition on nonlocal function is needed.

(2) Let us note that in [71], assumption (H_1), an invariance condition on the nonlinearity and some other conditions were used to derive existence of C^0-solutions for nonlocal Cauchy problems (see [71, Theorem 4.1]). In the present result, the lack of invariance condition on the nonlinearity F enables us to find that assumption (H_1) is not enough to obtain the existence of mild solutions. Therefore, assumption (H_1) is replaced by the stronger one $(H_1)'$. In fact this condition ensures an invariance condition for a sufficiently large r_0.

Remark 5.14 Condition (5.8) plays a key role in the proof of the above theorem. If the evolution family U satisfies the original condition then one, along our line, has to take aim at the particular case of nonlinearity F, i.e., $F \equiv 0$.

5.4.2 Invariance of Reachability Set

In this subsection we study the control problem (5.3) in the case of single-valued nonlinearity G.

The set

$$K_G = \{u(b, w, G) : w \in L^2([0, b], V)\}$$

is called the reachability set of the control problem (5.3). By K_0 we denote the reachability set for the corresponding linear problem ($G \equiv 0$).

We refer the reader to [40, 100] for the basic notions and facts of control problems.

Definition 5.1 The control problem (5.3) is said to be approximately controllable on $[0, b]$ if $\overline{K_G} = X$, where $\overline{K_G}$ denotes the closure of K_G.

Let Λ be the multivalued mapping appearing in Theorem 5.4. We first present the following result.

Lemma 5.7 *Let the hypothesis* (H_2) *be satisfied. Suppose in addition that the evolution family U is compact and G is continuous. Then* $\Lambda : L^2([0, b], V) \to P(C([-\tau, b], X))$ *is an* R_δ-*mapping.*

Proof From Theorem 5.4 it follows that the solution set $\Lambda(w)$ is an R_δ-set for every $w \in L^2([0, b], V)$. Therefore, to prove the assertion of this lemma, it suffices to show that Λ is u.s.c. To this aim, let $\mathscr{D} \subset L^2([0, b], V)$ be a compact set and $\{y_n\} \subset \Lambda(\mathscr{D})$. Then one can take a sequence $\{w_n\} \subset \mathscr{D}$ such that $\Lambda(w_n) = y_n$, $n \geq 1$. Noting that $\{w_n\}$ is bounded in $L^2([0, b], V)$, it follows from Lemma 5.6 that $\Lambda\{w_n\}$ is relatively compact in $C([-\tau, b], X)$. Hence, $\{y_n\}$ is relatively compact in $C([-\tau, b], X)$. This implies that Λ is quasicompact. Next, thanks to Lemma 1.10, it suffices to show that Λ is closed. Let $w_n \to w$ in $L^2([0, b], V)$ and $u_n \in \Lambda(w_n)$, $u_n \to u$ in $C([-\tau, b], X)$. Noting that u_n verify the following integral equations

$$u_n(t) = \begin{cases} \phi(t), & t \in [-\tau, 0), \\ U(t, 0)\phi(0) + \displaystyle\int_0^t U(t, s)\big(G(s, u_{ns}) + Bw_n(s)\big)ds, & t \in [0, b], \ n \geq 1, \end{cases}$$

$G(s, u_{ns}) \to G(s, u_s)$ for all $s \in [0, b]$ and $G(\cdot, \{u_{n\cdot}\})$ and $\{Bw_n\}$ are L^2-integrably bounded, we conclude, in view of Lebesgue's dominated convergence theorem, that $u \in \Lambda(w)$, as desired.

We are in a position to prove that the reachability set is invariant under nonlinear perturbations.

Theorem 5.6 *Assume that hypotheses* (H_2) *and* (S) *are satisfied. Suppose in addition that the evolution family U is compact and G is continuous. Then there exists $r_0 > 0$ such that $K_G = K_0$ provided $\|B\|_{\mathcal{L}(V,X)} < r_0$.*

Proof We proceed in two steps.

Step 1. Let $\mathscr{L}$ be the operator in Lemma 5.3. Denote by $N_G : C([-\tau, b], X) \to L^2([0, b], X)$ the Nemytskii operator corresponding to the nonlinearity G, i.e.,

$$(N_G u)(t) = G(t, u_t)$$

for each $t \in [0, b]$, $u \in C([-\tau, b], X)$. It is clear that $\mathscr{L}$, N_G are R_δ-mappings.
Let $w_0 \in L^2([0, b], V)$ be given. Consider the following multivalued mapping:

$$\mathcal{K}_{w_0} : L^2([0, b], V) \rightarrow P(L^2([0, b], V)),$$
$$\mathcal{K}_{w_0} w = \mathscr{L} \circ N_G \circ \Lambda(w_0 + w), \quad w \in L^2([0, b], V),$$

where $\Lambda : L^2([0, b], V) \rightarrow P(C([-\tau, b], X))$ is the R_δ-mapping due to Lemma 5.7.
In what follows, we shall prove $\mathrm{Fix}(\mathcal{K}_{w_0}) \neq \emptyset$, where $\mathrm{Fix}(\mathcal{K}_{w_0})$ denotes the set of all
fixed points of $\mathcal{K}_{w_0}$.
For every $r > 0$ put

$$W_r = \{w \in L^2([0, b], V) : \|w\|_{L^2([0,b],V)} \leq r\}.$$

Let $w \in L^2([0, b], V)$ and $\tilde{u} \in \Lambda(w_0 + w)$. Then it follows from (H_2) that

$$|\tilde{u}(t)| \leq |\phi|_0 + \int_0^t \gamma(s)(1 + |\tilde{u}_s|_0)ds + \sqrt{b}\|B(w_0 + w)\|_{L^2([0,b],X)}$$

for all $t \in [0, b]$ and $|\tilde{u}(t)| \leq |\phi|_0$ for all $t \in [-\tau, 0]$, which, in view of the generalized Gronwall-Bellman's inequality, implies that

$$\|\tilde{u}\|_{C([-\tau,b],X)} \leq c_0 \left(1 + \|\gamma\|_{L([0,b],\mathbb{R}^+)}e^{\|\gamma\|_{L([0,b],\mathbb{R}^+)}}\right).$$

where $c_0 = |\phi|_0 + \|\gamma\|_{L([0,b],\mathbb{R}^+)} + \sqrt{b}\|B\|_{\mathcal{L}(V,X)}\|(w_0 + w)\|_{L^2([0,b],V)}$. Therefore, if
$\tilde{w} \in \mathcal{K}_{w_0} w$, then by assumption (S), we obtain

$$
\begin{aligned}
\|\tilde{w}\|_{L^2([0,T],V)} &= \|\mathscr{L}N_G(\tilde{u})\|_{L^2([0,b],V)} \\
&\leq M_0\|N_G(\tilde{u})\|_{L^2([0,b],X)} \\
&\leq M_0\|\gamma\|_{L^2([0,b],\mathbb{R}^+)}\left(1 + \|\tilde{u}\|_{C([-\tau,b],X)}\right) \\
&\leq M_0(1 + c_0)\|\gamma\|_{L^2([0,b],\mathbb{R}^+)}\left(1 + \|\gamma\|_{L([0,b],\mathbb{R}^+)}e^{\|\gamma\|_{L([0,b],\mathbb{R}^+)}}\right).
\end{aligned}
$$

Then there exists $r_0 > 0$ such that one can find $\tilde{r} > 0$ such that $\mathcal{K}_{w_0}(W_{\tilde{r}}) \subset W_{\tilde{r}}$
provided $\|B\|_{\mathcal{L}(V,X)} < r_0$. In the sequel, by the compactness of $\Lambda(w_0 + W_{\tilde{r}})$ one
finds that $\mathcal{K}_{w_0}(W_{\tilde{r}})$ is compact. Accordingly, due to Theorem 1.16 $\mathcal{K}_{w_0}$ admits at least
one fixed point. It thus proves that $\mathrm{Fix}(\mathcal{K}_{w_0}) \neq \emptyset$.

Step 2. We show $K_G = K_0$. To this aim, let $u(b, \tilde{w}_0, 0) \in K_0$ with $\tilde{w}_0 \in L^2([0, b]$,
$V)$. As in Step 1 one has $\mathrm{Fix}(\mathcal{K}_{\tilde{w}_0}) \neq \emptyset$. Therefore, letting $w^* \in \mathrm{Fix}(\mathcal{K}_{\tilde{w}_0})$ and $u^* \in$
$\Lambda(\tilde{w}_0 + w^*)$ it follows from Lemma 5.3 that

$$u(b, \widetilde{w}_0 + w^*, G) = U(b, 0)\phi(0) + (T(B\widetilde{w}_0))(b) + T(N_G(u^*) + Bw^*)(b)$$
$$= u(b, \widetilde{w}_0, 0) + T(N_G(u^*))(b) + T B \mathscr{L} N_G(u^*)(b)$$
$$= x(b, \widetilde{w}_0, 0).$$

Accordingly, we obtain $K_0 \subset K_G$.

Next, to show $K_0 \supset K_G$, we let $u(b, \widetilde{w}, G) \in K_G$ with $\widetilde{w} \in L^2([0, b], V)$. Then, again by Lemma 5.3, we get

$$u(b, \widetilde{w}, G) = U(b, 0)\phi(0) + (T(N_G(u(b, \widetilde{w}, G)) + B\widetilde{w}))(b)$$
$$- (TN_G(u(b, \widetilde{w}, G)) + T B \mathscr{L} N_G(u(b, \widetilde{w}, G)))(b)$$
$$= U(b, 0)\phi(0) + (T(B\widetilde{w} - B \mathscr{L} N_G(u(b, \widetilde{w}, G))))(b)$$
$$= u(b, \widetilde{w} - \mathscr{L} N_G(u(b, \widetilde{w}, G)), 0),$$

which yields $K_0 \supset K_G$, proving our assertion.

Remark 5.15 There naturally arises a question: "Is there any chance to obtain the invariance of a reachability set to control problem (5.3) under nonlinear perturbations when the nonlinearity G is a "genuine" multivalued function. We look forward to seeing an affirmative answer to this question.

As a by-product of Theorem 5.6, we can conclude that the control problem (5.3) is approximately controllable under some supplementary conditions. To illustrate this, let us define the operator $T^* : X \to X$ by

$$T^* = \int_0^T U(t, s) B B^* U^*(t, s) ds,$$

where B^* and $U^*(t, s)$ stand for the adjoints of B and $U(t, s)$, respectively. Note that T^* is a linear bounded operator. Assuming that X, V are Hilbert spaces and $\lambda(\lambda I + T^*)^{-1} \to 0$ as $\lambda \to 0^+$ in the strong operator topology, from [142, Theorem 2] (see also [76]) we see that the corresponding linear control system is approximately controllable, i.e., $\overline{K_0} = X$, which together with Theorem 5.6 enables us to obtain $\overline{K_G} = X$.

5.5 Examples

In this section, we present two examples showing how to apply our abstract results to specific problems. The first example is inspired directly from the work of [71], [192, Example 5.1] (see also [197]). As we will see later, some known results cannot be applied to our examples.

Example 5.1 Take $X = L^2([0, \pi], \mathbb{R})$ and denote its norm by $|\cdot|$ and inner product by $(\cdot, \cdot)$. Let $c_1, c_2 > 0$ and $d : \mathbb{R}^+ \to [c_1, c_2]$ be Hölder continuous with exponent

γ_1. The operator $A' : D(A') \subset X \to X$ is defined as

$$A'x = x_{\xi\xi}, \quad x \in D(A'), \quad D(A') = H^2([0, \pi], \mathbb{R}) \cap H_0^1([0, \pi], \mathbb{R}).$$

As in Pazy [167], A' has a discrete spectrum and its eigenvalues are $-n^2$, $n \in \mathbb{N}^+$ with the corresponding normalized eigenvectors $x_n(\xi) = \sqrt{\frac{2}{\pi}} \sin(n\xi)$. We also let the functions

$$f_i : \mathbb{R}^+ \times [0, \pi] \times \mathbb{R} \times C([-\tau, 0], X) \to \mathbb{R}, \quad i = 1, 2$$

be such that
(F_1) $f_1(t, \xi, u, v) \le f_2(t, \xi, u, v)$ for   each $(t, \xi, u, v) \in \mathbb{R}^+ \times [0, \pi] \times \mathbb{R} \times C([-\tau, 0], X)$,
(F_2) f_1 is l.s.c. and f_2 is u.s.c.

Consider the system governed by a non-autonomous diffusion inclusion with time delay in the form

$$\begin{cases} D_t u(t, \xi) - d(t)u_{\xi\xi}(t, \xi) \in F(t, \xi, u(t, \xi), (u_t)(\xi)), & (t, \xi) \in \mathbb{R}^+ \times [0, \pi], \\ u(t, 0) = u(t, \pi) = 0, & t \in \mathbb{R}^+, \\ u(t, \xi) = \phi(t, \xi), & (t, \xi) \in [-\tau, 0] \times [0, \pi] \end{cases}$$
$$(5.11)$$

in X, where $u_t(\theta, \xi) = u(t + \theta, \xi)$,

$$F(t, \xi, u, v) = [f_1(t, \xi, u, v), f_2(t, \xi, u, v)]$$

is a closed interval for each $(t, \xi, u, v) \in \mathbb{R}^+ \times [0, \pi] \times \mathbb{R} \times C([-\tau, 0], X)$ and $\phi \in C([-\tau, 0], X)$.

Firstly, we are specially interested in R_δ-structure of mild solutions to system (5.11).

Theorem 5.7 *Let (F_1), (F_2) and the following hypothesis be satisfied:*
(F_3) there exist functions $\mu_1, \mu_2 \in L^\infty(\mathbb{R}^+, \mathbb{R}^+)$ such that

$$f_i(t, \xi, u, v)| \le \mu_1(t)(|u| + |v|_0) + \mu_2(t), \quad i = 1, 2,$$

for each $(t, \xi, u, v) \in \mathbb{R}^+ \times [0, \pi] \times \mathbb{R} \times C([-\tau, 0], X)$.
Then the set of mild solutions to system (5.11) is a compact R_δ-set. In particular, it is connected.

Proof Write

$$D(A(t)) = D(A'), \quad t \in \mathbb{R}^+,$$
$$A(t)x = d(t)A'x, \quad x \in D(A(t)).$$

It follows from [167, Lemma 6.1 in Chap. 7] that there are constants $\theta \in (\frac{\pi}{2}, \pi)$ and $K_1 \geq 0$ such that $A(t)$ satisfies

$$\Sigma_\theta \cup \{0\} \subset \rho(A(t)), \quad \text{where } \Sigma_\theta = \{\lambda \in \mathbb{C}\backslash\{0\} : |\lambda| \leq \theta\},$$

and

$$\|R(\lambda, A(t))\|_{\mathcal{L}(X)} \leq \frac{K_1}{1 + |\lambda|}.$$

for all $\lambda \in \Sigma_\theta \cup \{0\}$ and $t \in \mathbb{R}^+$. Moreover, we note that for $\lambda \in \Sigma_\theta \cup \{0\}$, $t, s \in \mathbb{R}^+$, $x \in X$,

$$R(\lambda, A(t))x = \sum_{n=1}^{\infty} \frac{1}{\lambda + n^2 d(t)}(x, x_n)x_n$$

and

$$A(t)R(\lambda, A(t))[R(0, A(t)) - R(0, A(s))]x = \sum_{n=1}^{\infty} \frac{1}{\lambda + n^2 d(t)} \frac{d(t) - d(s)}{d(s)}(x, x_n)x_n,$$

where from we see that for $\lambda \in \Sigma_\theta \cup \{0\}$, $t, s \in \mathbb{R}^+$,

$$\|A(t)R(\lambda, A(t))[R(0, A(t)) - R(0, A(s))]\|_{\mathcal{L}(X)} \leq \frac{\|R(\lambda, A(t))\|_{\mathcal{L}(X)} \cdot |d(t) - d(s)|}{d(s)}.$$

Accordingly, $A(t)$ satisfy the known Acquistapace-Terreni conditions (cf. [2]). Thus, the family $\{A(t)\}_{t \in \mathbb{R}^+}$ generates an evolution family $U = \{U(t, s)\}_{t \geq s \geq 0}$:

$$U(t, s)x = \sum_{n=1}^{\infty} e^{-n^2 \int_s^t d(\tau)d\tau}(x, x_n)x_n \quad \text{for } 0 \leq s \leq t < \infty, \ x \in X.$$

A direct calculation gives

$$\|U(t, s)\|_{\mathcal{L}(X)} \leq e^{-c_1(t-s)} \leq 1 \quad \text{for } 0 \leq s \leq t < +\infty.$$

Moreover, U is compact due to the boundedness of $A(t)U(t, s)$ for $0 \leq s < t < +\infty$ (cf. [167, Theorem 6.1 in Chap. 5]).

Next, let us define the multivalued function $F : \mathbb{R}^+ \times X \times C([-\tau, 0], X) \to P(X)$ as

$$F(t, u, v) = \{x \in X : x(\xi) \in [f_1(t, \xi, u(\xi), v), f_2(t, \xi, u(\xi), v)] \text{ a.e. in } [0, \pi]\}$$

for each $(t, u, v) \in \mathbb{R}^+ \times X \times C([-\tau, 0], X)$. Noticing (F_3), we readily see from Lemma 2.10 (see also [192, Lemma 5.1, Theorem 5.1]) that F has nonempty, convex and closed values, $F(\cdot, u, v)$ has a strongly measurable selection for every $(u, v) \in X \times C([-\tau, 0], X)$, and $F(t, \cdot, \cdot)$ is weakly u.s.c. for each $t \in \mathbb{R}^+$. Moreover,

$$|F(t, u, v)| \leq \sqrt{\pi} \max\{\mu_1(t), \mu_2(t)\}(1 + |u| + |v|_0)$$

for a.e. $t \in \mathbb{R}^+$, each $u \in X$ and $v \in C([-\tau, 0], X)$, which justify hypotheses (H_0), (H_1) with

$$\mu(t) = \sqrt{\pi} \max\{\mu_1(t), \mu_2(t)\}.$$

Then, system (5.11) can be rewritten as an abstract Cauchy problem of the form (5.1). Accordingly, Theorem 5.1 can apply to the situation, and hence we assert that the set of mild solutions to system (5.11) is an R_δ-set. The proof is complete.

In the sequel, with the help of Theorem 5.7 we show the existence of mild solutions to the following system with nonlocal initial condition

$$\begin{cases} D_t u(t, \xi) - d(t)u_{\xi\xi}(t, \xi) \in F(t, \xi, u(t, \xi), (u_t)(\xi)), & (t, \xi) \in \mathbb{R}^+ \times [0, \pi], \\ u(t, 0) = u(t, \pi) = 0, & t \in \mathbb{R}^+, \\ u(t, \xi) = \int_\tau^\infty \mathscr{G}(u(t + \theta, \xi))d\widetilde{\mu}(\theta), & (t, \xi) \in [-\tau, 0] \times [0, \pi] \end{cases}$$

$$(5.12)$$

in X, where $\mathscr{G} : \mathbb{R} \to \mathbb{R}$ is a continuous function such that either for some $C_1, C_2 \geq 0$ and $b \in [0, 1)$,

$$|\mathscr{G}(u)| \leq C_1 + C_2|u|^b \quad \text{for all } u \in \mathbb{R},$$

or

$$|\mathscr{G}(u)| \leq |u| \quad \text{for all } u \in \mathbb{R},$$

and $\widetilde{\mu}$ is a σ-finite and complete measure on $[\tau, \infty)$ such that

$$\widetilde{\mu}([\tau, \infty)) = 1, \quad \lim_{s \to \tau^+} \widetilde{\mu}([\tau, s]) = 0.$$

Theorem 5.8 *Let (F_1), (F_2) and the following hypothesis be satisfied:*
$(F_3)'$ *there exists a function $\mu_2 \in L^1(\mathbb{R}^+, \mathbb{R}^+) \cap L^\infty(\mathbb{R}^+, \mathbb{R}^+)$ such that*

$$|f_i(t, \xi, u, v)| \leq \mu_2(t), \quad i = 1, 2,$$

for each $(t, \xi, u, v) \in \mathbb{R}^+ \times [0, \pi] \times \mathbb{R} \times C([-\tau, 0], X)$.
Then the system (5.12) has at least one mild solution.

Proof We use the setting as in the proof of Theorem 5.7. Then from the assumption $(F_3)'$ one finds that F has nonempty, convex and closed values, $F(\cdot, u, v)$ has a strongly measurable selection for every $(u, v) \in X \times C([-\tau, 0], X)$, and $F(t, \cdot, \cdot)$ is weakly u.s.c. for each $t \in \mathbb{R}^+$. In addition,

$$|F(t, u, v)| \leq \sqrt{\pi}\mu_2(t)$$

for a.e. $t \in \mathbb{R}^+$, each $u \in X$ and $v \in C([-\tau, 0], X)$, which implies that hypotheses (H_0), $(H_1)'$ hold with

$$\mu(t) = \sqrt{\pi}\mu_2(t).$$

Let us define the Nemytskii operator $\mathscr{G}$ mapping from X into itself by

$$\mathscr{G}(x)(\xi) = \mathscr{G}(x(\xi)) \text{ for each } x \in X.$$

Then it is easy to see that $\mathscr{G}$ is continuous on X. Consider the operator

$$H : \widetilde{C}_b([-\tau, \infty), X) \to C([-\tau, 0], X)$$

defined as

$$H(u)(t) = \int_\tau^\infty \mathscr{G}(u(t+\theta))d\widetilde{\mu}(\theta), \quad u \in \widetilde{C}_b([-\tau, \infty), X).$$

Therefore, it follows that H is continuous and $|H(u)|_0 \leq r$ for all $u \in \Omega_r$ (given $r > 0$) (see, e.g., [71, Remark 5.1, the proof of Theorem 5.1] for more details), which justifies the assumption (H_a) in Theorem 5.5. Moreover, we see that for each $\mathscr{D} \subset \Omega_r$ which restricted to $[\eta, +\infty)$ is relatively compact in $\widetilde{C}_b([\eta, +\infty), X)$ for each $\eta > 0$, $H(\mathscr{D})$ is relatively compact in $C([-\tau, 0], X)$. This implies that the assumption (H_b) in Theorem 5.5 holds true. Finally, applying Theorem 5.5 we get what we wanted. This completes the proof.

Remark 5.16 Let us note that the existence results established recently in Vrabie [192, Theorem 3.1], Wang and Zhu [197, Theorems 3.3, 3.4] cannot be applied to system (5.12), since the nonlocal function H would not necessarily satisfy nonexpansive condition.

Remark 5.17 The key assumption (F_4) used in Vrabie [192, Theorem 5.1] (see also [71, the assumption (A_2) of Theorems 5.2]) is not needed in the above result.

Example 5.2 Let us consider the control system of non-autonomous diffusion equation with time delay

$$\begin{cases} D_t u(t, \xi) - d(t)u_{\xi\xi}(t, \xi) = g(t, \xi, (u_t)(\xi)) + \kappa w(t, \xi), & (t, \xi) \in [0, b] \times [0, \pi], \\ u(t, 0) = u(t, \pi) = 0, & t \in [0, b], \\ u(t, \xi) = \phi(t, \xi), & (t, \xi) \in [-\tau, 0] \times [0, \pi], \end{cases}$$

$$(5.13)$$

where the controller κ is a real parameter. We assume that:

(F_g) $g : [0, b] \times [0, \pi] \times C([-\tau, 0], L^2([0, \pi], \mathbb{R})) \to \mathbb{R}$ is continuous, and there exists a function $\gamma \in L^2([0, b], \mathbb{R}^+)$ such that

$$|g(t, \xi, v)| \leq \gamma(t)(1 + |v|_0)$$

for each $(t, \xi, v) \in [0, b] \times [0, \pi] \times C([-\tau, 0], L^2([0, \pi], \mathbb{R}))$.

We continue to use the setting as in Example 5.1. Let $V = X = L^2([0, \pi], \mathbb{R})$ and let us define

$$G(t, v)(\xi) = g(t, \xi, v), \quad (t, v) \in [0, b] \times C([-\tau, 0], X).$$

It is easy to see that G is continuous from $[0, b] \times C([-\tau, 0], X)$ to X. Furthermore

$$|G(t, v)| \leq \sqrt{\pi}\gamma(t)(1 + |v|_0)$$

for each $t \in [0, b]$ and $v \in C([-\tau, 0], X)$. In addition, we consider the bounded linear operator $B : L^2([0, b], X) \to L^2([0, b], X)$ given by $(Bw)(t) = \kappa w(t)$ for each $w(\cdot) \in L^2([0, b], X)$. It is clear that the assumption (S) is satisfied. Hence, all the assumptions in Theorem 5.6 are satisfied. Now we can rewrite our system as an abstract control problem (5.3) in X. Therefore, applying Theorem 5.6 we conclude that when $|\kappa|$ is sufficiently small, the reachability set of the control system (5.13) is invariant under nonlinear perturbations.

Chapter 6
Neutral Functional Evolution Inclusions

Abstract This chapter deals with functional evolution inclusions of neutral type in Banach space when the semigroup is compact as well as noncompact. The topological properties of the solution set is investigated. It is shown that the solution set is nonempty, compact and an R_δ-set which means that the solution set may not be a singleton but, from the point of view of algebraic topology, it is equivalent to a point, in the sense that it has the same homology group as one-point space. As a sample of application, we consider a partial differential inclusion.

6.1 Introduction

In this chapter, we study the following functional evolution inclusion of neutral type

$$
\begin{cases}
\dfrac{d}{dt}\big(u(t) - h(t, u_t)\big) \in Au(t) + F(t, u_t), & t \in [0, b], \\
u(t) = \phi(t), & t \in [-\tau, 0],
\end{cases}
\tag{6.1}
$$

where the state $u(\cdot)$ takes values in a Banach space X with the norm $|\cdot|$, F is a multimap defined on a subset of $[0, b] \times X$, A is the infinitesimal generator of an analytic semigroup $\{T(t)\}_{t\geq 0}$. For any continuous function u defined on $[-\tau, b]$ and any $t \in [0, b]$, we denote by u_t the element of $C([-\tau, 0], X)$ defined by $u_t(\theta) = u(t + \theta)$, $\theta \in [-\tau, 0]$. Here, $u_t(\cdot)$ represents the history of the state from time $t - \tau$, up to the present time t. For any $v \in C([-\tau, 0], X)$ the norm of c is defined by $|v|_0 = \sup_{\theta \in [-\tau, 0]} |c(\theta)|$.

The study of (6.1) is justified and motivated by a neutral partial differential inclusion of parabolic type

$$
\begin{cases}
\dfrac{\partial}{\partial t}\left(u(t, \xi) - \displaystyle\int_0^\pi U(\xi, \zeta)u_t(\theta, \zeta)d\zeta\right) \in \dfrac{\partial^2}{\partial \xi^2}u(t, \xi) + F(t, \xi, u_t(\theta, \xi)), \\
\hspace{6cm} t \in [0, 1],\ \xi \in [0, \pi], \\
u(t, 0) = u(t, \pi) = 0, \hspace{2.6cm} t \in [0, 1], \\
x(\theta, \xi) = \phi(\theta)(\xi), \hspace{2.3cm} \theta \in [-\tau, 0],\ \xi \in [0, \pi],
\end{cases}
$$

© Springer Nature Singapore Pte Ltd. 2017
Y. Zhou et al., *Topological Structure of the Solution Set for Evolution Inclusions*,
Developments in Mathematics 51, https://doi.org/10.1007/978-981-10-6656-6_6

where the functions U and ϕ satisfy appropriate conditions, $F : [0, 1] \times [0, \pi] \times C([-\tau, 0], L^2([0, \pi], \mathbb{R})) \to P(\mathbb{R})$ is with closed convex values.

The researches on the theory for nonlinear evolution inclusions of neutral type are only on their initial stage of development, see [34, 155, 199]. However, to the best of our knowledge, nothing has been done with the topological properties of solution sets for nonlinear evolution inclusions of neutral type. Our purpose in this chapter is to study the topological structure of the solution set for inclusion (6.1).

This chapter is organized as follows. Section 6.2.1 is devoted to proving that the solution set for inclusion (6.1) is nonempty compact in the case that the semigroup is compact, then proceed to study the R_δ-set. Section 6.2.2 provides that the solution set for inclusion (6.1) is nonempty compact in the case that the semigroup is noncompact, then proceed to study the R_δ-structure of the solution set of inclusion (6.1).

Throughout this chapter, $\mathcal{L}(X)$ stands for the space of all linear bounded operators on Banach space X with the norm $\| \cdot \|_{\mathcal{L}(X)}$. We denote by $C([-\tau, b], X)$ the Banach space consisting of continuous functions from $[-\tau, b]$ to X equipped with the norm

$$\|u\|_{[-\tau, b]} = \max_{t \in [-\tau, b]} |u(t)|.$$

We assume that A is the infinitesimal generator of an analytic semigroup $\{T(t)\}_{t \geq 0}$ of uniformly bounded linear operators on X. Let $0 \in \rho(A)$, where $\rho(A)$ is the resolvent set of A. Then for $0 < \beta \leq 1$, it is possible to define the fractional power A^β as a closed linear operator on its domain $D(A^\beta)$. For analytic semigroup $\{T(t)\}_{t \geq 0}$, the following properties will be used:

(i) there is a $M \geq 1$ such that $M := \sup_{t \geq 0} \|T(t)\|_{\mathcal{L}(X)} < \infty$;
(ii) for any $\beta \in (0, 1]$, there exists a positive constant C_β such that

$$\|A^\beta T(t)\|_{\mathcal{L}(X)} \leq \frac{C_\beta}{t^\beta}, \; 0 < t \leq b.$$

It is clear that $A^\beta T(t)u = T(t)A^\beta u$ for $u \in D(A^\beta)$. Then $AT(t)u = A^{1-\beta} T(t) A^\beta u$ for $u \in D(A^\beta)$.

The results in this chapter are taken from Zhou and Peng [209].

6.2 Topological Structure of Solution Set

In this section, let X be reflexive. We study the topological structure of the solution set in cases that $T(t)$ is compact and noncompact, respectively. Before stating and proving the main results, we introduce the following hypotheses:

(H_1) the multivalued nonlinearity $F : [0, b] \times C([-\tau, 0], X) \to P_{cl,cv}(X)$ satisfies

(i) $F(t, \cdot)$ is weakly u.s.c. for a.e. $t \in [0, b]$, and the multimap $F(\cdot, v)$ has a strongly measurable selection for every $v \in C([-\tau, 0], X)$;

(ii) there exists a function $\alpha \in L^1([0, b], \mathbb{R}^+)$ such that

$$|F(t, v)| \leq \alpha(t)(1 + |v|_0) \text{ for a.e. } t \in [0, b] \text{ and } v \in C([-\tau, 0], X).$$

(H_2) The function $h : [0, b] \times C([-\tau, 0], X) \to X$ is continuous and there exists a constant $\beta \in (0, 1)$ and $d, d_1 > 0$ with $d\left(\|A^{-\beta}\|_{\mathcal{L}(X)} + \frac{C_{1-\beta}b^\beta}{\beta}\right) < \frac{1}{2}$, such that $h \in D(A^\beta)$ and for any $v_1, v_2 \in C([-\tau, 0], X)$, the function $A^\beta h(t, \cdot)$ is strongly measurable and $A^\beta h(t, \cdot)$ satisfies the Lipschitz condition

$$|A^\beta h(t, v_1) - A^\beta h(t, v_2)| \leq d|v_1 - v_2|_0$$

and the inequality

$$|A^\beta h(t, v_1)| \leq d_1(1 + |v_1|_0) \text{ for every } t \in [0, b].$$

Given $u \in C([-\tau, b], X)$, let us denote

$$Sel_F(u) = \{f \in L^1([0, b], X) : f(t) \in F(t, u_t) \text{ for a.e. } t \in [0, b]\}.$$

The set $Sel_F(u)$ is always nonempty as Lemma 6.1 below shows.

Lemma 6.1 ([71] (see also [45])) *Let the condition* (H_1) *be satisfied. Then* $Sel_F(u)$: $C([-\tau, b], X) \to P(L^1([0, b], X))$ *is weakly u.s.c. with nonempty, convex and weakly compact values.*

Definition 6.1 A continuous function $u : [-\tau, b] \to X$ is said to be a mild solution of inclusion (6.1) if $u(t) = \phi(t)$ for $t \in [-\tau, 0]$ and there exists $f \in L^1([0, b], X)$ such that $f(t) \in F(t, u_t)$, and u satisfies the following integral equation

$$u(t) = T(t)\big(\phi(0) - h(0, \phi)\big) + h(t, u_t) + \int_0^t AT(t - s)h(s, u_s)ds$$

$$+ \int_0^t T(t - s)f(s)ds \text{ for } t \in [0, b]. \tag{6.2}$$

Remark 6.1 For any $u \in C([-\tau, b], X)$, now define a solution multioperator $\mathscr{F}$: $C([-\tau, b], X) \to P(C([-\tau, b], X))$ as follows:

$$\mathscr{F}(u) = \Gamma_1(u) + \Gamma_2(u),$$

where

$$\Gamma_1(u)(t) = \begin{cases} -T(t)h(0, \phi) + h(t, u_t) + \int_0^t AT(t - s)h(s, u_s)ds, & t \in [0, b], \\ 0, & t \in [-\tau, 0], \end{cases}$$

and

$$\Gamma_2(u)(t) = \left\{ y(t) \in C([-\tau, b], X) : y(t) = \begin{cases} S(f)(t), & f \in Sel_F(u), \quad t \in [0, b], \\ \phi(t), & t \in [-\tau, 0], \end{cases} \right\},$$

here, the operator $S : L^1([0, b], X) \to C([0, b], X)$ is defined by

$$S(f)(t) = T(t)\phi(0) + \int_0^t T(t - s)f(s)ds.$$

It is easy to verify that the fixed points of the multioperator $\mathscr{F}$ are mild solutions of inclusion (6.1).

Lemma 6.2 ([71, Lemma 3.3]) *Let the hypothesis* (H_1) *be satisfied. Then there exists a sequence* $\{F_n\}$ *with* $F_n : [0, b] \times C([-\tau, 0], X) \to P_{cl,cv}(X)$ *such that*

(i) $F(t, v) \subset \cdots \subset F_{n+1}(t, v) \subset F_n(t, v) \subset \cdots \subset \overline{co}(F(t, B_{3^{1-n}}(v))), \; n \geq 1,$ *for each* $t \in [0, b]$ *and* $v \in C([-\tau, 0], X)$;

(ii) $|F_n(t, v)| \leq \alpha(t)(2 + |v|_0), \; n \geq 1,$ *for a.e.* $t \in [0, b]$ *and each* $v \in C([-\tau, 0], X)$;

(iii) *there exists* $E \subset [0, b]$ *with* $mes(E) = 0$ *such that for each* $x^* \in X^*, \; \varepsilon > 0$ *and* $(t, v) \in [0, b] \setminus E \times C([-\tau, 0], X),$ *there exists* $N > 0$ *such that for all* $n \geq N,$

$$x^*(F_n(t, v)) \subset x^*(F(t, v)) + (-\varepsilon, \varepsilon);$$

(iv) $F_n(t, \cdot) : C([-\tau, 0], X) \to P_{cl,cv}(X)$ *is continuous for a.e.* $t \in [0, b]$ *with respect to Hausdorff metric for each* $n \geq 1$;

(v) *for each* $n \geq 1,$ *there exists a selection* $g_n : [0, b] \times C([-\tau, 0], X) \to X$ *of* F_n *such that* $g_n(\cdot, v)$ *is measurable for each* $v \in C([-\tau, 0], X)$ *and for any compact subset* $\mathscr{D} \subset C([-\tau, 0], X)$ *there exist constants* $C_V > 0$ *and* $\delta > 0$ *for which the estimate*

$$|g_n(t, v_1) - g_n(t, v_2)| \leq C_V \alpha(t)|v_1 - v_2|_0$$

holds for a.e. $t \in [0, b]$ *and each* $v_1, v_2 \in C([-\tau, 0], X)$ *with* $V := \mathscr{D} + B_\delta(0)$;

(vi) F_n *verifies the condition* (H_1) *with* F_n *instead of* F *for each* $n \geq 1,$ *provided that* X *is reflexive.*

6.2.1　Compact Semigroup Case

The following compactness characterizations of the solution set to inclusion (6.1) will be useful.

Lemma 6.3 *Suppose that $\{T(t)\}_{t>0}$ is compact and there exists $\varpi \in L^1([0, b], \mathbb{R}^+)$ such that*

$$|F(t, u)| \leq \varpi(t) \text{ for a.e. } t \in [0, b] \text{ and } u \in C([-\tau, 0], X).$$

Then the multimap Γ_2 is compact in $C([-\tau, b], X)$.

Proof Let D be a bounded set of $C([-\tau, b], X)$. We will prove that for each $t \in [-\tau, b]$, $V(t) = \{\Gamma_2(u)(t) : u \in D\}$ is relatively compact in X.

Obviously, for $t \in [-\tau, 0]$, $V(t) = \{\phi(t)\}$ is relatively compact in X. Let $t \in [0, b]$ be fixed, for $u \in D$ and $v(t) \in V(t)$, there exists $f \in Sel_F(u)$ such that

$$v(t) = T(t)\phi(0) + \int_0^t T(t-s)f(s)ds.$$

For arbitrary $\varepsilon \in (0, t)$, define an operator $J_\varepsilon : V(t) \to X$ by

$$J_\varepsilon v(t) = T(t)\phi(0) + T(\varepsilon)\int_0^{t-\varepsilon} T(t-\varepsilon-s)f(s)ds.$$

From the compactness of $T(t)$, $t > 0$, we get that the set $V_\varepsilon(t) = \{J_\varepsilon v(t) : v(t) \in V(t)\}$ is relatively compact in X for each $\varepsilon \in (0, t)$. Moreover, it follows

$$|v(t) - J_\varepsilon v(t)| \leq \left| \int_{t-\varepsilon}^t T(t-s)f(s)ds \right| \leq M \int_{t-\varepsilon}^t \varpi(s)ds.$$

Therefore, there is a relatively compact set arbitrarily close to the set $V(t)$. Thus the set $V(t)$ is also relatively compact in X, which yields that $V(t) = \{\Gamma_2(u)(t) : u \in D\}$ is relatively compact in X for each $t \in [-\tau, b]$.

We proceed to verify that the set $\{\Gamma_2(u) : u \in D\}$ is equicontinuous on $[-\tau, b]$. Taking $0 < t_1 < t_2 \leq b$ and $\delta > 0$ small enough, for any $v(t) \in \Gamma_2(u)(t)$, we obtain

$$|v(t_2) - v(t_1)|$$

$$\leq \|T(t_2) - T(t_1)\|_{\mathcal{L}(X)}|\phi(0)| + \left| \int_{t_1}^{t_2} T(t_2-s)f(s)ds \right|$$

$$+ \left| \int_0^{t_1-\delta} (T(t_2-s) - T(t_1-s))f(s)ds \right| + \left| \int_{t_1-\delta}^{t_1} (T(t_2-s) - T(t_1-s))f(s)ds \right|$$

$$\leq \|T(t_2) - T(t_1)\|_{\mathcal{L}(X)}|\phi|_0 + M \int_{t_1}^{t_2} \varpi(s)ds$$

$$+ \sup_{s \in [0, t_1-\delta]} \|T(t_2-s) - T(t_1-s)\|_{\mathcal{L}(X)} \int_0^{t_1-\delta} \varpi(s)ds + 2M \int_{t_1-\delta}^{t_1} \varpi(s)ds.$$

The right-hand side tends to zero as $t_2 - t_1 \to 0$, since $T(t)$ is strongly continuous, and the compactness of $T(t)$ $(t > 0)$, implies the continuity in the uniform operator topology.

For $-\tau \le t_1 < 0 < t_2 \le b$, we have

$$|v(t_2) - v(t_1)| \le |T(t_2)\phi(0) - \phi(t_1)| + \left| \int_0^{t_2} T(t_2 - s)f(s)ds \right|$$

$$\le \|T(t_2) - I\|_{\mathcal{L}(X)} |\phi|_0 + |\phi(t_1) - \phi(0)| + M \int_0^{t_2} \varpi(s)ds.$$

The right-hand side tends to zero as $t_2 - t_1 \to 0$ $(t_2 \to 0^+, \ t_1 \to 0^-)$, since ϕ is continuous. Note that for $t_1, t_2 \in [-\tau, 0]$, $|v(t_2) - v(t_1)| = |\phi(t_2) - \phi(t_1)| \to 0$ as $t_2 - t_1 \to 0$. Thus $\{\Gamma_2(u) : u \in D\}$ is equicontinuous as well. Thus, an application of Arzela-Ascoli's theorem justifies that $\{\Gamma_2(u) : u \in D\}$ is relatively compact in $C([-\tau, b], X)$. Hence Γ_2 is compact in $C([-\tau, b], X)$. This completes the proof.

Let $a \in [0, b)$ and $\varphi \in C([-\tau, a], X)$. Consider the integral equation of the form

$$u(t) = \begin{cases} T(t - a)\big(\varphi(a) - h(a, \varphi)\big) + h(t, u_t) + \displaystyle\int_a^t AT(t - s)h(s, u_s)ds \\[2mm] + \displaystyle\int_a^t T(t - s)g(s, u_s)ds, \qquad\qquad t \in [a, b], \\[2mm] \varphi(t), \qquad\qquad\qquad\qquad\qquad\qquad t \in [-\tau, a]. \end{cases} \qquad (6.3)$$

Lemma 6.4 *Assume that for every* $v \in C([a - \tau, a], X)$, $g(\cdot, v)$ *is* L^1-*integrable,* $\{T(t)\}_{t>0}$ *is compact and* (H_2) *holds. Suppose in addition that*

(i) *for any compact subset* $K \subset C([a - \tau, a], X)$, *there exist* $\delta > 0$ *and* $L_K \in L^1([a, b], \mathbb{R}^+)$ *such that*

$$|g(t, v_1) - g(t, v_2)| \le L_K(t)|v_1 - v_2|_0 \text{ for a.e. } t \in [a, b] \text{ and each } v_1, v_2 \in B_\delta(K);$$

(ii) *there exists* $\varpi_1(t) \in L^1([a, b], \mathbb{R}^+)$ *such that* $|g(t, v)| \le \varpi_1(t)(c' + |v|_0)$ *for a.e.* $t \in [a, b]$ *and every* $v \in C([a - \tau, a], X)$, *where* c' *is arbitrary, but fixed.*

If $d_1 \|A^{-\beta}\| < 1$, *then integral equation* (6.3) *admits a unique solution for every* $\varphi(t) \in C([-\tau, a], X)$. *Moreover, the solution of* (6.3) *depends continuously on* φ.

Proof Step 1. (Priori estimate). We have

$$|u(t)| \le |A^{-\beta}A^\beta h(t, u_t)| + \left| \int_a^t A^{1-\beta}T(t - s)A^\beta h(s, u_s)ds \right|$$

$$+ |T(t - a)\big(\varphi(a) - h(a, \varphi)\big)| + \left| \int_a^t T(t - s)g(s, u_s)ds \right|$$

$$\leq d_1 \|A^{-\beta}\|_{\mathcal{L}(X)}(1 + |u_t|_0) + d_1 C_{1-\beta} \int_a^t (t-s)^{\beta-1}(1 + |u_s|_0)ds$$

$$+ M(\|\varphi\| + d_1(1 + \|\varphi\|)) + M \int_a^t \varpi_1(s)(c' + |u_s|_0)ds$$

$$\leq d_1 \|A^{-\beta}\|_{\mathcal{L}(X)}(1 + \|u\|_{[a-\tau,t]}) + d_1 C_{1-\beta} \int_a^t (t-s)^{\beta-1}(1 + \|u\|_{[a-\tau,s]})ds$$

$$+ M(\|\varphi\| + d_1(1 + \|\varphi\|)) + M \int_a^t \varpi_1(s)(c' + \|u\|_{[a-\tau,s]})ds$$

for $t \in [a, b]$, and notice that $|u(t)| = |\varphi(t)|$ for $t \in [-\tau, a]$. Then

$$\|u\|_{[a-\tau,t]} \leq \frac{1}{1 - d_1\|A^{-\beta}\|_{\mathcal{L}(X)}}\left(c(b) + M(\|\varphi\| + d_1(1 + \|\varphi\|)) + c'M\|\varpi_1\|_{L([a,b],\mathbb{R}^+)}\right)$$

$$+ \frac{1}{1 - d_1\|A^{-\beta}\|_{\mathcal{L}(X)}} \int_a^t \left(d_1 C_{1-\beta}(t-s)^{\beta-1} + M\varpi_1(s)\right)\|u\|_{[a-\tau,s]}ds,$$

where $c(b) = d_1\|A^{-\beta}\|_{\mathcal{L}(X)} + \frac{d_1 C_{1-\beta} b^\beta}{\beta}$. By Gronwall's inequality, we get that there exists $M_1 > 0$ such that $\|u\|_{[-\tau,b]} \leq M_1$.

Step 2. Let $\varphi \in C([-\tau, a], X)$ be fixed. From $d_1\|A^{-\beta}\|_{\mathcal{L}(X)} < 1$, we can find one $\xi < 1 + a$ arbitrarily close to a such that

$$d_1\left(\|A^{-\beta}\|_{\mathcal{L}(X)} + \frac{C_{1-\beta}(\xi - a)^\beta}{\beta}\right) + M\|\varpi_1\|_{L([a,\xi],\mathbb{R}^+)} < 1.$$

Then for one of such ξ, we choose one ρ satisfying

$$\rho \geq \frac{d_1\left(\|A^{-\beta}\|_{\mathcal{L}(X)} + \frac{C_{1-\beta}(\xi-a)^\beta}{\beta}\right) + M(\|\varphi\| + d_1(1 + \|\varphi\|)) + Mc'\|\varpi_1\|_{L([a,\xi],\mathbb{R}^+)}}{1 - d_1\left(\|A^{-\beta}\|_{\mathcal{L}(X)} + \frac{C_{1-\beta}(\xi-a)^\beta}{\beta}\right) - M\|\varpi_1\|_{L([a,\xi],\mathbb{R}^+)}},$$

that is,

$$c(\xi - a)(1 + \rho) + M(\|\varphi\| + d_1(1 + \|\varphi\|)) + M(c' + \rho)\|\varpi_1\|_{L([a,\xi],\mathbb{R}^+)} \leq \rho.$$

Write

$$B_\rho(\xi) = \{u \in C([-\tau, \xi], X) : \max_{t\in[a-\tau,\xi]} |u(t)| \leq \rho \text{ and } u(t) = \varphi(t) \text{ for } t \in [-\tau, a]\}.$$

Let us define the operator W:

$$Wu(t) = W_1 u(t) + W_2 u(t), \tag{6.4}$$

where

$$
W_1 u(t) = \begin{cases} h(t, u_t) + \displaystyle\int_a^t AT(t-s)h(s, u_s)ds, & t \in [a, b], \\ 0, & t \in [-\tau, a], \end{cases}
$$

and

$$
W_2 u(t) = \begin{cases} T(t-a)\big(\varphi(a) - h(a, \varphi)\big) + \displaystyle\int_a^t T(t-s)g(s, u_s)ds, & t \in [a, b], \\ \varphi(t), & t \in [-\tau, a]. \end{cases}
$$

For $u \in B_\rho(\xi)$, we have

$$
|W_1 u(t) + W_2 u(t)|
$$

$$
\leq |A^{-\beta} A^{\beta} h(t, u_t)| + \left| \int_a^t A^{1-\beta} T(t-s) A^{\beta} h(s, u_s)ds \right|
$$

$$
+ |T(t-a)\big(\varphi(a) - h(a, \varphi)\big)| + \left| \int_a^t T(t-s)g(s, u_s)ds \right|
$$

$$
\leq d_1 \|A^{-\beta}\|_{\mathcal{L}(X)} \big(1 + \|u\|_{[a-\tau, t]}\big) + d_1 C_{1-\beta} \int_a^t (t-s)^{\beta-1}\big(1 + \|u\|_{[a-\tau, s]}\big)ds
$$

$$
+ M\big(\|\varphi\| + d_1(1 + \|\varphi\|)\big) + M \int_a^t \varpi_1(s)\big(c' + \|u\|_{[a-\tau, s]}\big)ds
$$

$$
\leq c(\xi - a)(1 + \rho) + M\big(\|\varphi\| + d_1(1 + \|\varphi\|)\big) + M(c' + \rho)\|\varpi_1\|_{L([a,b],\mathbb{R}^+)}
$$

$$
\leq \rho
$$

for $t \in [a, b]$. Obviously, W maps $B_\rho(\xi)$ into itself.

For any $u, v \in B_\rho(\xi)$ and $t \in [a, b]$, we have

$$
|W_1 u(t) - W_1 v(t)|
$$

$$
\leq |h(t, u_t) - h(t, v_t)| + \left| \int_a^t AT(t-s)\big(h(s, u_s) - h(s, v_s)\big)ds \right|
$$

$$
= \left| A^{-\beta} A^{\beta}\big(h(t, u_t) - h(t, v_t)\big) \right| + \left| \int_a^t A^{1-\beta} T(t-s) A^{\beta}\big(h(s, u_s) - h(s, v_s)\big)ds \right|
$$

$$
\leq d\|A^{-\beta}\|_{\mathcal{L}(X)}|u_t - v_t|_0 + dC_{1-\beta} \int_a^t (t-s)^{\beta-1}|u_s - v_s|_0 ds
$$

$$
\leq d\|A^{-\beta}\|_{\mathcal{L}(X)}\|u - v\|_{[a-\tau, t]} + dC_{1-\beta} \int_a^t (t-s)^{\beta-1}\|u - v\|_{[a-\tau, s]}ds
$$

$$
\leq d\left(\|A^{-\beta}\|_{\mathcal{L}(X)} + \frac{C_{1-\beta}(\xi - a)^{\beta}}{\beta} \right)\|u - v\|_{[a-\tau, \xi]}.
$$

Noting that $W_1 u(t) = 0$ for $t \in [-\tau, a]$, which implies that

$$\|W_1 u - W_1 v\|_{[a-\tau,\xi]} \leq d\left(\|A^{-\beta}\|_{\mathcal{L}(X)} + \frac{C_{1-\beta}(\xi - a)^\beta}{\beta}\right)\|u - v\|_{[a-\tau,\xi]},$$

we get that W_1 is a contraction.

Next, we will prove that W_2 is continuous on $B_\rho(\xi)$. Let $u^n, u \in B_\rho(\xi)$ with $u^n \to u$ on $B_\rho(\xi)$. Then by (i) and the fact that $u_t^n \to u_t$ for $t \in [a, \xi]$, we have

$$g(s, u_s^n) \to g(s, u_s) \text{ for a.e. } s \in [a, \xi] \text{ as } n \to \infty.$$

Noting that $|g(s, u_s^n) - g(s, u_s)| \leq 2\varpi_1(t)(c' + \rho)$, by Lebesgue's dominated convergence theorem, we have

$$|W_2 u^n(t) - W_2 u(t)| \leq M \int_a^t |g(s, u_s^n) - g(s, u_s)| ds \to 0 \text{ as } n \to \infty.$$

Moreover, from the proof of Lemma 6.3 we see that W_2 is a compact operator. Thus, W_2 is a completely continuous operator. Hence, Krasnoselskii's fixed point theorem shows that there is a fixed point of W, denoted by y, which is a local solution to equation (6.3).

Step 3. We prove that this solution is unique. In fact, let z be another local solution of equation (6.3). According to the condition (i), we obtain

$$|y(t) - z(t)|$$

$$\leq |h(t, y_t) - h(t, z_t)| + \left|\int_a^t AT(t - s)(h(s, y_s) - h(s, z_s))ds\right|$$

$$+ \left|\int_a^t T(t - s)(g(s, y_s) - g(s, z_s))ds\right|$$

$$\leq d\|A^{-\beta}\|_{\mathcal{L}(X)}|y_t - z_t|_0 + dC_{1-\beta}\int_a^t (t - s)^{\beta-1}|y_s - z_s|_0 ds + M\int_a^t L_K(s)|y_s - z_s|_0 ds$$

$$\leq d\|A^{-\beta}\|_{\mathcal{L}(X)}\|y - z\|_{[a-\tau,t]} + \int_a^t \left(dC_{1-\beta}(t - s)^{\beta-1} + ML_K(s)\right)\|y - z\|_{[a-\tau,s]}ds$$

for $t \in [a, \xi]$, and $|y(t) - z(t)| = 0$ for $t \in [-\tau, a]$. It follows that

$$\|y - z\|_{[a-\tau,t]} \leq \frac{1}{1 - d\|A^{-\beta}\|_{\mathcal{L}(X)}} \int_a^t \left(dC_{1-\beta}(t - s)^{\beta-1} + ML_K(s)\right)\|y - z\|_{[a-\tau,s]}ds.$$

Applying Gronwall's inequality, we get $\|y - z\|_{[a-\tau,t]} = 0$, which implies $y(t) = z(t)$ for $t \in [-\tau, \xi]$.

If $\xi = b$, the proof is complete. If this is not the case, we can repeat the previous argument on $[-\tau, \xi_1]$, where $\xi_1 = \min\{\xi + h_0, b\}$, with the initial condition

$u(t) = y(t)$ for $t \in [\xi - \tau, \xi]$; here $h_0 > 0$ is such that $d_1 \left(\|A^{-\beta}\|_{\mathcal{L}(X)} + \frac{C_{1-\beta} h_0^{\beta}}{\beta} \right) +$
$M \|\varpi_1\|_{L([\xi, \xi + h_0], \mathbb{R}^+)} < 1$. In this case, it is easy to see that, for $t \in [\xi, \xi_1]$, we have

$$u(t) = T(t - \xi)\big(y(t) - h(t, y_t)\big) + h(t, u_t)$$
$$+ \int_{\xi}^{t} AT(t - s)h(s, u_s)ds + \int_{\xi}^{t} T(t - s)g(s, u_s)ds.$$

We remark that

$$d_1 \left(\|A^{-\beta}\|_{\mathcal{L}(X)} + \frac{C_{1-\beta}(\xi_1 - \xi)^{\beta}}{\beta} \right) + M \|\varpi_1\|_{L([\xi, \xi_1], \mathbb{R}^+)} < 1.$$

We obtain in such a way a mild solution on $[-\tau, \xi_1]$. If $\xi_1 < b$, we can repeat the previous argument. At the end, we obtain a mild solution of equation (6.3) defined on $[-\tau, b]$ by the priori estimate.

Step 4. Let $\varphi^n \to \varphi^0$ in $C([-\tau, b], X)$ as $n \to \infty$, and u^n be the solution of equation (6.3) with the perturbation φ^n, i.e.,

$$u^n(t) = T(t - a)\big(\varphi^n(a) - h(a, \varphi^n)\big) + h(t, u_t^n) + \int_{a}^{t} AT(t - s)h(s, u_s^n)ds$$
$$+ \int_{a}^{t} T(t - s)g(s, u_s^n)ds$$

$$(6.5)$$

for $t \in [a, b]$ and $u^n(t) = \varphi^n(t)$ for $t \in [-\tau, a]$. It is clear that $\lim_{n \to \infty} u^n$ exists in $C([-\tau, a], X)$. From the condition (ii) and the compactness of $T(t)$ for $t > 0$ it follows that the set

$$\left\{ \int_{a}^{t} T(t - s)g(s, u_s^n)ds \right\}_{n \geq 1}$$

is relatively compact in $C([a, b], X)$. This gives that the family

$$\left\{ u^n(t) - h(t, u_t^n) - \int_{a}^{t} AT(t - s)h(s, u_s^n)ds \right\}_{n \geq 1}$$

is relatively compact in $C([a, b], X)$. We prove that $\lim_{n \to \infty} u^n$ exists in $C([a, b], X)$. On the contrary, if $\lim_{n \to \infty} u^n$ does not exist in $C([a, b], X)$, then for any $n \in \mathbb{N}$, we have n_1, n_2 with $n_1, n_2 > n$ such that $\|u^{n_1} - u^{n_2}\|_{[a,b]} > \varepsilon_0$ ($\varepsilon_0 > 0$ is a constant), that is, there exists t^* such that

$$|u^{n_1}(t^*) - u^{n_2}(t^*)| = \|u^{n_1} - u^{n_2}\|_{[a,b]} > \varepsilon_0.$$

Let $z^n(t) = u^n(t) - h(t, u_t^n) - \int_a^t AT(t-s)h(s, u_s^n)ds$. Using (H_2), we estimate

$$|z^{n_1}(t^*) - z^{n_2}(t^*)| \geq |u^{n_1}(t^*) - u^{n_2}(t^*)| - |h(t^*, x_{t^*}^{n_1}) - h(t^*, x_{t^*}^{n_2})|$$

$$- \left| \int_a^{t^*} AT(t^* - s)[h(s, u_s^{n_1}) - h(s, u_s^{n_2})]ds \right|$$

$$\geq |u^{n_1}(t^*) - u^{n_2}(t^*)| - d\|A^{-\beta}\|_{\mathcal{L}(X)}\|u_{t^*}^{n_1} - u_{t^*}^{n_2}\|_*$$

$$- dC_{1-\beta} \int_a^{t^*} (t^* - s)^{\beta-1}\|u_s^{n_1} - u_s^{n_2}\|_* ds$$

$$\geq |u^{n_1}(t^*) - u^{n_2}(t^*)| - d\left(\|A^{-\beta}\|_{\mathcal{L}(X)} + \frac{C_{1-\beta}b^\beta}{\beta}\right)\|u^{n_1} - u^{n_2}\|_{[a,b]}$$

$$= \left[1 - d\left(\|A^{-\beta}\|_{\mathcal{L}(X)} + \frac{C_{1-\beta}b^\beta}{\beta}\right)\right]\varepsilon_0,$$

which contracts with the compactness of z^n in $C([a, b], X)$. Hence, $\{u^n\}$ converges in $C([-\tau, b], X)$, the limit is denoted by u. Therefore, taking the limit in (6.5) as $n \to \infty$, one finds, again by (H_2) and Lebesgue's dominated convergence theorem, that u is the solution of equation (6.3) with the perturbation φ^0. This completes the proof.

For convenience, define

$$\tilde{d} = d_1\left(\|A^{-\beta}\|_{\mathcal{L}(X)} + \frac{C_{1-\beta}b^\beta}{\beta}\right) + M\|\alpha\|_{L([0,b],\mathbb{R}^+)}.$$

Theorem 6.1 *Let X be reflexive and assume that (H_1) and (H_2) hold. In addition, suppose that $\{T(t)\}_{t>0}$ is compact in X. If $\tilde{d} < 1$, then the solution set of the inclusion (6.1) is a nonempty compact subset of $C([-\tau, b], X)$ for each $\phi \in C([-\tau, 0], X)$.*

Proof Step 1. Let $\phi \in C([-\tau, 0], X)$ be fixed. Consider the set

$$B_R(b) = \{u \in C([-\tau, b], X) : \max_{t\in[-\tau,b]} |u(t)| \leq R\},$$

where

$$R > \frac{|\phi|_0 + M[|\phi|_0 + d_1\|A^{-\beta}\|_{\mathcal{L}(X)}(1 + |\phi|_0)] + \tilde{d}}{1 - \tilde{d}}.$$

It is clear that $B_R(b)$ is a bounded, closed and convex set of $C([-\tau, b], X)$. We first show that $\Gamma_1(B_R(b)) + \Gamma_2(B_R(b)) \subset B_R(b)$. Indeed, taking $u \in B_R(b)$ and $z(t) \in \Gamma_2(u)$, there exists $f \in Sel_F(u)$ such that

$$|\Gamma_1 u(t)| \leq |T(t)h(0, \phi)]| + |A^{-\beta}A^\beta h(t, u_t)| + \left|\int_0^t A^{1-\beta}T(t-s)A^\beta h(s, u_s)ds\right|$$

$$\leq M|A^{-\beta}A^\beta h(0, \phi)| + d_1\|A^{-\beta}\|_{\mathcal{L}(X)}(1 + |u_t|_0)$$

$$+ d_1 C_{1-\beta} \int_0^t (t-s)^{\beta-1}(1+|u_s|_0)ds$$

$$\leq M d_1 \|A^{-\beta}\|_{\mathcal{L}(X)}(1+|\phi|_0) + d_1 \|A^{-\beta}\|_{\mathcal{L}(X)}\big(1+\|x\|_{[-\tau,t]}\big)$$

$$+ d_1 C_{1-\beta} \int_0^t (t-s)^{\beta-1}\big(1+\|x\|_{[-\tau,s]}\big)ds$$

$$\leq M d_1 \|A^{-\beta}\|_{\mathcal{L}(X)}(1+|\phi|_0) + d_1 \|A^{-\beta}\|_{\mathcal{L}(X)}(1+R) + d_1(1+R)\frac{C_{1-\beta}b^{\beta}}{\beta},$$

and

$$|z(t)| \leq |T(t)\phi(0)| + \left| \int_0^t T(t-s)f(s)ds \right|$$

$$\leq M|\phi|_0 + M \int_0^t \alpha(s)(1+|u_s|_0)ds$$

$$\leq M|\phi|_0 + M(1+R)\|\alpha\|_{L([0,b],\mathbb{R}^+)},$$

it follows that

$$|\Gamma_1 u(t) + z(t)| \leq M[|\phi|_0 + d_1\|A^{-\beta}\|_{\mathcal{L}(X)}(1+|\phi|_0)] + \tilde{d}(1+R)$$

for $t \in [0, b]$. From $\Gamma_1 u(t) + z(t) = \phi(t)$ for $t \in [-\tau, 0]$, we know

$$|\Gamma_1 u(t) + \Gamma_2 u(t)| \leq |\phi|_0 + M[|\phi|_0 + d_1\|A^{-\beta}\|_{\mathcal{L}(X)}(1+|\phi|_0)] + \tilde{d}(1+R) \leq R$$

for $t \in [-\tau, b]$,

Step 2. We show that Γ_1 is a contraction on $C([-\tau, b], X)$. Let $u, v \in C([-\tau, b], X)$. Then

$$|\Gamma_1 u(t) - \Gamma_1 v(t)| \leq |h(t, u_t) - h(t, v_t)| + \left| \int_0^t AT(t-s)\big(h(s, u_s) - h(s, v_s)\big)ds \right|$$

$$\leq d\|A^{-\beta}\|_{\mathcal{L}(X)}|u_t - v_t|_0 + dC_{1-\beta} \int_0^t (t-s)^{\beta-1}|u_s - v_s|_0 ds$$

$$\leq d\|A^{-\beta}\|_{\mathcal{L}(X)}\|u - v\|_{[-\tau,t]} + dC_{1-\beta} \int_0^t (t-s)^{\beta-1}\|u - v\|_{[-\tau,s]}ds$$

$$\leq d\left(\|A^{-\beta}\|_{\mathcal{L}(X)} + \frac{C_{1-\beta}b^{\beta}}{\beta}\right)\|u - v\|_{[-\tau,b]}.$$

Noting that $\Gamma_1 u(t) = 0$ for $t \in [-\tau, 0]$, which implies that

$$\|\Gamma_1 u - \Gamma_1 v\|_{[-\tau,b]} \leq d\left(\|A^{-\beta}\|_{\mathcal{L}(X)} + \frac{C_{1-\beta}b^{\beta}}{\beta}\right)\|u - v\|_{[-\tau,b]}.$$

This shows that Γ_1 is a contraction, since $d\left(\|A^{-\beta}\|_{\mathcal{L}(X)} + \frac{C_{1-\beta}b^{\beta}}{\beta}\right) < \frac{1}{2}$.

Step 3. An application of Lemma 6.3 enables us to find that Γ_2 is compact on $B_R(b)$. We only show that Γ_2 is u.s.c.

By Lemma 1.9, it suffices to show that Γ_2 has a closed graph (and therefore has closed values). Let $u^n \subset B_R(b)$ with $u^n \to u$ and $z^n \in \Gamma_2(u^n)$ with $z^n \to z$. We shall prove that $z \in \Gamma_2(u)$. By the definition of Γ_2, there exist $f^n \in Sel_F(u^n)$ such that

$$z^n(t) = \begin{cases} T(t)\phi(0) + \displaystyle\int_0^t T(t-s)f^n(s)ds, & \text{for } t \in [0, b], \\ \phi(t), & \text{for } t \in [-\tau, 0]. \end{cases}$$

We need to prove that there exists $f \in Sel_F(u)$ such that

$$z(t) = \begin{cases} T(t)\phi(0) + \displaystyle\int_0^t T(t-s)f(s)ds, & \text{for } t \in [0, b], \\ \phi(t), & \text{for } t \in [-\tau, 0]. \end{cases} \tag{6.6}$$

By (H_1)(ii), noticing that $Sel_F(u)$ is weakly u.s.c. with weakly compact and convex values due to Lemma 6.1, an application of Lemma 1.7 yields that there exists $f \in Sel_F(u)$ and a subsequence of f^n, still denoted by f^n, such that $f^n \rightharpoonup f$ in $L^1([0, b], X)$. From this and Lemma 6.3, we see

$$z^n(t) = T(t)\phi(0) + \int_0^t T(t-s)f^n(s)ds \to T(t)\phi(0)$$
$$+ \int_0^t T(t-s)f(s)ds \text{ as } n \to \infty.$$

By the uniqueness of the limit, (6.6) holds and $z \in \Gamma_2(u)$. It follows that Γ_2 is closed and therefore has compact values.

Therefore, the operators Γ_1 and Γ_2 satisfy all conditions of Theorem 1.13, thus the fixed points set of the operator $\Gamma_1 + \Gamma_2$ is a nonempty compact subset of $C([-\tau, b], X)$.

Now, let $\Theta(\phi)$ denote the set of all mild solutions of inclusion (6.1).

Theorem 6.2 *Under the conditions in Theorem 6.1, the solution set of (6.1) is an R_δ-set.*

Proof To this aim, let us consider the following semilinear evolution inclusion

$$\begin{cases} \dfrac{d}{dt}[u(t) - h(t, u_t)] \in Au(t) + F_n(t, u_t), & t \in [0, b], \\ u(t) = \phi(t), & t \in [-\tau, 0], \end{cases} \tag{6.7}$$

where multivalued functions $F_n : [0, b] \times C([-\tau, 0], X) \to P_{cl, cv}(X)$ are established in Lemma 6.2. Let $\Theta_n(\phi)$ denote the set of all mild solutions of inclusion (6.7).

From Lemma 6.2 (ii) and (vi), it follows that $\{F_n\}$ verifies the condition (H_1) for each $n \geq 1$. Then from Lemma 6.1 one finds that Sel_{F_n} is weakly u.s.c. with convex and weakly compact values. Moreover, one can see from Theorem 6.1 that each set $\Theta_n(\phi)$ is nonempty and compact in $C([-\tau, b], X)$ for each $n \geq 1$.

We show that the set $\Theta_n(\phi)$ is contractible for each $n \geq 1$. In fact, let $u \in \Theta_n(\phi)$. For any $\lambda \in [0, 1]$, we consider the Cauchy problem of the form

$$\begin{cases} \dfrac{d}{dt}[v(t) - h(t, v_t)] = Av(t) + g_n(t, v_t), & t \in [\lambda b, b], \\ v(\lambda b + t) = u_{\lambda b}(t), & t \in [-\tau, 0], \end{cases} \tag{6.8}$$

where g_n is the selection of F_n. Since the functions g_n satisfy the conditions of Lemma 6.4 due to Lemma 6.2 (ii) and (v), by Lemma 6.4, we know that equation (6.8) has a unique solution for every $u(t) \in C([-\tau, \lambda b], X)$. Moreover, the solution of (6.8) depends continuously on (λ, u), denoted by $v(t, \lambda b, u)$.

Define the function $H : [0, 1] \times \Theta_n(\phi) \to \Theta_n(\phi)$ by the formula

$$H(\lambda, u) = \begin{cases} u(t), & t \in [-\tau, \lambda b], \\ v(t, \lambda b, u), & t \in (\lambda b, b]. \end{cases}$$

Clearly $H(\lambda, u) \in \Theta_n(\phi)$. In fact, for each $u \in \Theta_n(\phi)$, there exists $\tilde{g} \in Sel_{F_n}(u)$ such that $u = \Gamma_1(u) + S(\tilde{g})$. Put

$$\hat{g}(t) = \tilde{g}(t)\chi_{[0,\lambda b]}(t) + g_n(t)\chi_{[\lambda b, b]}(t) \ \text{ for each } t \in [0, b].$$

It is clear that $\hat{g} \in Sel_{F_n}(H)$. Also, it is readily checked that $\Gamma_1(H(\lambda, u)) + S(\hat{g})(t) = u(t)$ for all $t \in [-\tau, \lambda b]$ and $\Gamma_1(H(\lambda, u)) + S(\hat{g})(t) = v(t, \lambda b, u)$ for all $t \in [\lambda b, b]$, which gives $\Gamma_1(H(\lambda, u)) + S(\hat{g}) = H(\lambda, u)$ and hence $H(\lambda, u) \in \Theta_n(\phi)$.

To show that h is a continuous homotopy, let $(\lambda^m, u^m) \in [0, 1] \times \Theta_n(\phi)$ be such that $(\lambda^m, u^m) \to (\lambda, u)$ as $m \to \infty$. Then

$$H(\lambda^m, u^m) = \begin{cases} u^m, & t \in [-\tau, \lambda_m b], \\ v(t, \lambda^m b, u^m), & t \in (\lambda_m b, b]. \end{cases}$$

We shall prove that $H(\lambda^m, u^m) \to H(\lambda, u)$ as $m \to \infty$. Without loss of generality we assume that $\lambda^m \leq \lambda$. If $t \in [-\tau, \lambda^m b]$, then

$$|H(\lambda^m, u^m)(t) - H(\lambda, u)(t)| = |u^m(t) - u(t)| \to 0 \ \text{ as } m \to \infty.$$

If $t \in [\lambda b, b]$, then

$$\|H(\lambda^m, u^m) - H(\lambda, u)\|_{[\lambda b, b]} = \sup_{t \in [\lambda b, b]} |v(t, \lambda^m b, u^m) - v(t, \lambda, u)|,$$

which tends to 0 as $m \to \infty$, since $v(t, \lambda b, u)$ depends continuously on (λ, u). If $t \in [\lambda^m b, \lambda b]$, then

$$
\begin{aligned}
|H(\lambda^m, u^m)(t) - H(\lambda, u)(t)| &= |v(t, \lambda^m b, u^m) - u(t)| \\
&\le |v(t, \lambda^m b, u^m) - u^m(t)| + |u^m(t) - u(t)| \\
&\to 0, \quad \text{as } m \to \infty
\end{aligned}
$$

due to $v(t, \lambda^m b, u^m) \to u^m(t)$ $(t \to \lambda_m b)$. But $H(0, \cdot) = H(t, 0, \phi)$ and $H(1, \cdot)$ is the identity, hence $\Theta_n(\phi)$ is contractible.

Finally, in view of Lemma 6.2 (i), it is easy to verify that $\Theta(\phi) \subset \cdots \subset \Theta_n(\phi) \cdots \subset \Theta_2(\phi) \subset \Theta_1(\phi)$, this implies that $\Theta(\phi) \subset \bigcap_{n \ge 1} \Theta_n(\phi)$. To prove the reverse inclusion, we take $u \in \bigcap_{n \ge 1} \Theta_n(\phi)$. Therefore, there exists a sequence $\{g_n\} \subset L^1([0, b], \mathbb{R}^+)$ such that $g_n \in Sel_{F_n}(u)$, $u = \Gamma_1(u) + S(g_n)$ and for $n \ge 1$,

$$
|g_n(t)| \le \alpha(t)(3 + |u_t|_0) \quad \text{for a.e. } t \in [0, b]
$$

in view of Lemma 6.2 (ii). According to the reflexivity of the space X and Lemma 1.24, we have the existence of a subsequence, denoted as the sequence, such that $g_n \rightharpoonup g \in L^1([0, b], X)$. By Mazur's theorem, we obtain a sequence $\tilde{g}_n \in co\{g_k : k \ge n\}$ for $n \ge 1$ such that $\tilde{g}_n \to g$ in $L^1([0, b], X)$ and, up to subsequence, $\tilde{g}_n(t) \to g(t)$ for a.e. $t \in [0, b]$ and $g_n(t) \in F_n(t, u_t)$ for all $n \ge 1$.

Denote by $\mathcal{T}$ the set of all $t \in [0, b]$ such that $\tilde{g}_n(t) \to g(t)$ in X and $g_n(t) \in F_n(t, u_t)$ for all $n \ge 1$. According to Lemma 6.2 (iii), we have that there exists $E \subset [0, b]$ with $\text{mes}(E) = 0$ such that for each $t \in ([0, b] \setminus E) \cap \mathcal{T}$ and $x^* \in X^*$, $\varepsilon > 0$,

$$
x^*(\tilde{g}_n(t)) \in co\{x^*(g_k(t)) : k \ge n\} \subset x^*(F_n(t, u_t)) \subset x^*(F(t, u_t)) + (-\varepsilon, \varepsilon).
$$

Therefore, we obtain that $x^*(g(t)) \in x^*(F(t, u_t))$ for each $x^* \in X^*$ and $t \in ([0, b] \setminus E) \cap \mathcal{T}$. Since F has convex and closed values, we conclude that $g(t) \in F(t, u_t)$ for each $t \in ([0, b] \setminus E) \cap \mathcal{T}$, which implies that $g \in Sel_F(u)$. Moreover,

$$
\begin{aligned}
u(t) = T(t)\big(\phi(0) - h(0, \phi)\big) + h(t, u_t) + \int_0^t AT(t - s)h(s, u_s)ds \\
+ \int_0^t T(t - s)g_n(s)ds.
\end{aligned}
$$

By Lemma 6.3, we know that $\int_0^t T(t - s)g_n(s)ds \to \int_0^t T(t - s)g(s)ds$, which implies that $u = \Gamma_1(u) + S(g)$. This proves that $u \in \bigcap_{n \ge 1} \Theta_n(\phi)$. We conclude that $\Theta(\phi) = \bigcap_{n \ge 1} \Theta_n(\phi)$. Consequently, we conclude that $\Theta(\phi)$ is an R_δ-set, completing this proof.

Example 6.1 Let $X = L^2([0, \pi], \mathbb{R})$, we consider the following partial differential inclusions of neutral type:

$$
\begin{cases}
\dfrac{\partial}{\partial t}\left(z(t, \xi) - \displaystyle\int_0^\pi U(\xi, y)z_t(\theta, y)dy\right) \in \dfrac{\partial^2}{\partial \xi^2}z(t, \xi) + G(t, z_t(\theta, \xi)), \\
\qquad\qquad\qquad\qquad\qquad\qquad\qquad t \in [0, b], \ \xi \in [0, \pi], \\
z(t, 0) = z(t, \pi) = 0, \qquad\qquad\qquad\quad t \in [0, b], \\
z(\theta, \xi) = \phi(\theta)(\xi), \qquad\qquad\qquad\quad \theta \in [-\tau, 0], \ \xi \in [0, \pi],
\end{cases}
\tag{6.9}
$$

where $\phi \in C([-\tau, 0], X)$, that is, $\phi(\theta) \in X$ and $z_t(\theta, \xi) = z(t + \theta, \xi)$, $t \in [0, b]$, $\theta \in [-\tau, 0]$.

We consider the operator $A : D(A) \subset X \to X$ defined as $Av = -v''$ with the domain

$$D(A) = \{v(\cdot) \in X : v, \ v' \text{ absolutely continuous}, \ v'' \in X \text{ and } v(0) = v(\pi) = 0\}.$$

Then A generates a strongly continuous semigroup $\{T(t)\}_{t \geq 0}$, which is compact, analytic and self-adjoint. Furthermore, A has a discrete spectrum, the eigenvalues are n^2 ($n \in \mathbb{N}$), with corresponding normalized eigenvectors $u_n(\xi) = \sqrt{\frac{2}{\pi}} \sin n\xi$. This implies that $\sup_{t \geq 0} \|T(t)\|_{\mathcal{L}(X)} < \infty$ (see [167]). We also use the following properties:

(i) for each $v \in X$, $T(t)v = \sum_{n=1}^\infty e^{-n^2 t}\langle v, u_n\rangle u_n$;
(ii) for each $v \in X$, $A^{-\frac{1}{2}}v = \sum_{n=1}^\infty \frac{1}{n}\langle v, u_n\rangle u_n$;
(iii) the operator $A^{\frac{1}{2}}$ is given by

$$A^{\frac{1}{2}}v = \sum_{n=1}^\infty n\langle v, u_n\rangle u_n$$

on the space $D(A^{\frac{1}{2}}) = \{v(\cdot) \in X : \sum_{n=1}^\infty n\langle v, u_n\rangle u_n \in X\}$.

Then system (6.9) can be reformulated as

$$
\begin{cases}
\dfrac{d}{dt}\left(u(t) - h(t, u_t)\right) \in Au(t) + F(t, u_t), & t \in [0, b], \\
u(t) = \phi(t), & t \in [-\tau, 0],
\end{cases}
$$

where $u(t)(\xi) = z(t, \xi)$, $u_t(\theta, \xi) = z_t(\theta, \xi)$, $F(t, u_t)(\xi) = G(t, z_t(\theta, \xi))$. The function $h(t, u_t) : [0, b] \times C([-\tau, 0], X) \to X$ is defined by

$$h(t, u_t) = \int_0^\pi U(\xi, y)z_t(\theta, y)dy.$$

Moreover, we assume that the following conditions hold:

(h_1) the function $U(\xi, y)$ is measurable and

$$\int_0^\pi \int_0^\pi U^2(\xi, y)dyd\xi < \infty;$$

(h_2) the function $\partial_\xi U(\xi, y)$ is measurable, $U(0, y) = U(\pi, y) = 0$, and let

$$\overline{H} = \left(\int_0^\pi \int_0^\pi (\partial_\xi U(\xi, y))^2 dyd\xi \right)^{\frac{1}{2}} < \infty.$$

Clearly, (H_2) is satisfied.

Let $F(t, z_t) = [f_1(t, z_t), f_2(t, z_t)]$. Now, we assume that:

$$f_i : [0, b] \times \mathbb{R} \to \mathbb{R}, \quad i = 1, 2,$$

satisfy

(F_1) f_1 is l.s.c. and f_2 is u.s.c.;
(F_2) $f_1(t, v) \le f_2(t, v)$ for each $(t, v) \in [0, b] \times C([-\tau, 0], X)$;
(F_3) there exists $\alpha_1, \alpha_2 \in L^\infty([0, b], \mathbb{R}^+)$ such that

$$|f_i(t, v)| \le \alpha_i(t)(1 + |v|_0), \quad i = 1, 2,$$

for each $(t, v) \in [0, b] \times C([-\tau, 0], X)$.

From our assumptions on (F_1)–(F_3), it follows readily that the multivalued function $F(\cdot, \cdot) : [0, b] \times C([-\tau, 0], X) \to P(\mathbb{R})$ satisfies (H_1).

Thus, all the assumptions in Theorems 6.1 and 6.2 are satisfied, our results can be used to problem (6.9).

6.2.2 Noncompact Semigroup Case

We study the semilinear differential inclusion (6.1) under the following assumptions:
(H_2)$'$ h satisfies (H_2) with

$$|A^\beta h(t, v_1) - A^\beta h(s, v_2)| \le d|v_1 - v_2|_0 \text{ for } t, s \in [0, b]$$

instead of

$$|A^\beta h(t, v_1) - A^\beta h(t, v_2)| \le d|v_1 - v_2|_0;$$

(H_3) for every $\varepsilon > 0$ and every bounded set $D \subset C([-\tau, 0], X)$ there exists $\delta > 0$ and a function $k \in L^1([0, b], \mathbb{R}^+)$ such that

$$\gamma(F(t, B_\delta(D))) \le k(t) \sup_{-\tau \le \theta \le 0} \gamma(B_\varepsilon(D(\theta))) \text{ for a.e. } t \in [0, b],$$

where γ denotes Hausdorff MNC on X and $B_\delta(D)$ denotes a δ-neighborhood of D defined as

$$B_\delta(D) := \{z \in C([-\tau, 0], X) : \text{dist}(z, D) < \delta\}.$$

The assumption (H_3) was introduced and used in [107] and it implies the compactness of values of F.

Theorem 6.3 *Let X be reflexive and conditions (H_1), $(H_2)'$ and (H_3) be satisfied. If $\tilde{d} < 1$, then the solution set of inclusion (6.1) is a nonempty compact subset of $C([-\tau, b], X)$ for each $\phi \in C([-\tau, 0], X)$.*

Proof For the same $B_R(b)$, as the reason for Theorem 6.1, we see that $B_R(b)$ is a closed and convex subset of $C([-\tau, b], X)$.

Claim 1. The multimap $\mathscr{F}$ has a closed graph with compact values. Let $u^n \subset B_R(b)$ with $u^n \to u$ and $z^n \in \mathscr{F}(u^n)$ with $z^n \to z$. We shall prove that $z \in \mathscr{F}(u)$. By the definition of $\mathscr{F}$, there exist $f_n \in Sel_F(u^n)$ such that

$$z^n(t) = \begin{cases} \Gamma_1(u^n)(t) + S(f_n)(t), & t \in [0, b], \\ \phi(t), & t \in [-\tau, 0]. \end{cases}$$

The operator S satisfies the properties (L_1) and (L_2) of Property 1.6, since $T(t)$ is a strongly continuous operator. In view of (H_1)(ii), we have that $\{f_n\}$ is integrably bounded, and the condition (H_3) implies

$$\gamma(\{f_n(t)\}) \le \gamma(F(t, u_t^n)) \le k(t) \sup_{-\tau \le \theta \le 0} \gamma(u_t^n(\theta)) \le k(t) \sup_{-\tau \le \theta \le t} \gamma(u^n(s)) = 0.$$

Then $\{f_n\}$ is a semicompact sequence. Consequently, $\{f_n\}$ is weakly compact in $L^1([0, b], X)$, we may assume, without loss of generality, $f_n \rightharpoonup f$ in $L^1([0, b], X)$. By Property 1.6 (ii), one obtains that $S(f_n) \to S(f)$ in $C([0, b], X)$. Since Sel_F is weakly u.s.c. with weakly compact and convex values (see Lemma 6.1), from Lemma 1.7, we have that $f \in Sel_F(u)$.

On the other hand, we have the inequalities:

$$|\Gamma_1(u^n)(t) - \Gamma_1(u)(t)|$$

$$\le |h(t, u_t^n) - h(t, u_t)| + \left| \int_0^t AT(t - s)\big(h(s, u_s^n) - h(s, u_s)\big)ds \right|$$

$$\le d\|A^{-\beta}\|_{\mathcal{L}(X)}\|u^n - u\|_{[-\tau, t]} + dC_{1-\beta} \int_0^t (t - s)^{\beta-1}\|u^n - u\|_{[-\tau, s]}ds$$

$$\le d\left(\|A^{-\beta}\|_{\mathcal{L}(X)} + \frac{C_{1-\beta}b^\beta}{\beta}\right)\|u^n - u\|_{[-\tau, b]}$$

for $t \in [0, b]$. For $t \in [-\tau, 0]$, we have

$$|\Gamma_1(u^n)(t) - \Gamma_1(u)(t)| = 0.$$

Then

$$\|\Gamma_1(u^n) - \Gamma_1(u)\|_{[-\tau, b]} \leq d\left(\|A^{-\beta}\|_{\mathcal{L}(X)} + \frac{C_{1-\beta} b^\beta}{\beta}\right) \|u^n - u\|_{[-\tau, b]} \to 0 \text{ as } n \to \infty.$$

It follows immediately that $z^n \to z$ with

$$z(t) = \begin{cases} \Gamma_1(u)(t) + S(f)(t), & t \in [0, b], \\ \phi(t), & t \in [-\tau, 0], \end{cases}$$

where $f \in Sel_F(u)$ and $z \in \Gamma(u)$. Hence, $\mathscr{F}$ is closed.

It remains to show that, for $u \in B_R(b)$ and $\{f_n\}$ chosen in $Sel_F(u)$, the sequence $\{S(f_n)\}$ is relatively compact in $C([-\tau, b], X)$. Hypotheses (H_1)(ii) and (H_3) imply that $\{f_n\}$ is semicompact. Using Property 1.6 (ii), we obtain that $\{S(f_n)\}$ is relatively compact in $C([0, b], X)$. Thus $\mathscr{F}(u)$ is relatively compact in $C([-\tau, b], X)$, together with the closeness of $\mathscr{F}$, then $\mathscr{F}$ has compact values.

Claim 2. The multioperator $\mathscr{F}$ is u.s.c. In view of Lemma 1.9, it suffices to check that $\mathscr{F}$ is a quasicompact multimap. Let Q be a compact set. We prove that $\mathscr{F}(Q)$ is a relatively compact subset of $C([-\tau, b], X)$. Assume that $\{z^n\} \subset \mathscr{F}(Q)$. Then

$$z^n(t) = \begin{cases} \Gamma_1(u^n)(t) + S(f_n)(t), & t \in [0, b], \\ \phi(t), & t \in [-\tau, 0], \end{cases}$$

where $\{f_n\} \in Sel_F(u_t^n)$ for a certain sequence $\{u^n\} \subset Q$. Hypotheses (H_1)(ii) and (H_3) yield the fact that $\{f_n\}$ is semicompact and then it is a weakly compact sequence in $L^1([0, b], X)$. Similar arguments as in the previous proof of closeness imply that $\{\Gamma_1(u^n)\}$ and $\{S(f_n)\}$ are relatively compact in $C([0, b], X)$. Thus, $\{z^n\}$ converges in $C([-\tau, b], X)$, so the multioperator $\mathscr{F}$ is u.s.c.

Claim 3. The multioperator $\mathscr{F}$ is a condensing multioperator. Now in the space $C([-\tau, b], X)$ we consider the measure of noncompactness ν defined as: for a bounded subset $\Omega \subset B_R(b)$, let $\mathrm{mod}_C(\Omega)$ be the modulus of equicontinuity of the set of functions Ω given by

$$\mathrm{mod}_C(\Omega) = \lim_{\delta \to 0} \sup_{u \in \Omega} \max_{|t_2 - t_1| < \delta} |u(t_2) - u(t_1)|.$$

Given Hausdorff MNC γ, let χ be the real MNC defined on bounded set $D \subset C([-\tau, b], X)$ by

$$\chi(D) = \sup_{t \in [0, b]} e^{-Lt} \gamma(D(s)).$$

Here, the constant L is chosen such that

$$l := d\|A^{-\beta}\|_{\mathcal{L}(X)} + \sup_{t\in[0,b]} \left(c_0 \int_0^t e^{-L(t-s)}k(s)ds + dC_{1-\beta} \int_0^t e^{-L(t-s)}(t-s)^{\beta-1}ds \right)$$
$$< \frac{1}{2},$$

where $k(t)$ is the function from the condition (H_3).

Consider the function $\nu(\Omega) = \max_{D\in\Delta(\Omega)} \left(\sup_{t\in[-\tau,0]} \gamma(D(t)|_{[-\tau,0]}), \chi(D), \mathrm{mod}_C(D) \right)$ in the space of $C([-\tau, b], X)$, where $\Delta(\Omega)$ is the collection of all countable subsets of Ω.

To show that $\mathcal{F}$ is ν-condensing, let $\Omega \subset B_R(b)$ be a bounded set in $B_R(b)$ such that

$$\nu(\Omega) \leq \nu(\mathcal{F}(\Omega)). \tag{6.10}$$

We will show that Ω is relatively compact. Let $\nu(\mathcal{F}(\Omega))$ be achieved on a sequence $\{z^n\} \subset \mathcal{F}(\Omega)$, i.e.,

$$\nu(\{z^n\}) = \left(\sup_{t\in[-\tau,0]} \gamma(\{z^n(t)\}|_{[-\tau,0]}), \chi(\{z^n\}), \mathrm{mod}_C(\{z^n\}) \right).$$

Then

$$z^n(t) = \begin{cases} \Gamma_1(u^n)(t) + S(f_n)(t), & t \in [0, b], \\ \phi(t), & t \in [-\tau, 0], \end{cases}$$

where $\{u^n\} \subset \Omega$ and $f_n \in Sel_F(u^n)$. From inequality (6.10), it follows that

$$\sup_{t\in[-\tau,0]} \gamma(\{u^n(t)\}) = 0.$$

Indeed, we have

$$\gamma(\{z^n\}|_{[-\tau,0]}) = \gamma(\{\phi(t) : t \in [-\tau, 0]\}) = 0 \geq \sup_{t\in[-\tau,0]} \gamma(\{u^n(t)\}) \geq 0.$$

Now we give an upper estimate of $\gamma(\{z^n(t)\})$ for any $t \in [0, b]$. Using (H_3), we have $\gamma(\{f_n(t)\}) \leq k(t) \sup_{-\tau\leq\theta\leq0} \gamma(\{u_t^n(\theta)\})$. Then

$$\gamma(\{f_n(t)\}) \leq k(t) \left(\sup_{s\in[-\tau,0]} \gamma(\{u^n(s)\}) + \sup_{s\in[0,t]} \gamma(\{u^n(s)\}) \right)$$
$$\leq e^{Lt}k(t) \left(\sup_{s\in[0,t]} e^{-Ls}\gamma(\{u^n(s)\}) \right)$$
$$\leq e^{Lt}k(t)\chi(\{u^n\}).$$

Then, from Property 1.6 (i) with $c_0 = M$, we get

$$e^{-Lt}\gamma(\{S(f_n)(t)\}) \leq 2Me^{-Lt} \int_0^t e^{Ls} k(s)ds \cdot \chi(\{u^n\})$$
$$\leq 2M \int_0^t e^{-L(t-s)} k(s)ds \cdot \chi(\{u^n\}). \tag{6.11}$$

Since the measure γ is monotone, from $(H_2)'$, for $t \in [0, b]$ we get

$$\gamma\left(\{A^\beta h(t, u_t^n)\}\right) \leq \mu\left(\{A^\beta h(t, u_t^n)\}\right)$$
$$\leq d\left(\sup_{\theta \in [-\tau, 0]} \mu(\{u_t^n\})\right)$$
$$\leq 2d\left(\sup_{\theta \in [-\tau, 0]} \gamma(\{u_t^n\})\right),$$

where μ denotes Kuratowski measure of noncompactness. Then

$$e^{-Lt}\gamma(\{h(t, u_t^n)\}) \leq e^{-Lt}\gamma\left(\{A^{-\beta} A^\beta h(t, u_t^n)\}\right)$$
$$\leq 2d\|A^{-\beta}\|_{\mathcal{L}(X)} e^{-Lt}\left(\sup_{\theta \in [-\tau, 0]} \gamma(\{u_t^n\})\right)$$
$$\leq 2d\|A^{-\beta}\|_{\mathcal{L}(X)} e^{-Lt}\left(\sup_{s \in [0, t]} \gamma(\{u^n(s)\})\right) \tag{6.12}$$
$$\leq 2d\|A^{-\beta}\|_{\mathcal{L}(X)}\left(\sup_{s \in [0, t]} e^{-Ls}\gamma(\{u^n(s)\})\right)$$
$$\leq 2d\|A^{-\beta}\|_{\mathcal{L}(X)}\chi(\{u^n\}).$$

Let $t \in [0, b]$ and $s \in [0, t]$. Clearly, the function $G : s \mapsto AT(t-s)h(s, u_s^n)$ is integrable and integrably bounded. Since

$$\gamma(\{AT(t-s)h(s, u_s^n)\}) = \gamma\left(\{A^{1-\beta}T(t-s)A^\beta h(s, u_s^n)\}\right)$$
$$\leq \|A^{1-\beta}T(t-s)\|\gamma\left(\{A^\beta h(s, u_s^n)\}\right)$$
$$\leq 2dC_{1-\beta}(t-s)^{\beta-1}e^{Ls}\left(\sup_{s_1 \in [0, s]} e^{-Ls_1}\gamma(\{u^n(s_1)\})\right)$$
$$\leq 2dC_{1-\beta}(t-s)^{\beta-1}e^{Ls}\chi(\{u^n\}),$$

by Lemma 1.7, one obtains

$$e^{-Lt}\int_0^t \gamma(\{AT(t-s)h(s, u_s^n)\})ds \leq 2dC_{1-\beta}\chi(\{u^n\})\int_0^t e^{-L(t-s)}(t-s)^{\beta-1}ds$$
$$\leq 2dC_{1-\beta}\chi(\{u^n\})\sup_{t \in [0, b]}\int_0^t e^{-L(t-s)}(t-s)^{\beta-1}ds. \tag{6.13}$$

From (6.11)–(6.13), and the fact that $d\|A^{-\beta}\|_{\mathcal{L}(X)} < 1$, it follows

$$\chi(\{z^n\}) = \sup_{t\in[0,b]} e^{-Lt}\gamma(\{z^n(t)\})$$

$$\leq \sup_{t\in[0,b]} e^{-Lt}\gamma\left(\left\{S(f_n)(t) + h(t, u_t^n) + \int_0^t AT(t-s)h(s, u_s^n)ds\right\}\right)$$

$$\leq 2\left[\sup_{t\in[0,b]}\left(M\int_0^t e^{-L(t-s)}k(s)ds + dC_{1-\beta}\int_0^t e^{-L(t-s)}(t-s)^{\beta-1}ds\right)\right.$$

$$\left. + d\|A^{-\beta}\|_{\mathcal{L}(X)}\right]\chi(\{u^n\})$$

$$\leq 2l\chi(\{u^n\}).$$

But (6.10) implies

$$\chi(\{z^n\}) \geq \chi(\{u^n\}),$$

and consequently, $\chi(\{u^n\}) = 0$. This implies that $\gamma(\{u^n(t)\}) = 0$.

Using (H_1)(ii) and (H_3) again, one gets that $\{f_n\}$ is a semicompact sequence. Then, Property 1.6 (ii) ensures that $\{S(f_n)\}$ is relatively compact in $C([0, b], X)$. Hence, $\text{mod}_C(\{S(f_n)\}) = 0$.

Now we will show that the set $\{\Gamma_1(u^n)\}$ is equicontinuous on $C([-\tau, b], X)$. For $-\tau \leq t_1 < t_2 \leq 0$, we have

$$|\Gamma_1(u^n)(t_2) - \Gamma_1(u^n)(t_1)| = |\phi(t_2) - \phi(t_1)| \to 0 \text{ as } |t_1 - t_2| \to 0.$$

For $0 < t_1 < t_2 \leq b$, we obtain

$$|\Gamma_1(u^n)(t_2) - \Gamma_1(u^n)(t_1)|$$

$$\leq \|T(t_2) - T(t_1)\|_{\mathcal{L}(X)}|h(0, \phi)| + |h(t_2, u_{t_2}^n) - h(t_1, u_{t_1}^n)|$$

$$+ \left|\int_{t_1}^{t_2} AT(t_2 - s)h(s, u_s^n)ds\right| + \left|\int_0^{t_1} A(T(t_2 - s) - T(t_1 - s))h(s, u_s^n)ds\right|$$

$$\leq \|T(t_2) - T(t_1)\|_{\mathcal{L}(X)}|h(0, \phi)| + d\|A^{-\beta}\|_{\mathcal{L}(X)}|u_{t_2}^n - u_{t_1}^n|_0$$

$$+ \frac{d_1 C_{1-\beta}(t_2 - t_1)^\beta}{\beta}\left(1 + \|u^n\|_{[t_1-\tau,t_2]}\right)$$

$$+ \left|(T(t_2 - t_1) - I)\int_0^{t_1} AT(t_1 - s)h(s, u_s^n)ds\right|.$$

Since $\gamma\left(\int_0^t AT(t-s)h(s, u_s^n)ds\right) = 0$ for all $t \in [0, b]$, the last term on the right-hand side converges to zero when $t_2 - t_1$ tends uniformly to 0. From the inequality

$$\text{mod}_C(\{z^n\}) \leq \text{mod}_C(\{\Gamma_1(u^n)\}) + \text{mod}_C(\{S(f_n)\}),$$

we get

$$\mathrm{mod}_C(\{z^n\}) \le d\|A^{-\beta}\|_{\mathcal{L}(X)}|u^n_{t_2} - u^n_{t_1}|_0 \le d\|A^{-\beta}\|_{\mathcal{L}(X)}\mathrm{mod}_C(\{u^n\}).$$

In view of $d\|A^{-\beta}\|_{\mathcal{L}(X)} < 1$, from the last inequality and inequality (6.10) follows $\mathrm{mod}_C(\{z^n\}) = 0$, it implies that $\mathrm{mod}_C(\{u^n\}) = 0$. Hence, the subset $\{u^n\}$ is relatively compact, thus $\nu(\{u^n\}) = 0$, and so the map $\mathscr{F}$ is ν-condensing.

From Theorem 1.12, we deduce that the fixed point set $\mathrm{Fix}\mathscr{F}$ is a nonempty compact set.

Before proving the main result of this subsection, we give an important lemma to prove the contractibility of the solution set.

Lemma 6.5 *Under the conditions in Lemma 6.4 except that $\{T(t)\}_{t>0}$ is compact, We assume $(H_2)'$ instead of (H_2). Then integral equation (6.3) admits a unique solution for every $\varphi(t) \in C([-\tau, a], X)$. Moreover, the solution of (6.3) depends continuously on φ.*

Proof Let $\varphi \in C([-\tau, a], X)$ be fixed. From $d_1\|A^{-\beta}\|_{\mathcal{L}(X)} < 1$, we can find one ξ arbitrarily close to a such that

$$d_1\left(\|A^{-\beta}\|_{\mathcal{L}(X)} + \frac{C_{1-\beta}(\xi - a)^\beta}{\beta}\right) + M\|\varpi_1\|_{L([a,\xi],\mathbb{R}^+)} < 1$$

$$d\left(\|A^{-\beta}\|_{\mathcal{L}(X)} + \frac{C_{1-\beta}(\xi - a)^\beta}{\beta}\right) + M\|L_{B_\rho}\|_{L([a,\xi],\mathbb{R}^+)} < 1.$$

Then for one of such ξ, we choose one ρ satisfying

$$\rho \ge \frac{d_1\left(\|A^{-\beta}\|_{\mathcal{L}(X)} + \frac{C_{1-\beta}(\xi-a)^\beta}{\beta}\right) + M(\|\varphi\| + d_1(1 + \|\varphi\|)) + Mc'\|\varpi_1\|_{L([a,\xi],\mathbb{R}^+)}}{1 - d_1\left(\|A^{-\beta}\|_{\mathcal{L}(X)} + \frac{C_{1-\beta}(\xi-a)^\beta}{\beta}\right) - M\|\varpi_1\|_{L([a,\xi],\mathbb{R}^+)}},$$

For the same $B_\rho(\xi)$ and the operator W as defined in Lemma 6.4, it is easy to know that W maps $B_\rho(\xi)$ into itself.

For any $u, v \in B_\rho(\xi)$ and $t \in [a, b]$, we have

$$|Wu(t) - Wv(t)|$$

$$\le \left|A^{-\beta}A^\beta[h(t, u_t) - h(t, v_t)]\right| + \left|\int_a^t A^{1-\beta}T(t - s)A^\beta[h(s, u_s) - h(s, v_s)]ds\right|$$

$$+ \left|\int_a^t T(t - s)[g(s, u_s) - g(s, v_s)]ds\right|$$

$$\le d\|A^{-\beta}\|_{\mathcal{L}(X)}\|u_t - v_t\|_* + dC_{1-\beta}\int_a^t (t - s)^{\beta-1}\|u_s - v_s\|_*ds$$

$$+ M \int_a^t L_{B_\rho}(s) \|u_s - v_s\|_* ds$$

$$\leq d\|A^{-\beta}\|_{\mathcal{L}(X)} \|u - v\|_{[a-\tau,t]} + dC_{1-\beta} \int_a^t (t-s)^{\beta-1} \|u - v\|_{[a-\tau,s]} ds$$

$$+ M \int_a^t L_{B_\rho}(s) \|u - v\|_{[a-\tau,s]} ds$$

$$\leq \left(d\|A^{-\beta}\|_{\mathcal{L}(X)} + \frac{dC_{1-\beta}(\xi - a)^\beta}{\beta} + M\|L_{B_\rho}\|_{L([a,\xi],\mathbb{R}^+)} \right) \|u - v\|_{[a-\tau,\xi]}.$$

Noting that $Wu(t) - Wv(t) = 0$ for $t \in [-\tau, a]$, which implies that

$$\|Wu - Wv\|_{[a-\tau,\xi]}$$

$$\leq \left(d\|A^{-\beta}\|_{\mathcal{L}(X)} + \frac{dC_{1-\beta}(\xi - a)^\beta}{\beta} + M\|L_{B_\rho}\|_{L([a,\xi],\mathbb{R}^+)} \right) \|u - v\|_{[a-\tau,\xi]},$$

we get that W is a strict contraction. Hence, Banach's fixed point theorem shows that W has a unique fixed point, which implies that Eq. (6.3) has a local solution on $[-\tau, \xi]$. According to Lemma 6.4, we also obtain that Eq. (6.3) has a solution on $[-\tau, b]$. Therefore, the first part of this lemma is proved.

We only prove that the solution of equation (6.3) depends continuously on φ. Let $\varphi^n \to \varphi^0$ in $C([-\tau, b], X)$ as $n \to \infty$, and u^n be the solution of equation (6.3) with the perturbation φ^n, i.e.,

$$u^n(t) = T(t - a)(\varphi^n(a) - h(a, \varphi^n)) + h(t, u_t^n) + \int_a^t AT(t - s)h(s, u_s^n) ds$$

$$+ \int_a^t T(t - s)g(s, u_s^n) ds$$

for $t \in [a, b]$ and $u^n(t) = \varphi^n(t)$ for $t \in [-\tau, a]$. We prove that $\{u^n\}$ is a convergent sequence in $C([-\tau, b], X)$. It is clear that $\{u^n\}$ is a convergent sequence in $C([-\tau, a], X)$. we only consider $\{u^n\}$ in $C([a, b], X)$.

By Lemma 6.4, we obtain the bound of $\{u^n\}$. Thus for $t \in [a, b]$,

$$|u^n(t) - u^m(t)|$$

$$\leq M|\varphi^n(a) - \varphi^m(a) + h(a, \varphi^m) - h(a, \varphi^n)| + |h(t, u_t^n) - h(t, u_t^m)|$$

$$+ \left| \int_a^t AT(t - s)[h(s, u_s^n) - h(s, u_s^m)] ds \right| + \left| \int_a^t T(t - s)[g(s, u_s^n) - g(s, u_s^m)] ds \right|$$

$$\leq M(1 + d\|A^{-\beta}\|_{\mathcal{L}(X)}) \|\varphi^n - \varphi^m\| + d\|A^{-\beta}\|_{\mathcal{L}(X)} \|u_t^n - u_t^m\|_*$$

$$+ dC_{1-\beta} \int_a^t (t - s)^{\beta-1} \|u_s^n - u_s^m\|_* ds + M \int_a^t L_{M_1}(s) \|u_s^n - u_s^m\|_* ds$$

$$\leq \|\varphi^n - \varphi^m\| \left(M + (1 + M)d\|A^{-\beta}\|_{\mathcal{L}(X)} + \frac{dC_{1-\beta}(b - a)^\beta}{\beta} + M\|L_{M_1}\|_{L([a,b],\mathbb{R}^+)} \right)$$

$$+ d\|A^{-\beta}\|_{\mathcal{L}(X)}\|u^n - u^m\|_{[a,t]} + \int_a^t [dC_{1-\beta}(t-s)^{\beta-1} + ML_{M_1}(s)]\|u^n - u^m\|_{[a,s]}ds.$$

Let

$$M_2 = \frac{M + (1+M)d\|A^{-\beta}\|_{\mathcal{L}(X)} + \frac{dC_{1-\beta}(b-a)^\beta}{\beta} + M\|L_{M_1}\|_{L([a,b],\mathbb{R}^+)}}{1 - d\|A^{-\beta}\|_{\mathcal{L}(X)}}.$$

It follows that

$$\|u^n - u^m\|_{[a,t]} \leq M_2\|\varphi^n - \varphi^m\|$$
$$+ \frac{1}{1 - d\|A^{-\beta}\|_{\mathcal{L}(X)}} \int_a^t [dC_{1-\beta}(t-s)^{\beta-1}$$
$$+ ML_{M_1}(s)]\|u^n - u^m\|_{[a,s]}ds.$$

Applying Gronwall's inequality, we get

$$\|u^n - u^m\|_{[a,t]}$$
$$\leq M_2\|\varphi^n - \varphi^m\| \exp\left[\frac{1}{1 - d\|A^{-\beta}\|_{\mathcal{L}(X)}}\left(\frac{dC_{1-\beta}(t-a)^\beta}{\beta} + M\|L_{M_1}\|_{L([a,t],\mathbb{R}^+)}\right)\right]$$
$$\to 0 \quad \text{as } n, m \to \infty,$$

which implies that $\{u^n\}$ is a Cauchy sequence in $C([a, b], X)$. Thus $\{u^n\}$ is relatively compact in $C([a, b], X)$. Therefore, taking the limit in (6.5) as $n \to \infty$, one finds, again by $(H_2)'$ and Lebesgue's dominated convergence theorem, that u is the solution of equation (6.3) with the perturbation φ^0. The proof is completed.

Theorem 6.4 *Under the conditions in Theorem 6.3, the solution set of (6.1) is an R_δ-set.*

Proof We also consider inclusion (6.7), where the multivalued functions F_n : $[0, b] \times C([-\tau, 0], X) \to P_{cl,cv}(X)$ are established in view of Lemma 6.2, and F_n satisfy the condition (H_1) for each $n \geq 1$.

Let $\Theta_n(\phi)$ denote the set of all mild solutions of inclusion (6.7). Clearly, each set $\Theta_n(\phi)$ is nonempty in $C([-\tau, b], X)$ for each $n \geq 1$.

We show that each sequence $\{u^n\}$ such that $u^n \in \Theta_n(\phi)$ for all $n \geq 1$ has a convergent subsequence $u^{n_k} \to u \in \Theta(\phi)$. At first we notice

$$u^n(t) = T(t)(\phi(0) - h(0, \phi)) + h(t, u_t^n) + \int_0^t AT(t-s)h(s, u_s^n)ds$$
$$+ \int_0^t T(t-s)g_n(s)ds, \quad g_n(s) \in F_n(s, u_s^n)$$

for $t \in [0, b]$, and $u^n(t) = \phi(t)$ for $t \in [-\tau, 0]$. It is obvious that $\gamma(\{u^n(t)\}) = 0$ for $t \in [-\tau, 0]$. By (H_3), for any $\varepsilon > 0$ there exists δ such that

$$\gamma(\{F_{B_\delta}(s, u^n_s)\}_{n \geq 1}) \leq k(s)\left(\sup_{-\tau \leq \theta \leq 0} \gamma(\{u^n_s\}_{n \geq 1}) + \varepsilon \right).$$

Take some $N \in \mathbb{N}$ with $3^{1-N} < \delta$. Then

$$\begin{aligned}
\gamma(\{g_n(s)\}_{n \geq 1}) = \gamma(\{g_n(s)\}_{n \geq N}) &\leq \gamma(\{F_N(s, u^n_s)\}_{n \geq N}) \\
&\leq \gamma\left(\overline{co}F(t, B_{3^{1-N}}(\{u^n_s\}_{n \geq N}))\right) \\
&\leq k(s)\left(\sup_{-\tau \leq \theta \leq 0} \gamma(\{u^n_s\}_{n \geq N}) + \varepsilon \right) \qquad (6.14) \\
&\leq k(s)\left(\sup_{\theta \in [0,s]} \gamma(\{u^n(\theta)\}_{n \geq 1}) + \varepsilon \right).
\end{aligned}$$

Therefore,

$$\gamma(\{g_n(s)\}_{n \geq 1}) \leq k(s) \sup_{\theta \in [0,s]} \gamma(\{u^n(\theta)\}_{n \geq 1}).$$

This, together with (6.12) and (6.13), implies

$$\begin{aligned}
\chi(\{u^n\}) = \sup_{t \in [0,b]} e^{-Lt}\gamma(\{u^n(t)\}) \\
\leq \sup_{t \in [0,b]} e^{-Lt}\gamma(\{h(t, u^n_t)\}) + \sup_{t \in [0,b]} e^{-Lt}\gamma\left(\left\{ \int_0^t AT(t-s)h(s, u^n_s)ds \right\}\right) \\
+ \sup_{t \in [0,b]} e^{-Lt}\gamma\left(\left\{ \int_0^t T(t-s)g_n(s)ds \right\}\right) \\
\leq 2\left[d\|A^{-\beta}\|_{\mathcal{L}(X)} + \sup_{t \in [0,b]} \left(2M \int_0^t e^{-L(t-s)}k(s)ds \right.\right. \\
\left.\left. + dC_{1-\beta} \int_0^t e^{-L(t-s)}(t-s)^{\beta-1}ds \right) \right]\chi(\{u^n\}) \\
\leq 2l\chi(\{u^n\}) \\
< \chi(\{u^n\}).
\end{aligned}$$

Thus $\chi(\{u^n\}) = 0$, then $\gamma(\{u^n(t)\}) = 0$ for $t \in [0, b]$. From (6.14), we get that $\gamma(\{g_n(s)\}) = 0$. This, together with the equicontinuity of $\{u^n\}$ is proved in Theorem 6.3, implies the existence of a subsequence $\{u^{n_k}\}$ which is convergent on $[-\tau, b]$. Denote the limit by u.

Since $\gamma(\{g_n(s)\}) = 0$, we can assume, up to subsequence, that $g_n(s) \to g(s)$ in X for $s \in [0, t]$. Together with the above discussion, we have

$$u(t) = T(t)\big(\phi(0) - h(0, \phi)\big) + h(t, u_t) + \int_0^t AT(t - s)h(s, u_s)ds$$

$$+ \int_0^t T(t - s)g(s)ds,$$

for $t \in [0, b]$ and $u(t) = \phi(t)$ for $t \in [-\tau, 0]$. As the reason for Theorem 6.2, the fact that F_n satisfy the condition (H_1) shows that $g(t) \in F(t, u_t)$ for a.e. $t \in [0, b]$.

It follows that $\sup\{\text{dist}(u, \Theta(\phi)) : u \in \Theta_n(\phi)\} \to 0$ (an easy proof by contradiction). Therefore $\sup\{\text{dist}(u, \Theta(\phi)) : u \in \overline{\Theta_n(\phi)}\} \to 0$ as well. Since $\Theta(\phi)$ is compact and $\Theta_{n+1}(\phi) \subset \Theta_n(\phi)$, $\gamma(\Theta_n(\phi)) = \gamma(\overline{\Theta_n(\phi)}) \to 0$ as $n \to \infty$ and $\Theta(\phi) = \bigcap_{n=1}^{\infty} \overline{\Theta_n(\phi)}$.

By the same methods as in Theorem 6.1, together with Lemma 6.5, we know that $\Theta_n(\phi)$ is contractible for all $n \geq 1$. Consequently, we conclude that $\Theta(\phi)$ is an R_δ-set. The proof is completed.

Chapter 7
Impulsive Evolution Inclusions

Abstract In this chapter, the existence of mild solutions for impulsive differential inclusions in a reflexive Banach space is obtained. Weakly compact valued nonlinear terms are considered, combined with strongly continuous evolution operators generated by the linear part. A continuation principle or a fixed point theorem are used, according to the various regularity and growth conditions assumed. Secondly, a topological structure of the set of solutions to impulsive functional differential inclusions on the half-line is investigated. It is shown that the solution set is nonempty, compact and, moreover, an R_δ-set. It is proved on compact intervals and then, using the inverse limit method, obtained on the half-line.

7.1 Introduction

For the first time differential equations with impulses were investigated by Milman and Myshkis [148]. Impulsive differential equations and inclusions have applications in biology, economics, medicine, physics and other fields. These problems often describe phenomenas in which states are changing rapidly. One of the example is the motion of an elastic ball bouncing vertically on a surface. The moments of the impulses are in the time when the ball touches the surface and rapidly its velocity is changed. The moments of the impulse effects for the impulsive problems can be chosen in various ways: randomly, fixed beforehand, determined by the state of a system. For some recent works on impulsive differential problems, concerning the aspects we deal with, we refer to [37, 57, 87, 88, 156].

In Sect. 7.2, we consider the impulsive evolution inclusions in a reflexive Banach space of the form

$$
\begin{cases}
u'(t) \in A(t)u(t) + F(t, u(t)), & \text{a.e. } t \in J = [0, b], \ t \neq t_k, \ k = 1, 2, \ldots, N, \\
u(0) = u_0, \\
u(t_k^+) = u(t_k) + I_k(u(t_k)), & k = 1, 2, \ldots, N,
\end{cases}
$$

$$(7.1)$$

© Springer Nature Singapore Pte Ltd. 2017
Y. Zhou et al., *Topological Structure of the Solution Set for Evolution Inclusions*,
Developments in Mathematics 51, https://doi.org/10.1007/978-981-10-6656-6_7

where $\{A(t)\}_{t\in[0,b]}$ is in X generating an evolution operator $U : \Delta \to \mathcal{L}(X)$, where $\Delta = \{(t, s) \in J \times J : 0 \le s \le t\}$ and $\mathcal{L}(X)$ is the space of all bounded linear operators in X; u_0 is an element of the Banach space X and $u(t^+) = \lim_{s \to t^+} u(s)$ $0 < t_0 < t_1 < t_2 < \cdots < t_N < b$, $F : J \times X \to P(X)$ is a multivalued map, and I_k are some operators.

Two different sets of regularity and growth assumptions on F are assumed, which cause the use of different techniques for studying (7.1). Firstly we treat the case when the evolution operator $U(t, s)$ is compact for $t > s$ and the nonlinearity F has a superlinear growth, we make use of a classical continuation principle for compact multivalued fields. Secondly we allow $U(t, s)$ to be non-compact and are not posing any conditions on the multivalued nonlinearity expressed in terms of measures of noncompactness. We use the regularity of the nonlinear part with respect to the weak topology.

In Sect. 7.3, we study a topological structure of the solution set to the impulsive Cauchy problem governed by a semilinear differential inclusion on noncompact intervals.

For a fixed $\tau > 0$ and a given piecewise continuous function $x : [-\tau, 0] \to X$, where X is a Banach space, the problem we deal with is

$$\begin{cases} u'(t) \in A(t)u(t) + F(t, u_t), & \text{for a.a. } t \in [0, \infty), \ t \ne t_k, \ k \in \mathbb{N}, \\ u(t) = x(t), & \text{for a.a. } t \in [-\tau, 0], \\ u(t_k^+) = u(t_k) + I_k(u_{t_k}), & \text{for } k \in \mathbb{N}^+, \end{cases} \qquad (7.2)$$

where F is an upper-Carathéodory map; $u_t(\theta) = u(t + \theta)$, $\theta \in [-\tau, 0]$; $k \in \mathbb{N}$, and the time sequence $\{t_k\}_{k\in\mathbb{N}}$ is an increasing sequence of given points in $[0, \infty)$ without accumulation points. Hence $u_t(\cdot)$ represents the history of the state from $t - \tau$ to the present time t.

The solution sets for differential problems often correspond with fixed point sets of multivalued operators in function spaces. In this section we use the inverse system method, which, in studying the topological structure of fixed point sets of operators in function spaces, was initiated in [104]. This method was developed in [12] and then also in [14]. It is observed that differential problems on noncompact intervals can be reformulated as fixed point problems in Fréchet (function) spaces which are inverse limits of Banach spaces that appear when we consider these differential problems on compact intervals. Some interesting properties of fixed point sets of limit maps become very useful.

The existence of mild solutions for problem (7.2) has been obtained in [37]. We state and prove the compactness of the solution set for this problem. Next we prove that the set of solutions to problem (7.2) is an R_δ-set.

This gives an important information from the topological point of view. The translation operator along trajectories which is often used to detect, for instance, periodic solutions, being R_δ-valued can be checked to be an admissible (in the sense of Górniewicz) multivalued operator, and the fixed point methods can be applied.

The chapter is organized as follows. In Subsect. 7.2.1 we obtain that the solution set for problem (7.1) is nonempty and compact when the family $\{A(t)\}_{t \in J}$ generates a compact evolution operator. Subsect. 7.2.2 is devoted to the weak compactness of the solution set for problem (7.1) when we drop the condition that the family $\{A(t)\}_{t \in J}$ generates a compact evolution operator and F is locally compact. In Sect. 7.3.1 we obtain the compactness of solution sets on compact intervals, then the R_δ-structure of the solution sets for the impulsive problem on compact intervals is shown. Note that in [87] it was shown that the solution set for the impulsive problem on compact intervals is an R_δ-set if F is a σ-Ca-selectionable multivalued map and $A(t) = A$ is the infinitesimal generator of a C_0-semigroup. The problem is that it is not clear if a sufficiently good σ-Ca-selectionability is possible in infinite dimensional spaces. In fact, as we show in the proof of Theorem 7.6, we can approximate the right-hand side of the inclusion by maps which have noncompact values and which are not k-set contractions. Therefore, we propose different arguments to avoid the obstacles and to prove the R_δ-structure on compact intervals. In Sect. 7.3.2, the compactness of the solution set is proved on the half-line. The result improves Theorem 4.2 in [37], where only the existence of solutions was shown. Our proof is essentially shorter and it shows how one can effectively use the inverse limit method. Finally we combine the information on a topological structure of solution sets on compact intervals with the inverse limit method to obtain an R_δ-structure on the half-line in Theorem 7.8. In this way we essentially develop some recent results in [88], where an R_δ-structure of the solution set for the multivalued impulsive differential inclusion on the half-line is shown only in a finite dimensional case, where the compactness properties become much easier, and for the problem without any retard.

The results in Sect. 7.2 are the extensions of Benedetti et al. [35], and the results in Sect. 7.3 are taken from Benedetti and Rubbioni [37], Gabor and Grudzka [106, 107].

7.2 Existence and Weak Compactness

In this section we study the solution set of problem (7.1). Let $(X, |\cdot|)$ be a reflexive Banach space. We denote with B the closed unit ball of X centered at 0. Given the measure space (S, Σ, μ) and the Banach space X, we denote with $\|\cdot\|_{L^p(S,X)}$ the norm of the Lebesgue space $L^p(S, X)$.

We denote by $PC([0, b], X)$ the space of piecewise continuous functions $c : [0, b] \to X$ with finite number of discontinuity points $\{t_*\}$ and such that

$$c(t_*^+) = \lim_{h \to 0^+} c(t_* + h) \quad \text{and} \quad \lim_{h \to 0^-} c(t_* + h) = c(t_*)$$

are finite. The space $PC([0, b], X)$ is a Banach space with the norm

$$\|c\|_{PC} = \sup\{|c(t)| : t \in [0, b]\}$$

and a space of continuous functions $C([0, b], X)$ is a closed subspace of it.

Let $\Delta = \{(t, s) \in J \times J : 0 \le s \le t\}$ and the evolution system $\{U(t, s)\}_{(t,s) \in \Delta}$. Since the evolution operator U is strongly continuous on the compact set, by the uniform boundedness theorem there exists a constant $\widetilde{M} = \widetilde{M}_\Delta > 0$ such that

$$\|U(t, s)\|_{\mathcal{L}(X)} \le \widetilde{M}, \quad (t, s) \in \Delta. \tag{7.3}$$

We study the compactness of the solution set of problem (7.1) under the following assumptions:

(H_1) $F(\cdot, u) : J \to P(X)$ has a measurable selection for any $u \in X$ and $F(t, u)$ is nonempty, convex and weakly compact for any $t \in J$ and $u \in X$;

(H_2) $F(t, \cdot) : X \to P(X_w)$ is u.s.c. for a.a. $t \in J$;

$(H_2)'$ $F(t, \cdot) : X \to P(X)$ is weakly sequentially closed for a.a. $t \in J$, i.e., it has a weakly sequentially closed graph;

(H_3) $\sup_{u \in \Omega} |F(t, u)| \le \eta_\Omega(t)$ for a.a. $t \in J$, with $\Omega \subset X$ bounded and $\eta_\Omega \in L^1(J, \mathbb{R})$;

$(H_3)'$ $|F(t, u)| \le \alpha(t)(1 + |u|)$ for a.a. $t \in J$, every $u \in X$ and some $\alpha \in L^1(J, \mathbb{R})$;

(H_4) let $I_k : X \to X$, $k = 1, 2, ..., N$ be a weakly continuous operator and there exist constants $l_k > 0$ such that

$$|I_k(u)| \le l_k |u| \quad \text{and} \quad \sum_{k=1}^{N} l_k < \frac{1}{\widetilde{M}}. \tag{7.4}$$

$(H_4)'$ let $I_k : X \to X$, $k = 1, 2, \ldots, N$ be a weakly continuous operator and there exist non-decreasing functions $L_k : \mathbb{R}^+ \to \mathbb{R}^+$, $k = 1, 2, \ldots, N$, such that

$$|I_k(u)| \le L_k(|u|) \quad \text{and} \quad \liminf_{m \to +\infty} \frac{L_k(m)}{m} = 0, \quad k = 1, 2, \ldots, N.$$

Given $u \in PC(J, X)$, allow us to represent

$$Sel_F(u) = \{f \in L^1(J, X) : f(t) \in F(t, u(t)) \text{ for a.a. } t \in J\}.$$

Lemma 7.1 ([35]) *Assume that the multimap F satisfies conditions (H_1), (H_2) and (H_3). Then the set $Sel_F(u)$ is nonempty for any $u \in PC(J, X)$.*

Proof Let $u \in PC(J, X)$; by the uniform continuity of u there exists a sequence $\{u_m\}$ of step functions, $u_m : [0, b] \to X$ such that

$$\sup_{t \in [0,b]} |u_m(t) - u(t)| \to 0 \text{ for } n \to \infty. \tag{7.5}$$

Hence, by (H_1), there exists a sequence of functions $\{f_m\}$ such that $f_m(t) \in F(t, u_m(t))$ for a.a. $t \in J$ and $f_m : J \to X$ is measurable for any $m \in \mathbb{N}$. From (7.5) there exists a bounded set $E \subset X$ such that $u_m(t), u(t) \in E$ for any $t \in [0, b]$ and $m \in \mathbb{N}$ and by (H_3) there exists $\eta_E \in L^1(J, \mathbb{R})$ such that

$$|f_m(t)| \leq |F(t, u_m(t))| \leq \eta_E(t), \quad \forall\, m \in \mathbb{N}, \text{ and a.a. } t \in J.$$

Hence $\{f_m\} \subset L^1(J, X)$, $\{f_m\}$ is bounded and uniformly integrable and $\{f_m(t)\}$ is bounded in X for a.a. $t \in J$. According to the reflexivity of the space X and by Dunford-Pettis theorem, we have the existence of a subsequence, denoted as the sequence, such that

$$f_m \rightharpoonup f \in L^1(J, X).$$

By Mazur's theorem we obtain a sequence

$$\widetilde{f}_m = \sum_{i=0}^{k_m} \lambda_{m,i} f_{m+i}, \quad \lambda_{m,i} \geq 0, \quad \sum_{i=0}^{k_m} \lambda_{m,i} = 1$$

such that $\widetilde{f}_m \to f$ in $L^1(J, X)$ and, up to subsequence, $\widetilde{f}_m(t) \to f(t)$ for all $t \in J$.

To conclude we have only to prove that $f(t) \in F(t, u(t))$ for a.a. $t \in J$. Indeed, let N_0 with Lebesgue measure zero be such that $F(t, \cdot) : X \to P(X_w)$ is u.s.c., $f_m(t) \in F(t, u_m(t))$ and $\widetilde{f}_m(t) \to f(t)$ for all $t \in J \setminus N_0$ and $m \in \mathbb{N}$.

Fix $t_0 \notin N_0$ and assume that $f(t_0) \notin F(t_0, u(t_0))$. Since $F(t_0, u(t_0))$ is closed and convex, from Hahn-Banach theorem there is a weakly open convex set $V \supset F(t_0, u(t_0))$ satisfying $f(t_0) \notin \overline{V}$. Since $F(t_0, \cdot) : X \to P(X_w)$ is u.s.c., we can find a neighbourhood U of $u(t_0)$ such that $F(t_0, u) \subset V$ for all $u \in U$. The convergence $u_m(t_0) \to u(t_0)$ as $m \to \infty$ then implies the existence of $m_0 \in \mathbb{N}$ such that $u_m(t_0) \in U$ for all $m > m_0$. Therefore $f_m(t_0) \in F(t_0, u_m(t_0)) \subset V$ for all $m > m_0$. Since V is convex we also have that $\widetilde{f}_m(t_0) \in V$ for all $m > m_0$ and, by the convergence, we arrive at the contradictory conclusion that $f(t_0) \in \overline{V}$. We obtain that $f(t) \in F(t, u(t))$ for a.a. $t \in J$.

Definition 7.1 A function $u(\cdot) \in PC(J, X)$ is called a mild solution of problem (7.1) if there exists $f \in L^1(J, X)$, $f(t) \in F(t, u(t))$ for a.a. $t \in J$ such that

(a) $u(t) = U(t, 0)u_0 + \int_0^t U(t, s) f(s)ds + \sum_{0 < t_k < t} U(t, t_k) I_k(u(t_k)), \quad t \in J$;
(b) $u(0) = u_0$;
(c) $u(t_k^+) = u(t_k) + I_k(u(t_k)), \quad k = 1, ..., N$.

We assume here the existence of $R > |u_0|$ satisfying

$$\widetilde{M}\left(|u_0| + \|\eta_{RB \setminus |u_0|B}\|_{L(J, \mathbb{R})} + \sum_{k=1}^{N} l_k R\right) \leq R, \tag{7.6}$$

where η appears in (H_3) and l_k satisfies (7.4).

7.2.1   Compact Operator Case

In this subsection we assume that the family $\{A(t)\}_{t \in J}$ generates a compact evolution operator and that the nonlinear term F satisfies the regularity condition (H_2) and, when not explicitly mentioned, the growth condition (H_3).

With a specific end goal to show the fundamental consequence of this subsection, we have to examine the consequent assistant statement. Presently, we delineate the solution multioperator $\Upsilon : PC(J, X) \times [0, 1] \rightarrow P(PC(J, X))$ as

$$
\Upsilon(u, \lambda) = \left\{
\begin{array}{l}
v \in PC(J, X) : v(t) = U(t, 0)u_0 + \displaystyle\int_0^t U(t, s) f(s) ds \\
\qquad\qquad + \displaystyle\sum_{0 < t_k < t} U(t, t_k) I_k(u(t_k)), \quad f \in Sel_F(u)
\end{array}
\right\},
$$

(7.7)

which is well-defined according to Lemma 7.1 and we investigate its regularity properties.

Notice that the fixed points of $\Upsilon(u, 1)$ are mild solutions of problem (7.1).

Lemma 7.2 *The multioperator Υ has a closed graph.*

Proof Since $PC(J, X)$ is a metric space, it is sufficient to prove the sequential closure of the graph. Let $\{u_m\}, \{v_m\} \subset PC(J, X)$ and $\{\lambda_m\} \subset [0, 1]$ satisfying $v_m \in \Upsilon_m(u_m, \lambda_m)$ for all m and $u_m \rightarrow u$, $v_m \rightarrow v$ in $PC(J, X)$, $\lambda_m \rightarrow \lambda$ in $[0, 1]$. We prove that $v \in \Upsilon(u, \lambda)$.

The fact that $v_m \in \Upsilon_m(u_m, \lambda_m)$ means that there exists a sequence $\{f_m\}$, $f_m \in Sel_F(u_m)$, such that

$$
\begin{aligned}
v_m(t) = U(t, 0)u_0 + \lambda_m \int_0^t U(t, s) f_m(s) ds \\
+ \lambda_m \sum_{0 < t_k < t} U(t, t_k) I_k(u_m(t_k)), \quad \forall t \in J.
\end{aligned}
$$

(7.8)

Let $\Omega \subset X$ be such that $u_m(t), u(t) \in \Omega$ for all $t \in J$ and $m \in \mathbb{N}$. Since $u_m \rightarrow u$ in $PC(J, X)$, it follows that Ω is bounded and according to (H_3) there is $\eta_\Omega \in L^1(J, \mathbb{R})$ satisfying $|f_m(t)| \leq \eta_\Omega(t)$ for a.a. $t \in J$ and every $m \in \mathbb{N}$, implying that $\{f_m\}$ is bounded and uniformly integrable in $L^1(J, X)$ and $\{f_m(t)\}$ is bounded in X for a.a. $t \in J$. Hence, by the reflexivity of the space X and by Dunford–Pettis theorem, we have the existence of a subsequence, denoted as the sequence, and a function g such that $f_m \rightharpoonup g$ in $L^1(J, X)$. It is also easy to show that $U(t, \cdot) f_m \rightharpoonup U(t, \cdot) g$ in $L^1([0, t], X)$ for all $t \in J$. Since $\lambda_m \rightarrow \lambda$, we obtain that

$$v_m(t) \rightharpoonup v^*(t) := U(t,0)u_0 + \lambda \int_0^t U(t,s)g(s)ds + \lambda \sum_{0 < t_k < t} U(t,t_k)I_k(u(t_k))$$

$$(7.9)$$

for all $t \in J$. By the uniqueness of the weak limit in X, we get that $v^*(t) = v(t)$ for all $t \in J$. Finally, reasoning as in the second part of the proof of Lemma 7.1 it is possible to show that $g(t) \in F(t,u(t))$ for a.a. $t \in J$.

Lemma 7.3 $\Upsilon(D \times [0,1])$ is relatively compact, for every bounded $D \subset PC(J,X)$.

Proof Let $D \subset PC(J,X)$ be bounded. Since $PC(J,X)$ is a metric space it is sufficient to prove the relative sequential compactness of $\Upsilon(D \times [0,1])$. Consider $\{u_m\} \subset D$, $\{v_m\} \subset PC(J,X)$ and $\{\lambda_m\} \subset [0,1]$ satisfying $v_m \in \Upsilon_m(u_m,\lambda_m)$ for all $m \in \mathbb{N}$. By the definition of the multioperator Υ, there exists a sequence $\{f_m\}$, $f_m \in Sel_F(u_m)$, such that v_m satisfies (7.8). Let $\Omega \subset X$ be such that $u_m(t) \in \Omega$ for all $t \in J$ and $m \in \mathbb{N}$. Since D is bounded, we have that Ω is bounded too and according to (H_3) there exists $\eta_\Omega \in L^1(J,\mathbb{R})$ such that $|f_m(t)| \leq \eta_\Omega(t)$ for a.a. $t \in J$ and all $m \in \mathbb{N}$.

According to the compactness of the evolution operator $U(t,s)$, the sequence $\{U(t,\cdot)f_m\}$ is semicompact in $[0,t]$ for every fixed $t \in (0,b)$.

Since the operator $S : L^1([0,t],X) \to PC([0,t],X)$ defined by

$$Sf(\tau) = \int_0^\tau U(t,s)f(s)ds \text{ for } \tau \in [0,t]$$

satisfies the conditions of Property 1.6, we obtain that the sequence

$$\tau \mapsto \int_0^\tau U(t,s)f_m(s)ds, \quad \tau \in [0,t], \quad m \in \mathbb{N}$$

is relatively compact in $PC([0,t],X)$; in particular $\left\{ \int_0^t U(t,s)f_m(s)ds \right\}$ is a relatively compact set in X for all $t \in J$.

Now consider $0 < t_0 < t \leq b$. For every $\sigma \in (0,t_0)$ we have that

$$\left| \int_0^t U(t,s)f_m(s)ds - \int_0^{t_0} U(t_0,s)f_m(s)ds \right|$$

$$\leq \left| \int_0^{t_0-\sigma} \left(U(t,s) - U(t_0,s) \right) f_m(s)ds \right| + \left| \int_{t_0-\sigma}^{t_0} \left(U(t,s) - U(t_0,s) \right) f_m(s)ds \right|$$

$$+ \left| \int_{t_0}^t U(t,s)f_m(s)ds \right| + \left| \left(U(t,s) - U(t_0,s) \right) I_k(u_m(t_k)) \right|.$$

$$(7.10)$$

Since it is known that $t \mapsto U(t,s)$ is continuous in the operator norm topology, uniformly with respect to s such that $t - s$ is bounded away from zero, for each

$\varepsilon > 0$ there is $\delta \in (0, t_0)$ satisfying

$$\left| \int_0^{t_0-\sigma} \big(U(t, s) - U(t_0, s)\big) f_m(s)ds \right| \leq \varepsilon \int_0^{t_0-\sigma} \eta_\Omega(s)ds,$$

whenever $t - t_0 < \delta$; hence, according to (7.10), we obtain that

$$\left| \int_0^t U(t, s) f_m(s)ds - \int_0^{t_0} U(t_0, s) f_m(s)ds \right| \leq \varepsilon \int_0^{t_0-\sigma} \eta_\Omega(s)ds$$

$$+ 2\widetilde{M} \int_{t_0-\sigma}^{t_0} \eta_\Omega(s)ds + \varepsilon \sum_{k=1}^N l_k |u_m(t_k)|.$$

According to the absolute continuity of the integral function, it implies that the sequence $\left\{ \int_0^t U(t, s) f_m(s)ds \right\}$ is equicontinuous in J. Consequently, passing to a subsequence, denoted as the sequence, such that $\lambda_m \to \lambda \in [0, 1]$ and using Arzela–Ascoli's theorem, we obtain that $\{v_m\}$ is relatively compact in $PC(J, X)$ and the proof is completed.

Lemma 7.4 *The multioperator Υ has convex and compact values.*

Proof Fix $u \in PC(J, X)$ and $\lambda \in [0, 1]$, since F is convex valued, the set Υ is convex from the linearity of the integral and of the operator $U(t, s)$ for all $(t, s) \in \Delta$. The compactness of Υ follows by Lemmas 7.2 and 7.3.

Theorem 7.1 *If conditions (H_1), (H_2), (H_3), (H_4) hold and $\{A(t)\}_{t \in J}$ generates a compact evolution operator, then problem (7.1) has at least one solution.*

Proof Consider the set $Q = PC(J, RB)$ with R defined in (7.6). We show that the solution multioperator Υ defined in (7.7), when restricted to Q, satisfies the assumptions of Theorem 1.7. In fact Q is closed, convex, bounded and with a non-empty interior. According to Lemmas 7.2–7.4, Υ satisfies conditions (a) and (b) in Theorem 1.7.

Notice that $(\Upsilon(Q \times \{0\}) \subset D|u_0|B \subset \text{int } Q$, hence the condition (c) in Theorem 1.7 holds and $\Upsilon(\cdot, 0)$ is fixed point free on ∂Q. Let us now prove that Υ also satisfies the condition (d) for $\lambda \in (0, 1)$. Let $u \in Q$ and $\lambda \in (0, 1)$ be such that $u \in \Upsilon(u, \lambda)$ and assume, by contradiction, the existence of $t_0 \in (0, b]$ such that $u(t_0) \in \partial Q$ which is equivalent to $|u(t_0)| = R$. Since u is continuous and $u \in \Upsilon(u, \lambda)$, from $|u_0| < R$ it follows that there exist $\hat{t}_0, \hat{t}_1 \in (0, t_0]$ with $\hat{t}_0 < \hat{t}_1$ such that $|u(\hat{t}_0)| = |u_0|$, $|u_0| < |u(t)| < R$ for $t \in (\hat{t}_0, \hat{t}_1)$ and $|u(\hat{t}_1)| = R$. Moreover there exists $f \in Sel_F(u)$ such that

$$u(t) = U(t, \hat{t}_0)u(\hat{t}_0) + \lambda \int_{\hat{t}_0}^t U(t, s) f(s)ds + \lambda \sum_{\hat{t}_0 < t_k < t} U(t, t_k) I_k(u(t_k))$$

for $t \in [\hat{t}_0, \hat{t}_1]$. According to (H_3), $|f(t)| \leq \eta_{RB \setminus |u_0|B}(t)$ for $t \in (\hat{t}_0, \hat{t}_1)$, so we arrive to the contradiction

$$R = |u(\hat{t}_1)| \leq \tilde{M}\left(|u_0| + \|\eta_{RB\setminus|u_0|B}\|_{L(J,\mathbb{R})} + \sum_{k=1}^{N} l_k R\right) < R,$$

and the condition (d) in Theorem 1.7 is also satisfied.

Hence $\Upsilon(\cdot, 1)$ has a fixed point in Q which is a mild solution of problem (7.1).

When the nonlinear term F has an at most linear growth, i.e., when it satisfies $(H_3)'$ instead of condition (H_3), then the transversality condition (7.6) can be eliminated and the compactness of then the solution set can be obtained too.

Theorem 7.2 *If conditions (H_1), (H_2), $(H_3)'$, (H_4) hold and $\{A(t)\}_{t \in J}$ generates a compact evolution operator, then the solution set of problem (7.1) is nonempty and compact.*

Proof Consider the set Q defined as

$$Q = \{u \in PC(J, X) : |u(t)| \leq \phi(t) \text{ a.a } t \in J\},$$

where ϕ is a solution of the following equation

$$\phi(t) = \tilde{M}\left(|u_0| + \|\alpha\|_{L(J,\mathbb{R})} + \sum_{k=1}^{N} l_k \phi(t_k)\right) + \tilde{M}\int_0^t \alpha(s)\phi(s)ds,$$

and α is given in $(H_3)'$. Define the operator $\Gamma := \Upsilon(\cdot, 1)$. According to Lemmas 7.2–7.4, it is easy to see that Γ is locally compact with nonempty convex compact values and it has a closed graph. Hence it is also u.s.c. by Lemma 1.9. We prove now that Γ maps the set Q into itself.

Indeed if $u \in Q$ and $v \in \Gamma(u)$, there exists a function $f \in Sel_F(u)$ such that

$$v(t) = U(t, 0)u_0 + \int_0^t U(t, s)f(s)ds + \sum_{0 < t_k < t} U(t, t_k)I_k(u(t_k)).$$

By the hypothesis $(H_3)'$ we have that

$$|v(t)| = \left|U(t, 0)u_0 + \int_0^t U(t, s)f(s)ds + \sum_{0 < t_k < t} U(t, t_k)I_k(u(t_k))\right|$$

$$\leq \tilde{M}\left(|u_0| + \int_0^t \alpha(s)(1 + \phi(s))ds + \sum_{k=1}^{N} l_k \phi(t_k)\right)$$

$$\leq \tilde{M}\left(|u_0| + \|\alpha\|_{L(J,\mathbb{R})} + \sum_{k=1}^{N} l_k \phi(t_k)\right) + \tilde{M}\int_0^t \alpha(s)\phi(s)ds$$

$$= \phi(t).$$

Then $\Gamma(Q) \subseteq Q$. Let $V = \Gamma(Q)$ and $W = \overline{\mathrm{co}}(V)$. Since $\overline{V}$ is a compact set, W is compact too. Moreover from the fact that $\Gamma(Q) \subset Q$ and that Q is a convex closed set we have that $W \subset Q$ and hence

$$\Gamma(W) = \Gamma(\overline{\mathrm{co}}(\Gamma(Q))) \subseteq \Gamma(Q) = V \subset W.$$

Hence, according to Theorem 1.6, Γ has a fixed point, which is a solution of problem (7.1).

We prove now that then the solution set is compact. Indeed a solution of problem (7.1) is a fixed point of the operator Γ. If $u \in \Gamma(u)$, by the definition of Γ and $(H_3)'$ we have the existence of $f \in Sel_F(u)$ and reasoning as above

$$|u(t)| \leq |U(t, s)u_0| + \left| \int_0^t U(t, s)f(s)ds \right| + \left| \sum_{0 < t_k < t} U(t, t_k)I_k(u(t_k)) \right|$$

$$\leq \tilde{M}\left(|u_0| + \|\alpha\|_{L(J, \mathbb{R})} + \int_0^t \alpha(s)|u(s)|ds + \sum_{k=1}^N l_k|u(t_k)| \right).$$

By Gronwall's inequality it holds

$$|u(t)| \leq \tilde{M}\left(|u_0| + \|\alpha\|_{L(J, \mathbb{R})} + \sum_{k=1}^N l_k|u(t_k)| \right) e^{\tilde{M}\|\alpha\|_{L(J, \mathbb{R})}} := \overline{n}.$$

Hence $\mathrm{Fix}(\Gamma)$ is a bounded set and so $\Gamma(\mathrm{Fix}(\Gamma))$ is relatively compact. Since $\mathrm{Fix}(\Gamma) \subset (\mathrm{Fix}(\Gamma))$, then $\mathrm{Fix}(\Gamma)$ is relatively compact too. Finally, according to the closure of the graph of Γ, $\mathrm{Fix}(\Gamma)$ is also closed and hence compact.

7.2.2 Noncompact Operator Case

If we drop the assumption that the family $\{A(t)\}$ generates a compact evolution operator, we need stronger regularity hypotheses on F to consider the richer class of evolution operators which we discuss now. We take, precisely, F satisfying $(H_2)'$; moreover, when not explicitly mentioned, we always assume the growth restriction (H_3), and the impulse condition $(H_4)'$.

By (H_3), the multimap $F(t, \cdot)$ is locally weakly compact for a.a. $t \in J$, i.e., for a.a. t and every $u \in X$ there is a neighbourhood U_u of u such that the restriction of $F(t, \cdot)$ to U_u is weakly compact. Hence by $(H_2)'$ and the locally weak compactness, we easily get that $F(t, \cdot) : X_w \to X_w$ is u.s.c. for a.a. $t \in J$. Lemma 7.1 is still true under the condition $(H_2)'$. Hence the set $Sel_F(u) \neq \emptyset$ for any $u \in PC(J, X)$ and the solution operator $\Upsilon : PC(J, X) \times [0, 1] \to PC(J, X)$ can be defined as in (7.7) and it has nonempty convex values.

Fix $n \in \mathbb{N}$, consider $\mathcal{D}_n$ the closed ball of radius n in $PC(J, X)$ centered at the origin and designate by $\Upsilon_n = \Upsilon|_{\mathcal{D}_n} : \mathcal{D}_n \to P(PC(J, X))$ the restriction of the multioperator Υ on the set $\mathcal{D}_n$.

With a similar reasoning as in Lemma 7.2 it is also possible to prove that Υ has a weakly sequentially closed graph. Now we show that Υ is relatively weakly compact.

Lemma 7.5 Υ_n *is relatively weakly compact.*

Proof We first prove that $\Upsilon_n(\mathcal{D}_n)$ is relatively weakly sequentially compact.

Let $\{u_m\} \subset \mathcal{D}_n$ and $\{v_m\} \subset PC(J, X)$ satisfying $v_m \in \Upsilon_n(u_m)$ for all $m \in \mathbb{N}$. By the definition of the multioperator Υ_n, there exists a sequence $\{f_m\}$, $f_m \in Sel_F(u_m)$, such that for all $t \in [0, b]$,

$$v_m(t) = U(t, 0)u_0 + \int_0^t U(t, s)f_m(s)ds + \sum_{0 < t_k < t} U(t, t_k)I_k(u_m(t_k)).$$

The sequence $\{I_k(u_m(t_k))\}$ is bounded in X, we obtain that, up to subsequence, $I_k(u_m(t_k)) \rightharpoonup \bar{x}_k$ as $m \to \infty$ implying $\sum_{0 < t_k < t} U(t, t_k)I_k(u_m(t_k)) \rightharpoonup \sum_{0 < t_k < t} U(t, t_k)\bar{x}_k$ in X. Further, as the reason for Lemma 7.2, we receive

$$v_m(t) \rightharpoonup \bar{v}(t) = U(t, 0)u_0 + \int_0^t U(t, s)g(s)ds$$
$$+ \sum_{0 < t_k < t} U(t, t_k)\bar{x}_k, \quad \forall t \in J.$$

Furthermore, by (H_A), (H_3) and $(H_4)'$, we have

$$|v_m(t)| \leq \tilde{M}\left(|u_0| + \int_0^t \eta_{\mathcal{D}_n}(s)ds + \sum_{k=1}^N L_k(|u_m(t_k)|)\right)$$
$$\leq \tilde{M}\left(|u_0| + \|\eta_{\mathcal{D}_n}\|_{L(J,\mathbb{R})} + \sum_{k=1}^N L_k(n)\right)$$

for all $m \in \mathbb{N}$ and $t \in J$. It is simple to confirm that $v_m \rightharpoonup \bar{v}$ in $PC(J, X)$. Therefore $\Upsilon_n(\mathcal{D}_n)$ is relatively weakly sequentially compact. Hence relatively weakly compact by Theorem 1.26.

Remark 7.1 Notice that, since Υ has a weakly sequentially closed graph and according to Lemma 7.5, Υ has also weakly compact values.

Theorem 7.3 *Assume that conditions (H_1), $(H_2)'$, (H_3) and $(H_4)'$ hold. If*

$$\liminf_{n \to \infty} \frac{1}{n} \int_0^b \eta_{\mathcal{D}_n}(s)ds = 0, \tag{7.11}$$

then problem (7.1) has at least one solution.

Proof We show that there exists $n \in \mathbb{N}$ such that the operator Υ_n maps the ball $\mathcal{Q}_n$ into itself.

Assume to the contrary, that there exist sequences $\{u_n\}, \{z_n\}$ such that $u_n \in \mathcal{Q}_n$, $z_n \in \Upsilon_n(u_n)$ and $z_n \notin \mathcal{Q}_n, \forall n \in \mathbb{N}$. Then there exists a sequence $\{f_n\} \subset L^1(J, X)$, $f_n(s) \in F(s, u_n(s)), \forall n \in \mathbb{N}$ and a.e. $s \in J$ such that

$$z_n(t) = U(t, 0)u_0 + \int_0^t U(t, s)f_n(s)ds + \sum_{0 < t_k < t} U(t, t_k)I_k(u_n(t_k)), \quad \forall t \in J.$$

As the reason for Lemma 7.5, we have

$$\|z_n\|_{PC} \le \tilde{M}|u_0| + \tilde{M} \int_0^b \eta_{\mathcal{Q}_n}(s)ds + \tilde{M} \sum_{k=1}^N L_k(|u_n(t_k)|)$$

$$\le \tilde{M}\left(|u_0| + \sum_{k=1}^N L_k(n)\right) + \tilde{M} \int_0^b \eta_{\mathcal{Q}_n}(s)ds,$$

which suggests that

$$1 < \frac{\|z_n\|_{PC}}{n} \le \frac{1}{n}\left(|u_0| + \sum_{k=1}^N L_k(n)\right) + \frac{\tilde{M}}{n} \int_0^b \eta_{\mathcal{Q}_n}(s)ds, \quad n \in \mathbb{N},$$

which contradicts (7.11).

Now, choose $n \in \mathbb{N}$ such that $\Upsilon_n(\mathcal{Q}_n) \subseteq \mathcal{Q}_n$. By Lemma 7.5, the set $V_n = \overline{\Upsilon_n(\mathcal{Q}_n)}^w$ is weakly compact. Let now $\tilde{V}_n = \overline{co}(V_n)$, where $\overline{co}(V_n)$ denotes the closed convex hull of V_n. By Theorem 1.27, $\tilde{V}_n$ is a weakly compact set. Moreover from the fact that $\Upsilon_n(\mathcal{Q}_n) \subset \mathcal{Q}_n$ and that $\mathcal{Q}_n$ is a convex closed set we have that $\tilde{V}_n \subset \mathcal{Q}_n$ and hence

$$\Upsilon_n(\tilde{V}_n) = \Upsilon_n(\overline{co}(\Upsilon_n(\mathcal{Q}_n))) \subseteq \Upsilon_n(\mathcal{Q}_n) \subseteq \overline{\Upsilon_n(\mathcal{Q}_n)}^w = V_n \subset \tilde{V}_n.$$

It is clear that Υ_n has a weakly sequentially closed graph, hence, by Remark 7.1, it is weakly u.s.c. Thus from Theorem 1.6, problem (7.1) has a solution. The proof is completed.

Theorem 7.4 *Let assumptions (H_1), $(H_2)'$, $(H_3)'$ and (H_4) hold. If*

$$\tilde{M}\|\alpha\|_{L(J,\mathbb{R})} < 1, \tag{7.12}$$

then the solution set of problem (7.1) is nonempty and weakly compact.

Proof Considering as in Theorem 7.3 and expecting that there exist $\{u_n\}, \{z_n\}$ such that $u_n \in \mathcal{Q}_n, z_n \in \Upsilon_n(u_n)$ and $z_n \notin \mathcal{Q}_n, \forall n \in \mathbb{N}$, we get

$$n < \|z_n\|_{PC} \leq \widetilde{M}|u_0| + \widetilde{M} \int_0^b \alpha(s)(1 + |u_n(s)|)ds + \widetilde{M} \sum_{k=1}^N L_k(|u_n(t_k)|)$$

$$\leq \widetilde{M}\left(|u_0| + \|\alpha\|_{L(J,\mathbb{R})} + \sum_{k=1}^N L_k(n)\right) + \widetilde{M} \int_0^b \alpha(s)ds,$$

which implies that

$$1 < \frac{\|z_n\|_{PC}}{n} \leq \frac{\widetilde{M}}{n}\left(|u_0| + (1+n)\|\alpha\|_{L(J,\mathbb{R})} + \sum_{k=1}^N L_k(n)\right), \quad n \in \mathbb{N}. \quad (7.13)$$

By passing to the limit as $n \to +\infty$ in (7.13), we receive $\widetilde{M}\|\alpha\|_{L(J,\mathbb{R})} \geq 1$, which contradicts to (7.12). The conclusion then takes after by Theorem 1.6, as in Theorem 7.3.

7.3 Topological Structure of Solution Set

Throughout this section, let $(X, |\cdot|)$ be a Banach space, we denote by $C([-\tau, 0], X)$ the space of piecewise continuous functions $c : [-\tau, 0] \to X$ with finite number of discontinuity points $\{t_*\}$ such that $t_* \neq 0$ and all values

$$c(t_*^+) = \lim_{h \to 0^+} c(t_* + h) \text{ and } c(t_*^-) = \lim_{h \to 0^-} c(t_* + h)$$

are finite. We consider the space $\mathcal{C}([-\tau, 0], X)$ with the L^1-norm, i.e.,

$$\|c\|_{\mathcal{C}} = \int_{-\tau}^0 |c(t)|dt.$$

We do not consider the space $\mathcal{C}([-\tau, 0], X)$ with the uniform convergence norm, because it creates problems: the function $t \in [0, \infty) \mapsto u_t$ is not continuous, moreover, it is not necessarily measurable (see Example 3.1, [117]). As a consequence, the multivalued superposition operator, which we will define in Sect. 7.3.1, would not be well defined. This space of delays with the integral norm was considered in [38] (see also [156]).

We denote by $PC([a, \infty), X)$ the space of piecewise continuous functions $c : [a, \infty) \to X$ with infinite number of discontinuity points $t_1, t_2, \ldots$ such that $\lim_{n \to \infty} t_n = +\infty$. The values $c(t_i^-)$, $c(t_i^+)$ for $i = 1, 2, \ldots$ are finite and $c(t_i^-) = c(t_i)$. The space $PC([a, \infty), X)$ is a Fréchet space with the family of seminorms $\{p_n\}$ given by

$$p_n(c) = \|c|_{[0,t_n]}\|_{PC}.$$

and a metric:

$$d(c_1, c_2) = \sum_{n=1}^{\infty} \frac{1}{2^n} \frac{p_n(c_1 - c_2)}{1 + p_n(c_1 - c_2)}.$$

We consider the following problem on a compact interval

$$\begin{cases} u'(t) \in A(t)u(t) + F(t, u_t), & \text{for a.a. } t \in [0, t_m), \ t \neq t_k, \ k < m, \\ u(t) = x(t), & \text{for a.a. } t \in [-\tau, 0], \\ u(t_k^+) = u(t_k) + I_k(u_{t_k}), & \text{for } k < m. \end{cases} \quad (7.14)$$

Let $[0, t_m]$ be a fixed interval on the real line. Put $\Delta_m = \{(t, s) \in [0, t_m] \times [0, t_m] : s \leq t \leq t_m\}$. We study the compactness of the solution set of problem (7.14) under the following assumptions:

$(A)_m$ $\{A(t)\}_{t \in [0, t_m]}$ is a family of linear (not necessarily bounded) operators $(A(t) : D(A) \subset X \to X, t \in [0, t_m], D(A)$ a dense subset of X not depending on $t)$ generating an evolution operator $U : \Delta_m \to \mathcal{L}(X)$.

We observe that, since the evolution operator $U(t, s)$ is strongly continuous on the compact set Δ_m, there exists a constant $B_m = B_{\Delta_m} > 0$ such that

$$\|U(t, s)\|_{\mathcal{L}(X)} \leq B_m, \quad (t, s) \in \Delta_m. \quad (7.15)$$

Firstly we give the concept of mild solutions for problem (7.14).

Definition 7.2 A piecewise continuous function $u : [-\tau, t_m] \to X$ is a mild solution for the impulsive Cauchy problem (7.14) if there exists $f \in L^1([0, t_m], X)$, $f(s) \in F(s, u_s)$ for a.e. $s \in [0, t_m]$ such that

(a) $u(t) = U(t, 0)x(0) + \sum_{0 < t_k < t} U(t, t_k)I_k(u_{t_k}) + \int_0^t U(t, s)f(s)ds, \ t \in [0, t_m]$;
(b) $u(t) = x(t), \ t \in [-\tau, 0]$;
(c) $u(t_k^+) = u(t_k) + I_k(u_{t_k}), \ k < m$.

We assume that the map $F : [0, t_m] \times \mathcal{C}([-\tau, 0], X) \to P_{cp,cv}(X)$ satisfies:

$(F_1)_m$ $F(\cdot, c)$ has a strongly measurable selection for every $c \in \mathcal{C}([-\tau, 0], X)$;
$(F_2)_m$ $F(t, \cdot)$ is u.s.c. for a.e. $t \in [0, t_m]$;
$(F_3)_m$ F has at most a linear growth for a.e.

$t \in [0, t_m]$, i.e., there exists a function $\alpha \in L^1([0, t_m], \mathbb{R}^+)$ such that

$$|F(t, c)| \leq \alpha(t)(1 + \|c\|_{\mathcal{C}}) \ \text{for a.e. } t \in [0, t_m];$$

$(F_4)_m$ for every $\varepsilon > 0$ and every bounded set $D \subset \mathcal{C}([-\tau, 0], X)$ there exist $\delta > 0$ and a function $\mu \in L^1_{loc}([0, t_m], \mathbb{R}^+)$ such that

$$\beta(F(t, O_\delta(D))) \leq \mu(t) \sup_{-\tau \leq \theta \leq 0} \beta(O_\varepsilon(D(\theta))) \ \text{for a.e. } t \geq 0.$$

It is clear that $(F_4)_m$ can infer the condition:

$(F_4)'_m$ there exists a function $\mu \in L^1([0, t_m])$ such that

$$\beta(F(t, D)) \le \mu(t) \sup_{-\tau \le \theta \le 0} \beta(D(\theta)) \quad \text{for a.e. } t \in [0, t_m]$$

and for every bounded $D \subset C([-\tau, 0], \mathbb{R}^+)$, $D(\theta) = \{c(\theta) : c \in D\}$.

For $v \in PC([0, t_m], X)$ we can define $v_i \in C([t_i, t_{i+1}], X)$, $i = 0, 1, ..., m - 1$ as $v_i(t) = v(t)$ on $(t_i, t_{i+1}]$ and $v_i(t_i) = v(t_i^+)$. For every set $K \subset PC([0, t_m], X)$ we denote by K_i, $i = 0, 1, ..., m - 1$ the set $K_i = \{v_i : v \in K\}$. It is easy to see that

Lemma 7.6 *A set $K \in PC([0, t_m], X)$ is relatively compact in $PC([0, t_m], X)$ if and only if each set K_i, $i = 0, 1, ..., m - 1$ is relatively compact in $C([t_i, t_{i+1}], X)$.*

For any $v \in PC([0, t_m], X)$, or $v \in PC([0, \infty), X)$, such that $v(0) = x(0)$ we define the function $v[x] : [-\tau, t_m] \to X$ as

$$v[x] = \begin{cases} x(t), & t \in [-\tau, 0], \\ v(t), & t \in [0, t_m], \end{cases} \tag{7.16}$$

where $x : [-\tau, 0] \to X$ is the function from the initial condition in (7.2). We denote $\Omega[x] = \{v[x] : v \in \Omega\}$.

7.3.1 Compact Interval Case

For a given multivalued map $F : [0, t_m] \times C([-\tau, 0], X) \to P_{cp,cv}(X)$ satisfying $(F_1)_m$-$(F_4)_m$ we consider the multivalued superposition operator $Sel_F^{t_m} : PC([0, t_m], X) \to P(L^1([0, t_m], X))$ defined as

$$Sel_F^{t_m}(v) = \{f \in L^1([0, t_m], X) : f(s) \in F(s, v[x]_s) \text{ for a.e. } s \in [0, t_m]\}. \tag{7.17}$$

This multivalued superposition operator $Sel_F^{t_m}$ is well defined (see e.g. [130]). Notice that the function $s \in [0, t_m] \to v[x]_s \in C([-\tau, 0], X)$ is continuous.

Lemma 7.7 ([130, Lemma 5.1.1]) *Assume that multimap the F satisfies hypotheses $(F_1)_m$-$(F_3)_m$. If the sequences*

$$\{u_n\}_{n=1}^{\infty} \subset C([t_{k-1}, t_k], X), \quad \{f_n\}_{n=1}^{\infty} \subset L^1([t_{k-1}, t_k], X),$$

$f_n \in Sel_F^{t_m}(u_n)$, $n \ge 1$, *are such that* $u_n \to u_0$, $f_n \rightharpoonup f_0$, *then* $f_0 \in Sel_F^{t_m}(u_0)$.

Now, we prove the compactness of the solution set of problem (7.2).

Theorem 7.5 *Under assumptions $(A)_m$, $(F_1)_m$-$(F_3)_m$ and $(F_4)'_m$, problem (7.14) has at least one mild solution on $[-\tau, t_m]$. Moreover, if the impulse functions*

$$I_k : C([-\tau, 0], X) \to X, \quad k \in \mathbb{N}$$

are continuous, then the solutions set is a compact subset of $C([-\tau, t_m], X)$.

Proof We divide the proof in several steps.

Step 1. Given the non-impulsive Cauchy problem

$$\begin{cases} u'(t) \in A(t)u(t) + F(t, u_t), & \text{a.e. } t \in [0, t_1], \\ u(t) = x(t), & t \in [-\tau, 0], \end{cases} \tag{P}_1$$

we search for the existence of a mild solution $u \in C([-\tau, t_1], X)$ such that

(a) $u(t) = U(t, 0)x(0) + \int_0^t U(t, s) f(s) ds$, $t \in [0, t_1]$, where $f \in L^1([0, t_1], X)$, $f(s) \in F(s, u_s)$ for a.e. $s \in [0, t_1]$;
(b) $u(t) = x(t)$, $t \in [-\tau, 0]$.

To this aim, we first consider the integral multioperator $\Gamma^1 : C([0, t_1], X) \to P(C([0, t_1], X))$ defined as:

$$\Gamma^1(u) = \left\{ v \in C([0, t_1], X) : v(t) = U(t, 0)x(0) + \int_0^t U(t, s) f(s) ds, \right.$$

$$\left. f \in L^1([0, t_1], X), \ f(s) \in F(s, u[x]_s) \text{ a.e. } s \in [0, t_1] \right\}.$$

It is clear that if $v \in \text{Fix}(\Gamma^1)$ then $u[x]$, where $u[x]$ is from (7.16), is a mild solution of $(P)_1$ on the interval $[-\tau, t_1]$.

We claim that The multioperator Γ^1 is a closed multioperator with compact, convex values. Indeed, let $\{u_n\}_{n=1}^\infty$, $\{v_n\}_{n=1}^\infty \subset C([0, t_1], X)$ with $u_n \to u_0$, $v_n \in \Gamma^1(u_n)$, $n \geq 1$ and $v_n \to v_0$. Take a sequence $\{f_n\}_{n=1}^\infty \subset L^1([0, t_1], X)$ such that $f_n \in Sel_F^1(u_n)$, $n \geq 1$. From assumption $(F3)_m$ it follows that the sequence $\{f_n\}_{n=1}^\infty$ is integrably bounded. Moreover, the hypothesis $(F4)_m'$ implies that

$$\beta(\{f_n(t)\}_{n=1}^\infty) \leq \mu(t) \sup_{-\tau \leq t \leq 0} \beta(\{(u_n[x])_t\}_{n=1}^\infty) \leq \mu(t) \sup_{0 \leq \theta \leq t} \beta(\{u_n(\theta)\}_{n=1}^\infty) = 0$$

for a.e. $t \in [0, t_1]$, so the set $\{f_n(t)\}_{n=1}^\infty$ is relatively compact for almost every $t \in [0, t_1]$. Hence the set $\{f_n\}_{n=1}^\infty$ is semicompact and then it is also weakly compact in $L^1([0, t_1], X)$ by Lemma 1.25. So we can assume, without loss of generality, that $f_n \rightharpoonup f_0$ in $L^1([0, t_1], X)$.

Applying Property 1.6 (ii) and the uniqueness of the limit algorithm, we conclude that $v_n = G^1 f_n \to G^1 f_0 = v_0$, where for $f_0 \in L^1([0, t_1], X)$,

$$G^1(f) = U(t, 0)x(0) + \int_0^t U(t, s) f_0(s) ds.$$

Moreover, by Lemma 7.7, we have $f_0 \in Sel_F^{t_1}(u_0)$, therefore $v_0 \in G^1 \circ Sel_F^{t_1}(u_0) = \Gamma^1(u_0)$, demonstrating that the multioperator Γ^1 is closed.

Let us prove now that Γ^1 has compact values. To this aim, we consider an arbitrary $u \in C([0, t_1], X)$. Then, every sequence $\{f_n\}_{n=1}^\infty \in Sel_F^{t_1}(u)$ is semicompact. Therefore, with the same arguments as above, the set $\{G^1(f_n)\}_{n=1}^\infty$ is relatively compact. So the compactness of $\Gamma^1(u)$ follows from its closeness.

Further, the convexity of $\Gamma^1(u)$ is the consequence of the convexity of values of the multimap F and the linearity of G^1.

Consider now the function ν defined on bounded sets $\Omega \subset C([0, t_1], X)$ with values in $(\mathbb{R}_d^2, \geq)$ as:

$$\nu(\Omega) = \max_{D \in \mathcal{D}(\Omega)} (\gamma(D), \mathrm{mod}_C(D)),$$

where $\mathcal{D}(\Omega)$ is the collection of all denumerable subsets of Ω and, for a given constant $L > 0$,

$$\gamma(D) = \sup_{0 \leq t \leq t_1} e^{-Lt} \beta(D(t)).$$

Example 1.3 shows that ν is a monotone, non singular, regular MNC.

Let us prove that the multioperator Γ^1 is condensing on bounded subsets of $C([0, t_1], X)$ with respect to the MNC ν.

Let $\Omega \subset C([0, t_1], X)$ be a bounded set such that

$$\nu(\Gamma^1(\Omega)) \geq \nu(\Omega). \tag{7.18}$$

Let the maximum of $\nu(\Gamma'(\Omega))$ be achieved for the countable set $D' = \{g_n\}_{n=1}^\infty$, where $g_n = G^1(f_n)$, $f_n \in Sel_F^{t_1}(u_n)$, $n \geq 1$ and $\{v_n\}_{n=1}^\infty \subset \Omega$. From (7.18) we have

$$\gamma(\{g_n\}_{n=1}^\infty) \geq \gamma(\{u_n\}_{n=1}^\infty). \tag{7.19}$$

From $(F_4)'_m$ we have for $s \in [0, t_1]$,

$$\beta(\{f_n(s)\}_{n=1}^\infty) \leq \mu(s) \sup_{0 \leq \theta \leq s} \beta(\{u_n(\theta)\}_{n=1}^\infty)$$

$$= e^{Ls} \mu(s) e^{-Ls} \sup_{0 \leq \theta \leq s} \beta(\{u_n(\theta)\}_{n=1}^\infty)$$

$$\leq e^{Ls} \mu(s) \sup_{0 \leq \theta \leq t_1} e^{-L\theta} \beta(\{u_n(\theta)\}_{n=1}^\infty)$$

$$= e^{Ls} \mu(s) \gamma(\{u_n\}_{n=1}^\infty).$$

Moreover, from properties of MNC β we have for $t \in [0, t_1]$,

$$\beta(\{g_n(t)\}_{n=1}^{\infty}) = \beta\left(U(t,0)x(0) + \int_0^t U(t,s)\{f_n(s)\}_{n=1}^{\infty} ds\right)$$

$$\leq \beta(U(t,0)x(0)) + \beta\left(\int_0^t U(t,s)\{f_n(s)\}_{n=1}^{\infty} ds\right)$$

$$= \beta\left(\int_0^t U(t,s)\{f_n(s)\}_{n=1}^{\infty} ds\right).$$

Applying Property 1.6 (i) we have

$$e^{-Lt}\beta(\{g_n(t)\}_{n=1}^{\infty}) \leq e^{-Lt}2B_1 \int_0^t e^{Ls}\mu(s)ds\gamma(\{u_n\}_{n=1}^{\infty})$$

$$\leq 2B_1 \sup_{t\in[0,t_1]} e^{-Lt} \int_0^t e^{Ls}\mu(s)ds\gamma(\{u_n\}_{n=1}^{\infty}).$$

Then

$$\gamma(\{g_n(t)\}_{n=1}^{\infty}) \leq \sup_{t\in[0,t_1]} e^{-Lt}2B_1 \int_0^t e^{Ls}\mu(s)ds\gamma(\{u_n\}_{n=1}^{\infty})$$

$$= 2B_1\gamma(\{u_n\}_{n=1}^{\infty}) \sup_{t\in[0,t_1]} \int_0^t e^{-L(t-s)}\mu(s)ds.$$

We can choose the constant $L > 0$ so that

$$\sup_{t\in[0,t_1]} \left(2B_1 \int_0^t e^{-L(t-s)}\mu(s)ds\right) < 1,$$

so

$$\gamma(\{g_n\}_{n=1}^{\infty}) < \gamma(\{u_n\}_{n=1}^{\infty}). \tag{7.20}$$

Then from (7.19) and (7.20) we have $\gamma(\{g_n\}_{n=1}^{\infty}) = \gamma(\{u_n\}_{n=1}^{\infty}) = 0$ and hence $\beta(\{u_n(t)\}_{n=1}^{\infty}) = 0$ for all $t \in (0, t_1]$.

With the same arguments used before we obtain that $\{g_n\}_{n=1}^{\infty}$ is a relatively compact sequence, therefore $\text{mod}_C(\{g_n\}_{n=1}^{\infty}) = 0$, i.e., $\nu(\Gamma^1(\Omega)) = 0$ and from (7.18) we know $\nu(\Omega) = 0$, then Ω is a relatively compact set.

Let the function $\tilde{u}^0 \in C([0, t_1], X)$ be defined by $\tilde{u}^0(t) := U(t, 0)x(0)$, $t \in [0, t_1]$. Consider the family of multimaps $\Phi : C([0, t_1], X) \times [0, 1] \to P_{cp,cv}(C([0, t_1], X))$ given by:

$$\Phi(u, \lambda) = \left\{\tilde{u}^0(t) + \lambda \int_0^t U(t,s)f(s)ds, \ t \in [0, t_1] : f \in Sel_F^{t_1}(u)\right\}. \tag{7.21}$$

We will show that the set of fixed points of Φ, i.e., $\text{Fix}(\Phi) = \{u \in \Phi(u, \lambda)$ for some $\lambda \in [0, 1]\}$, is a priori bounded.

Let $v \in \text{Fix}(\varPhi)$. Then there exists $f \in Sel_F^{t_1}(u)$ such that by using (7.15) and $(F_3)_m$, for any $t \in [0, t_1]$ we have

$$
|v(t)| = \left| U(t,0)x(0) + \lambda \int_0^t U(t,s)f(s)ds \right|
$$

$$
\leq B_1|x(0)| + B_1 \int_0^t |f(s)|ds
$$

$$
\leq B_1|x(0)| + B_1 \int_0^t \alpha(s)(1 + \|u[x]_s\|_C)ds
$$

$$
\leq B_1|x(0)| + B_1\|\alpha\|_{L([0,t_1],\mathbb{R}^+)} + B_1 \int_0^t \alpha(s) \int_{-\tau}^0 |u[x]_s(\theta)|ds
$$

$$
\leq B_1|x(0)| + B_1\|\alpha\|_{L([0,t_1],\mathbb{R}^+)} + B_1 \int_0^t \alpha(s)\tau \cdot \sup_{-\tau \leq \theta \leq 0} |u[x](s + \theta)|ds
$$

$$
\leq B_1|x(0)| + B_1\|\alpha\|_{L([0,t_1],\mathbb{R}^+)}
$$

$$
+ \tau B_1 \int_0^t \alpha(s)\left(\sup_{-\tau \leq \theta \leq 0} |x(\theta)| + \sup_{0 < \theta \leq s} |u(\theta)| \right)ds
$$

$$
\leq B_1|x(0)| + B_1\|\alpha\|_{L([0,t_1],\mathbb{R}^+)} + \tau B_1\|\alpha\|_{L([0,t_1],\mathbb{R}^+)} \sup_{-\tau \leq \theta \leq 0} |x(\theta)|
$$

$$
+ \tau B_1 \int_0^t \alpha(s) \sup_{0 < \theta \leq s} |u(\theta)|ds
$$

$$
\leq B_1|x(0)| + B_1\|\alpha\|_{L([0,t_1],\mathbb{R}^+)}(1 + \tau N) + \tau B_1 \int_0^t \alpha(s) \sup_{0 < \theta \leq s} |u(\theta)|ds,
$$

where $N = \sup_{-\tau \leq t \leq 0} |x(t)|$. The right-hand side is an increasing function in t, so we have the same estimate for all $0 < r \leq t$, i.e.,

$$
\sup_{0 < r \leq t} |v(r)| \leq R + \tau B_1 \int_0^t \alpha(s) \sup_{0 < \theta \leq s} |v(\theta)|ds,
$$

where $R = B_1|x(0)| + B_1\|\alpha\|_{L([0,t_1],\mathbb{R}^+)}(1 + \tau N)$. Since z is continuous on $[0, t_1]$, the function $\psi(t) = \sup_{0 < r \leq t} |z(r)|$ is also continuous, so

$$
\psi(t) \leq R + \tau B_1 \int_0^t \alpha(s)\psi(s)ds,
$$

by Gronwall–Bellmann's inequality, we obtain

$$
\psi(t) \leq R \exp\left(\tau B_1 \int_0^t \alpha(s)ds \right) \leq R \exp\{\tau B_1\|\alpha\|_{L([0,t_1],\mathbb{R}^+)}\} := H.
$$

Using the same arguments as before we may verify that the family $\varPhi$ defined in (7.21) is ν-condensing on every bounded set $\varOmega \subset C([0, t_1], X)$.

Now we take an open ball $U \subset C([0, t_1], X)$ of radius greater than H and with center $\tilde{u}^0$ (then containing the set Fix(Φ)). The family Φ is fixed point free on the boundary ∂U and hence it determines an homotopy between the multifield $i - \Gamma^1$ and the multifield $i - \tilde{u}^0$. In this framework it is possible to apply the relative topological degree theory for condensing multifields developed in [130]. In this case we evaluate the degree with respect to the whole space X. Taking into account that $\tilde{u}^0 \in U$ and using the homotopy and normalization properties of the degree, we obtain that $\deg(i - \Gamma^1, \overline{U}) = \deg(i - \tilde{u}^0, \overline{U}) = 1$ and therefore (see [130, Theorem 3.3.1])

$$\emptyset \neq \text{Fix}(\Gamma^1) \subset U.$$

Then also the set Σ_x^1 of all mild solutions of problem $(P)_1$ is nonempty.

Now, we prove that it is compact. First of all, by applying Theorem 1.11 we can claim that the fixed points set of Γ^1 is compact. Let us take the function κ^1 : $C([0, t_1], X) \to C([-\tau, t_1], X)$ defined by

$$\kappa^1(u) = u[x],$$

where $u[x]$ is from (7.16). Since κ^1 is a continuous map, then the set $\kappa^1(\text{Fix}\,\Gamma^1)$ is compact. The equality $\kappa^1(\text{Fix}(\Gamma^1)) = \Sigma_x^1$ concludes the proof of the step.

Step 2. Let us fix $v^1 \in \Sigma_x^1$ and let us consider the non-impulsive Cauchy problem

$$\begin{cases} u'(t) \in A(t)u(t) + F(t, u_t) & \text{a.e. } t \in [t_1, t_2], \\ u(t) = v^1(t), & t \in [-\tau, t_1], \\ u(t_1) = v^1(t_1) + I_1(v_{t_1}^1). \end{cases} \qquad (P)_{2;v^1}$$

Of course, for this problem a mild solution is a function $u \in C([-\tau, t_2], X)$ such that

(a)' $u(t) = U(t, t_1)(v^1(t_1) + I_1(v_{t_1}^1)) + \int_{t_1}^t U(t, s)f(s)ds$, for $t \in [t_1, t_2]$, where $f \in L^1([t_1, t_2], X)$, $f(s) \in F(s, u_s)$ for a.e. $s \in [t_1, t_2]$;

(b)' $u(t) = v^1(t)$, $t \in [-\tau, t_1]$.

To provide such a solution for problem $(P)_{2;v^1}$, in analogy with the previous step we consider the integral multioperator $\Gamma^2 : C([t_1, t_2], X) \to P(C([t_1, t_2], X))$ defined as:

$$\Gamma^2(u) = \left\{ v \in C([t_1, t_2], X) : v(t) = U(t, t_1)(v^1(t_1) + I_1(v_{t_1}^1)) + \int_{t_1}^t U(t, s)f(s)ds \right.$$

$$\left. f \in L^1([t_1, t_2], X), \ f(s) \in F(s, u[v^1]_s) \text{ a.e. } s \in [t_1, t_2] \right\},$$

where $u[v^1]$ is defined by (7.16).

Also here, if $v \in \text{Fix}(\Gamma^2)$ then $v[v^1]$ is a mild solution of $(P)_{2;v^1}$ on the interval $[-\tau, t_2]$.

Moreover, by proceeding in the same way as in Step 1, we can claim that this problem has at least one mild solution and the solution set is a compact set, say $\Sigma^2_{v^1}$.

Of course, we can iterate this process till a problem $(P)_{m;v^1,\ldots,v^{m-1}}$ and obtain that also this problem has solutions and that these solutions form a compact set $\Sigma^m_{v^1,\ldots,v^{m-1}}$.

Now, every solution of $(P)_{m;v^1,\ldots,v^{m-1}}$ is a solution of (7.14) and the first part of the theorem is proved.

Step 3. It remains to prove that the set of all solutions of (7.14), i.e.,

$$\Theta_m = \bigcup \left\{ \Sigma^m_{v^1,\ldots,v^{m-1}} : v^1 \in \Sigma^1_x, \cdots, v^{m-1} \in \Sigma^{m-1}_{v^1,\ldots,v^{m-2}} \right\} \tag{7.22}$$

is compact.

To this aim, from now on we assume that the impulse functions I_k are continuous. First of all, we define the multifunction $H^1 : \Sigma^1_x \to P(C([-\tau, t_2], X))$ as

$$H^1(v^1) = \Sigma^2_{v^1}.$$

From Step 2 we know both that Σ^1_x is compact and that H^1 has compact values. Now, we prove that it is u.s.c. so that, by applying Lemma 1.11, we get the range $\bigcup_{v^1 \in \Sigma^1_x} \Sigma^2_{v^1}$ to be compact.

Note that defining the multifunction $Q^1 : \Sigma^1_x \to P_{cp}(C([t_1, t_2], X))$ as

$$Q^1(v^1) = \Sigma^2_{v^1|[t_1,t_2]},$$

the multifunction H^1 can be written as the composition of the multimap $P^1 : \Sigma^1_x \to P_{cp}(\Sigma^1_x \times C([t_1, t_2], X))$ defined by

$$P^1(v^1) = \{v^1\} \times Q^1(v^1)$$

with the continuous map $\eta^1 : P^1(\Sigma^1_x) \to C([-\tau, t_2], X)$ defined by

$$\eta^1(v^1, v) = v[v^1].$$

We first prove that the multifunction Q^1 is u.s.c. We assume to the contrary that there exists $\bar{v}^1 \in \Sigma^1_x$ such that Q^1 is not u.s.c. in $\bar{v}^1$. Therefore there exist $\bar{\varepsilon} > 0$ and two sequences $\{v^1_n\}^\infty_{n=1}, v^1_n \to \bar{v}^1$ in $C([-\tau, t_1], X)$, and $\{v^2_n\}^\infty_{n=1}, v^2_n \in \Sigma^2_{v^1_n|[t_1,t_2]}$, such that

$$v^2_n \notin B(\Sigma^2_{\bar{v}^1}|[t_1, t_2], \bar{\varepsilon}), \quad n \geq 1. \tag{7.23}$$

Since $\{v^2_n\}^\infty_{n=1}$ is a sequence of solutions, we have

$$v^2_n = U(t, t_1)[v^1_n(t_1) + I_1(v^1_{nt_1})] + \int_{t_1}^t U(t, s) f^2_n(s) ds, \quad t \in [t_1, t_2], \tag{7.24}$$

where $f_n^2 \in L^1([t_1, t_2], X)$, $f_n^2(s) \in F(s, (v_n^2)_s)$ for a.e. $s \in [t_1, t_2]$. Then, for $s \in [t_1, t]$, $(F_4)_m'$ yields

$$\beta(\{f_n^2(s)\}_{n=1}^\infty) \leq \mu(s) \sup_{-\tau \leq \theta \leq 0} \beta(\{v_n^2(s+\theta)\}_{n=1}^\infty)$$

$$\leq \mu(s)\left(\sup_{-\tau \leq \eta \leq 0} \beta(x(\eta)) + \sup_{0 \leq \eta \leq t_1} \beta(\{v_n^1(\eta)\}_{n=1}^\infty) \right.$$

$$\left. + \sup_{t_1 \leq \eta \leq s} \beta(\{v_n^2(\eta)\}_{n=1}^\infty) \right).$$

On account of $\beta(\{v_n^1(\eta)\}_{n=1}^\infty) = 0$, since $\{v_n^1\}_{n=1}^\infty$ is a converging sequence, then

$$\beta(\{f_n^2(s)\}_{n=1}^\infty) \leq e^{Ls} \mu(s) \sup_{t_1 \leq \theta \leq t_2} e^{-L\eta} \beta(\{v_n^2(\theta)\}_{n=1}^\infty).$$

Now, using similar arguments as in the proof of condensivity in Step 1, it is possible to prove that the set $\{v_n^2\}_{n=1}^\infty$ is relatively compact in $C([t_1, t_2], X)$. Therefore without loss of generality we can assume that there exists $\bar{v}^2 \in C([t_1, t_2], X)$ such that $v_n^2 \to \bar{v}^2$ in $C([t_1, t_2], X)$.

Now we prove that $\bar{v}^2 \in Q^1(\bar{v}^1)$. For every $n \geq 1$, we consider $v_n^2 \in \Sigma_{v^1|[t_1,t_2]}^2$ and the corresponding function f_n^2 from (7.24).

As in Step 1 it is possible to prove that there exists $\overline{f}^2 \in L^1([t_1, t_2], X)$ such that $f_n^2 \rightharpoonup \overline{f}^2 \in L^1([t_1, t_2], X)$. Now, by using Lemma 7.7, we have

$$\overline{f}^2 \in F(t, \bar{v}_t^2) \quad \text{a.e. } t \in [t_1, t_2].$$

From Property 1.6 (ii) and the fact that the function I_1 is continuous, by considering the limit in both sides of (7.24) we get

$$\bar{v}^2(t) = U(t, t_1)\left(v^1(t_1) + I_1(v_{t_1}^1)\right) + \int_{t_1}^t U(t, s)\overline{f}^2(s)ds, \quad t \in [t_1, t_2],$$

where $\overline{f}^2 \in L^1([t_1, t_2], X)$, $\overline{f}^2(s) \in F(s, v_s^2)$ for a.e. $s \in [t_1, t_2]$, that is, $\bar{v}^2 \in \Sigma_{v^1|[t_1,t_2]}^2 = Q^1(\bar{v}^1)$.

The fact that $v_n^2 \to \bar{v}^2 \in Q^1(\bar{v}^1)$ leads a contradiction with (7.23). Therefore, by applying well known results on composition and cartesian product of multimaps (see e.g. [130, Theorems 1.2.12 and 1.2.8]), we can conclude that H^1 is u.s.c.

By iterating this process we obtain the compactness of the solution set Θ_m on the interval $[-\tau, t_m]$.

Now we prove the results about an R_δ-structure of the solution set. At first we examine the case of problems on compact intervals.

Theorem 7.6 *Let X be a Banach space and let the hypothesis $(A)_m$ hold. Suppose that the multivalued map $F : [0, t_m] \times C([-\tau, 0], X) \to P_{cp,cv}(X)$ satisfies conditions $(F_1)_m$-$(F_4)_m$. Moreover, assume that the maps $I_k : C([-\tau, 0], X) \to X$, $k \in \mathbb{N}$, are continuous and there exist constants $r_k > 0$ such that*

$$\beta(I_k(D)) \leq r_k \sup_{-\tau \leq \theta \leq 0} \beta(D(\theta)) \tag{7.25}$$

for every bounded $D \subset C([-\tau, 0], X)$. Then the solution set for problem (7.14) is an R_δ-set in $PC([0, t_m], X)[x]$.

Note that, since the evolution operator T is strongly continuous on the compact set Δ_m, the number B_m is finite, that is, $B_m < \infty$.

Proof We will proceed in several steps.
Step 1. Consider the non-impulsive Cauchy problem

$$\begin{cases} u'(t) \in A(t)u(t) + F(t, u_t), & \text{for a.a. } t \in [0, t_1], \\ u(t) = x(t), & \text{for } t \in [-\tau, 0]. \end{cases} \tag{7.26}$$

According to Theorem 7.5, it is shown that solutions of (7.26) are bounded (by some $\bar{K}_1 \geq 0$). We show that solutions on the interval $[t_1, t_2]$ are bounded by $\bar{K}_2$. To do this we consider the non-impulsive Cauchy problems

$$\begin{cases} u'(t) \in A(t)u(t) + F(t, u_t), & \text{for a.a. } t \in [t_1, t_2], \\ u(t) = v^1(t), & \text{for } t \in [-\tau, t_1], \\ u(t_1^+) = v^1(t_1) + I_1(v_{t_1}^1). \end{cases} \tag{7.27}$$

Here the function v^1 is any solution on the $[-\tau, t_1]$. The mild solutions for (7.27) have the forms

$$u(t) = U(t, t_1)v^1(t_1) + U(t, t_1)I_1(v_{t_1}^1) + \int_{t_1}^t U(t, s)f(s)ds, \quad t \in [t_1, t_2],$$

where $f \in L^1([t_1, t_2], X)$, $f(s) \in F(s, u_s)$.
We have

$$|u(t)| \leq B_m|v^1(t_1)| + B_m|I_1(v_{t_1}^1)| + B_m \int_{t_1}^t |f(s)|ds$$

$$\leq B_m(\bar{K}_1 + \bar{R}) + B_m \int_{t_1}^t \alpha(s)(1 + \|u_t\|_C)ds$$

$$\leq B_m(\bar{K}_1 + \bar{R} + \|\alpha\|_{L([t_1,t_2],\mathbb{R}^+)}) + B_m \int_{t_1}^t \alpha(s)\left(\int_{-\tau}^0 |u_s(\theta)|d\theta\right)ds,$$

where $\bar{R}$ is a common upper bound for $|I_1(v_{t_1})|$, where v is any solution on $[-\tau, t_1]$, which exists because the solution set for (7.26) is compact and I_1 is continuous. So we have

$$
\begin{aligned}
|u(t)| &\le \bar{M}_2 + B_m \int_{t_1}^t \alpha(s)\Big(\int_{-\tau}^0 |u_s(\theta)|d\theta\Big)ds, \\
&\le \bar{M}_2 + B_m \int_{t_1}^t \alpha(s)\tau \cdot \sup_{-\tau \le \theta \le 0} |u_s(\theta)|ds \\
&\le \bar{M}_2 + B_m \int_{t_1}^t \alpha(s)\tau \cdot \sup_{-\tau \le \theta \le 0} |u(s+\theta)|ds \\
&\le \bar{M}_2 + B_m \int_{t_1}^t \alpha(s)\tau \cdot \sup_{s-\tau \le \theta \le s} |u(\theta)|ds \\
&\le \bar{M}_2 + B_m \int_{t_1}^t \alpha(s)\tau \cdot \sup_{-\tau \le \theta \le s} |u(\theta)|ds,
\end{aligned}
$$

where $\bar{M}_2 \ge B_m(\bar{K}_1 + \bar{R} + \|\alpha\|_{L([t_1,t_2],\mathbb{R}^+)})$.

The right-hand side is an increasing function in t, so we have the same estimate for all $t_1 < r \le t$. Therefore

$$
\sup_{t_1 < r \le t} |u(r)| \le \bar{M}_2 + \tau B_m \int_{t_1}^t \alpha(s) \sup_{-\tau \le \theta \le s} |u(\theta)|ds.
$$

Since $|u(s)| \le \bar{M}_2$ for $s < t_1$, we obtain

$$
\sup_{-\tau < r \le t} |u(r)| \le \bar{M}_2 + \tau B_m \int_{t_1}^t \alpha(s) \sup_{-\tau \le \theta \le s} |u(\theta)|ds.
$$

The function $\psi_2(t) = \sup_{-\tau < r \le t} |u(r)|$ is piecewise continuous. For the function

$$
z(t) = \int_{t_1}^t \alpha(s)\psi_2(s)ds,
$$

we have $z(t_1) = 0$ and $z'(t) = \alpha(t)\psi_2(t)$.

We obtained the estimate $\psi_2(t) \le \bar{M}_2 + \tau B_m z(t)$, so

$$
z'(t) \le \alpha(t)(\bar{M}_2 + \tau B_m z(t)).
$$

Next we multiply the above inequality by $e^{-\tau B_m \int_{t_1}^t \alpha(s)ds}$,

$$
z'(t)e^{-\tau B_m \int_{t_1}^t \alpha(s)ds} \le e^{-\tau B_m \int_{t_1}^t \alpha(s)ds}\alpha(t)(\bar{M}_2 + \tau B_m z(t)).
$$

Hence

$$(v(t)e^{-\tau B_m \int_{t_1}^{t} \alpha(s)ds})' \leq e^{-\tau B_m \int_{t_1}^{t} \alpha(s)ds} \alpha(t)\bar{M}_2.$$

We integrate both sides of this inequality from t_1 to t_2 and we obtain

$$z(t_2)e^{-\tau B_m \int_{t_1}^{t_2} \alpha(s)ds} \leq \bar{M}_2 \int_{t_1}^{t_2} \alpha(t)e^{-\tau B_m \int_{t_1}^{t} \alpha(s)ds} dt,$$

so

$$z(t_2) \leq K_2,$$

where $K_2 = \dfrac{\bar{M}_2 \int_{t_1}^{t_2} \alpha(t)e^{-\tau B_m \int_{t_1}^{t} \alpha(s)ds} dt}{e^{-\tau B_m \int_{t_1}^{t_2} \alpha(s)ds}}$. The function z is nondecreasing, so

$$z(t) \leq K_2 \quad \text{for all } t \in [t_1, t_2],$$

hence

$$|u(t)| \leq \bar{M}_2 + \tau B_m K_2 := \bar{K}_2.$$

Without any loss of generality we can assume that $\bar{K}_2 \geq \bar{K}_1$. So for every $k \geq 1$, every solution of the non-impulsive Cauchy problem

$$\begin{cases} u'(t) \in A(t)u(t) + F(t, u_t), & \text{for a.a. } t \in [0, t_k], \\ u(t) = x(t), & \text{for } t \in [-\tau, 0], \\ u(t_i^+) = u(t_i) + I_i(u_{t_i}). & \text{for } i < k, \end{cases} \quad (7.28)$$

is bounded by $\bar{K}_k \geq \bar{K}_{k-1}$. Let $\tilde{K}_m = \tau \bar{K}_m$. Then $\|u_t\|_C \leq \tilde{K}_m$ for every solution u to (7.14). We define a mapping $\tilde{F} : [0, t_m] \times C([-\tau, 0], X) \to P_{cp,cv}(X)$,

$$\tilde{F}(t, c) = \begin{cases} F(t, c), & \text{if } t \in [0, t_m] \text{ and } \|c\|_C \leq \tilde{K}_m, \\ F(t, \dfrac{\tilde{K}_m c}{\|c\|_C}), & \text{if } t \in [0, t_m] \text{ and } \|c\|_C > \tilde{K}_m. \end{cases} \quad (7.29)$$

The function $r : C([-\tau, 0], X) \to \overline{B(0, \tilde{K}_m)} \subset C([-\tau, 0], X)$ given by the formula $r(c) = \frac{\tilde{K}_m c}{\|c\|_C}$ for every $c \in C([-\tau, 0], X)$ with $\|c\|_C > \tilde{K}_m$ and $r(c) = c$ for c with $\|c\|_C \leq \tilde{K}_m$ is a continuous retraction of $C([-\tau, 0], X)$ onto a closed ball. Therefore $\tilde{F}(t, c) = F(t, r(c))$ and $\tilde{F}$ has the same measurability and continuity properties as F. For every $m \geq 1$ and $t \in [0, t_m]$ the following inequalities hold,

$$|\tilde{F}(t, c)| = |F(t, c)| \leq \alpha(t)(1 + \|c\|_C) \leq \alpha(t)(1 + \tilde{K}_m)$$

for every $\|c\|_C \leq \tilde{K}_m$, and

$$|\tilde{F}(t, c)| = \left| F(t, \frac{\tilde{K}_m c}{\|c\|_C}) \right|$$

$$\leq \alpha(t)\left(1 + \left\| \frac{\tilde{K}_m c}{\|c\|_C} \right\|_C\right) \leq \alpha(t)(1 + \tilde{K}_m)$$

for every $\|c\|_C > \tilde{K}_m$. So we have

$$|\tilde{F}(t, c)| \leq \alpha(t)(1 + \tilde{K}_m) := \psi_m(t) \in L^1([0, t_m], \mathbb{R}^+).$$

Now we consider an impulsive problem for fixed m with a multivalued map $\tilde{F}$

$$\begin{cases} u'(t) \in A(t)u(t) + \tilde{F}(t, u_t), & \text{for a.a. } t \in [0, t_m], \ t \neq t_k, \ k < m, \\ u(t) = x(t), & \text{for } t \in [-\tau, 0], \\ u(t_k^+) = u(t_k) + I_k(u_{t_k}), & \text{for } k < m. \end{cases} \tag{7.30}$$

Let $\tilde{\Theta}_m$ be the solution set of problem (7.30). If u is a solution of (7.14), i.e., $u \in \Theta_m$, then $\|u\|_{PC} \leq \tilde{K}_m$. F and $\tilde{F}$ coincide on $\overline{B}(0, \tilde{K}_m)$, so we have that $u \in \tilde{\Theta}_m$. If $u \in \tilde{\Theta}_m$, then we can easily see that $\|u\|_{PC} \leq \tilde{K}_m$ and $u'(t) \in A(t)u(t) + F(t, u_t)$ for a.e. $t \in [0, t_m]$, $t \neq t_k$, $k < m$ and that $u(t) = x(t)$ for $t \in [-\tau, 0]$, so $u \in \Theta_m$. We have $\Theta_m = \tilde{\Theta}_m$. Consequently, we can assume from now on, without any loss of generality, that

$(F_3)'_m$ $|F(t, c)| \leq \psi_m(t)$ for every $t \in [0, t_m]$, where $\psi_m(t) \in L^1([0, t_m], \mathbb{R}^+)$.

Step 2. Now we prove that there exists a sequence of multivalued maps $\{G_n\}_{n=1}^\infty$, $G_n : [0, t_m] \times \mathcal{C}([-\tau, 0], X) \to P_{cl,cv}(X)$ such that

(i) each multivalued map $G_n(t, \cdot) : \mathcal{C}([-\tau, 0], X) \to P_{cl,cv}(X)$, $n \geq 1$, is continuous for a.e. $t \in [0, t_m]$;
(ii) $F(t, c) \subset \cdots \subset G_{n+1}(t, c) \subset G_n(t, c) \subset \overline{co}F(t, B_{3d_n}(c))$, $n \geq 1$;
(iii) $F(t, c) = \bigcap_{n \geq 1} G_n(t, c)$;
(iv) for each $n \geq 1$ there exists a selection $g_n : [0, t_m] \times \mathcal{C}([-\tau, 0], X) \to X$ of G_n, such that $g_n(\cdot, c)$ is measurable and $g_n(t, \cdot)$ is locally Lipschitz.

Consider the sequence $d_n = \frac{1}{3^n}$, $n \geq 1$. Let us cover $\mathcal{C}([-\tau, 0], X)$ by the open balls $\{B_{d_n}(c)\}_{c \in PC([-\tau,0],X)}$. Since the space $\mathcal{C}([-\tau, 0], X)$ is metric, there exists a locally finite refinement $\{V_j\}_{j \in J}$ of the cover $\{B_{d_n}(c)\}_{c \in \mathcal{C}([-\tau,0],X)}$, Now, we can associate a locally Lipschitz partition of unity $\{p_j\}_{j \in J}$ subordinated to the open covering $\{V_j\}_{j \in J}$. For every $j \in J$ let c_j be such that $V_j \subset B_{d_n}(c_j)$ and define

$$G_n(t, c) = \sum_{j \in J} p_j(c) \cdot \overline{co}F(t, B_{2d_n}(c_j)).$$

To prove (ii) and (iii) note that $p_j(c) > 0$ implies that $c \in V_j \subset B_{d_n}(c_j)$, hence $B_{2d_n}(c_j) \subset B_{3d_n(c)}$ and therefore

$$F(t, c) \subset G_n(t, c) \subset \overline{co}F(t, B_{3d_n}(c)).$$

Now we prove that $F(t, c) \supset \bigcap_{n \geq 1} G_n(t, c)$.

Let O be an open and convex set such that $F(t, c) \subset O$. Then from u.s.c. there exists $\delta > 0$ such that, if $d(c, \bar{c}) < \delta$, then $F(t, \bar{c}) \subset O$. Hence, if $3d_n < \delta$ for a.e. n, then

$$F(t, B_{3d_n}(c)) \subset O \Rightarrow \overline{co}F(t, B_{3d_n}(c)) \subset \overline{co}O \subset \overline{O}.$$

Let n_0 be such that $3d_{n_0} < \delta$.

$$\bigcap_{n \geq 1} G_n(t, c) \subset G_{n_0}(t, c) \subseteq \overline{co}F(t, B_{3d_{n_0}}(c)) \subset \overline{O}.$$

Since O was arbitrary, we have $\bigcap_{n \geq 1} G_n(t, c) \subset \bigcap_{\mathbb{O}} \overline{O} = F(t, c)$, where $\mathbb{O}$ denotes a family of all open and convex subsets O such that $F(t, c) \subset O$. To prove (iv) we take, for every c_j, $j \in J$, a measurable selection g_j of the multivalued map $F(\cdot, c_j)$ and define $g_n : [0, t_m] \times C([-\tau, 0], X) \to X$ as

$$g_n(t, c) = \sum_{j \in J} p_j(c) \cdot g_j(t).$$

Step 3. We consider the differential problem

$$\begin{cases} u'(t) \in A(t)u(t) + G_n(t, u_t), & \text{for a.a. } t \in [0, t_m], \ t \neq t_k, \ k < m, \\ u(t) = x(t), & \text{for } t \in [-\tau, 0], \\ u(t_k^+) = u(t_k) + I_k(u_{t_k}), & \text{for } k < m, \end{cases} \tag{7.31}$$

and denote by Θ_m^n the solution set of problem (7.31). From condition (ii) it follows that there is a constant $K_m \geq 0$ such that $\|u\|_{PC} \leq K_m$ for every $u \in \Theta_m^n$.

We show that each sequence $\{u_n\}$ such that $u_n \in \Theta_m^n$ for all $n \geq 1$ has a convergent subsequence $u_{n_k} \to u \in \Theta_m$. Notice that $u_n(t) = x(t)$ for every $t \in [-\tau, 0]$.

Consider the sequence $\{u_n\}$ on $[0, t_1]$. Then

$$u_n(t) = U(t, 0)x(0) + \int_0^t U(t, s)f_n(s)ds \text{ for } t \in [0, t_1],$$

where $f_n \in L^1([0, t_1], X)$, $f_n(t) \in G_n(t, (u_n)_t)$ for a.e. $t \in [0, t_1]$.

For d_k, $k \geq 1$, from $(F_4)_m$ we get $\delta > 0$ such that

$$\beta(F(\{s\} \times O_\delta(\{(u_n)_s\}_{n\geq 1}))) \leq \mu(s)\left(\sup_{-\tau \leq \theta \leq 0} \beta(\{(u_n)_s(\theta)\}_{n\geq 1}) + d_k\right).$$

Take d_l, $l \geq k$, such that $3d_l < \delta$. Then

$$\beta(\{f_n(s)\}_{n\geq 1}) = \beta(\{f_n(s)\}_{n\geq l}) \leq \beta\left[F\left(\{s\} \times O_{3d_l}(\{(u_n)_s\}_{n\geq l})\right)\right]$$

$$\leq \mu(s)\left(\sup_{-\tau \leq \theta \leq 0} \beta(\{u_n(s+\theta)\}_{n\geq l}) + d_k\right)$$

$$\leq \mu(s)\left[\max\left(\sup_{-\tau \leq \theta \leq 0} \beta(\{x(\theta)\}),\ \sup_{0 \leq \theta \leq s} \beta(\{u_n(\theta)\}_{n\geq l})\right) + d_k\right]$$

$$= \mu(s)(\bar{\rho}(s) + d_k),$$

where $\bar{\rho}(s) := \sup_{0 \leq \theta \leq s} \beta(\{u_n(\theta)\}_{n\geq 1})$. This implies that

$$\beta(\{u_n(t)\}_{n\geq 1}) = \beta\left(\left\{U(t,0)x(0) + \int_0^t U(t,s)f_n(s)ds\right\}_{n\geq 1}\right)$$

$$\leq 2B_m \int_0^t \mu(s)(\bar{\rho}(s) + d_k)ds$$

for every $k \geq 1$. But $d_k \searrow 0$ when $k \to \infty$, so

$$\bar{\rho}(t) \leq 2B_m \int_0^t \mu(s)\bar{\rho}(s)ds.$$

From Gronwall's inequality we obtain $\bar{\rho}(t) = 0$, which means that, in the interval $[0, t_1]$,

$$\beta(\{u_n(t)\}_{n\geq 1}) = 0.$$

Hence, $\beta(\{f_n(t)\}_{n\geq 1}) = 0$ for $s \leq t_1$.

For $t = t_1$ one has

$$\beta(\{U(t,t_1)I_1((u_n)_{t_1})\}_{n\geq 1})$$
$$= \beta(\{U(t,t_1)I_1((u_n)_{t_1})\}_{n\geq l})$$
$$\leq \|U(t,t_1)\|_{\mathcal{L}(X)} \cdot \beta(\{I_1((u_n)_{t_1})\}_{n\geq l})$$
$$\leq B_m r_1 \sup_{-\tau \leq \theta \leq 0} \beta(\{(u_n)_{t_1}(\theta)\}_{n\geq l})$$
$$\leq B_m r_1 \cdot \max\left(\sup_{-\tau \leq \theta \leq 0} \beta(\{x(\theta)\}),\ \sup_{0 \leq \theta \leq t_1} \beta(\{u_n(\theta)\}_{n\geq l})\right)$$

$$= B_m r_1 \bar{\rho}(t_1) = 0.$$

Every solution u_n has the form

$$u_n(t) = U(t, t_1)u_n(t_1) + U(t, t_1)I_1((u_n)_{t_1}) + \int_{t_1}^t U(t, s)f_n(s)ds \ \text{ for } t \in [t_1, t_2],$$

where $f_n \in L^1([t_1, t_2], X)$, $f_n(t) \in G_n(t, (u_n)_t)$ for a.e. $t \in [t_1, t_2]$. Analogously as before one shows that $\beta(\{f_n(s)\}_{n\geq 1}) = 0$. Next, we consider $t = t_2$ and so on. Finally, $\beta(\{u_n(t)\}_{n\geq 1}) = 0$ for $t \in [0, t_m]$, and hence $\beta(\{f_n(t)\}_{n\geq 1}) = 0$.

For $t \leq t'$ in $[0, t_1]$ we have

$$|u_n(t') - u_n(t)| = \left| \int_t^{t'} u_n'(s)ds \right| \leq B_m \int_t^{t'} \psi_m(s)ds,$$

since $|G_n(t, c)| \leq \psi_m(t)$ for a.e. $t \in [0, t_m]$ from (ii) and $(F_3)'_m$. Therefore, the family $\{u_n\}$ is equicontinuous. This implies the existence of a convergent subsequence on $[0, t_1]$ denoted by $\{u_{n_{q_1}}^1\}_{q_1} \in Q_1$. Define

$$u_{n_{q_1},1} := \begin{cases} u_{n_{q_1}}^1(t), & \text{for } t \in [0, t_1], \\[2mm] u_{n_{q_1}}^1(t_1) + I_1((u_{n_{q_1}}^1)_{t_1}) + \displaystyle\sum_{t_1 < t_j < t} U(t, t_j)I_j((u_{n_{q_1}})_{t_j}) \\[4mm] \quad + \displaystyle\int_{t_1}^t U(t, s)f_{n_{q_1}}(s)ds, & \text{for } t > t_1. \end{cases}$$

Let $u_{n_{q_1}}^2 := u_{n_{q_1},1}|_{[t_1, t_2]}$. Notice that $u_{n_{q_1}}^2(t_1^+) = u_{n_{q_1}}^1(t_1^+)$ for every $k \geq 1$. For $t \leq t'$ in $[t_1, t_2]$ we have

$$|u_{n_{q_1}}^2(t') - u_{n_{q_1}}^2(t)| \leq B_m \int_t^{t'} \psi_m(s)ds,$$

so the family $\{u_{n_{q_1}}^2\}$ is equicontinuous. Thus it contains a convergent subsequence $\{u_{n_{q_2}}^2\}_{q_2 \in Q_2}$ on $[t_1, t_2]$, where $Q_2 \subset Q_1$. We take a concatenation

$$u_{n_{q_2},2} := \begin{cases} u_{n_{q_2}}^1(t), & \text{for } t \in [0, t_1], \\[2mm] u_{n_{q_2}}^2(t), & \text{for } t \in (t_1, t_2], \\[2mm] u_{n_{q_2}}^2(t) + I_2((u_{n_{q_2}}^2)_{t_2}) + \displaystyle\sum_{t_2 < t_j < t} U(t, t_j)I_j((u_{n_{q_2}})_{t_j}) \\[4mm] \quad + \displaystyle\int_{t_2}^t U(t, s)f_{n_{q_2}}(s)ds, & \text{for } t > t_2. \end{cases}$$

We proceed up to m and find a convergent subsequence $\{u_{n_{qm},m}\}_{q_m \in Q_m}$ of (u_n), where $Q_m \subset Q_{m-1}$. Thus we have

$$
u_{n_{qm},m} := \begin{cases}
u^1_{n_{qm}}(t), & \text{for } t \in [0, t_1], \\
u^2_{n_{qm}}(t), & \text{for } t \in (t_1, t_2], \\
\vdots \\
u^m_{n_{qm}}(t), & \text{for } t \in (t_{m-1}, t_m].
\end{cases}
$$

Let u be a limit of the sequence $\{u^m_{n_{qm}}\}_{q_m \in Q_m}$. Denote the sequence $\{f_{n_{qm}}\}_{q_m \in Q_m}$ by $\{f_{n_{qm}(l)}\}_{l \geq 1}$.

From $(F_3)'_m$, the equality $\beta(\{f_{n_{qm}(l)}(s)\}_{l \geq 1}) = 0$ and Dunford-Pettis theorem it follows that, up to subsequence, $f_{n_{qm}(l)} \rightharpoonup f_0 \in L^1([0, t_m], X)$. By the continuity of impulse functions I_j, we have

$$
u(t) = U(t, 0)x(0) + \sum_{0 < t_j < t} U(t, t_j)I_j(u_{t_j}) + \int_0^t U(t, s)f_0(s)ds.
$$

To prove that $f_0(s) \in F(s, (u[x])_s)$ for a.e. $s \in [0, t_m]$, it is sufficient to use the convexity and closedness of values of F, an upper semicontinuity of $F(t, \cdot)$ and some standard procedures based on Mazur's theorem.

Step 4. From Step 3 it follows that $\sup\{\text{dist}(v, \Theta_m) : v \in \Theta^n_m\} \to 0$ (an easy proof by contradiction). Therefore $\sup\{\text{dist}(v, \Theta_m) : v \in \overline{\Theta^n_m}\} \to 0$ as well. Hence, since Θ_m is compact and $\Theta^{n+1}_m \subset \Theta^n_m$, $\beta(\Theta^n_m) = \beta(\overline{\Theta^n_m}) \searrow 0$ as $n \to \infty$ and $\Theta_m = \bigcap_{n=1}^\infty \overline{\Theta^n_m}$.

Step 5. We show, what is sufficient to finish the proof, that $\overline{\Theta^n_m}$ is contractible for every $n \geq 1$.

Fix $\bar{u} \in \overline{\Theta^n_m}$. We divide the interval $[0, 1]$ on m parts, so we have $0 < \frac{1}{m} < \frac{2}{m} < \cdots < \frac{m-1}{m} < 1$. Let $r \in (0, \frac{1}{m}]$. We consider the problem

$$
\begin{cases}
u'(t) = A(t)u(t) + g_n(t, u_t), & \text{for a.e. } t \in [t_m - mr(t_m - t_{m-1}), t_m], \\
u(t) = \bar{u}(t), & \text{for } t \in [-\tau, t_m - mr(t_m - t_{m-1})], \\
u(t_{m-1}) = \bar{u}(t_{m-1}) + I_{m-1}(\bar{u}_{t_{m-1}}).
\end{cases}
\tag{7.32}
$$

Here g_n is a measurable-locally Lipschitz selection of G_n from Step 2.

Let $\tilde{u}^m_{n,r}$ denote the unique solution of this problem. Then the function $u^m_{n,r}$ defined as

$$
u^m_{n,r} = \begin{cases}
\bar{u}(t), & t \in [0, t_m - mr(t_m - t_{m-1})], \\
\tilde{u}^m_{n,r}(t), & t \in (t_m - mr(t_m - t_{m-1}), t_m]
\end{cases}
$$

satisfies $u^m_{n,r} \in \overline{\Theta^n_m}$.

Next for $r \in (\frac{1}{m}, \frac{2}{m}]$ we consider the problem

$$\begin{cases} u'(t) = A(t)u(t) + g_n(t, u_t), & \text{for a.e. } t \in \left[t_{m-1} - m\left(r - \frac{1}{m}\right)(t_{m-1} - t_{m-2}), t_m\right], \\ u(t) = \bar{u}(t), & \text{for } t \in \left[-\tau, t_{m-1} - m\left(r - \frac{1}{m}\right)(t_{m-1} - t_{m-2})\right], \\ u(t_k^+) = u(t_k) + I_k(u_{t_k}), & k = m - 1, \\ u(t_{m-2}) = \bar{u}(t_{m-2}) + I_{m-2}(\bar{u}_{t_{m-2}}). \end{cases}$$

$$(7.33)$$

Let $\tilde{u}_{n,r}^{m-1}$ denote the unique solution of this problem. Then we have $\tilde{u}_{n,r}^{m-1} \in \overline{\Theta_m^n}$, where

$$u_{n,r}^{m-1} = \begin{cases} \bar{u}(t), & t \in \left[0, t_{m-1} - m\left(r - \frac{1}{m}\right)(t_{m-1} - t_{m-2})\right], \\ \tilde{u}_{n,r}^{m-1}(t), & t \in \left(t_{m-1} - m\left(r - \frac{1}{m}\right)(t_{m-1} - t_{m-2}), t_m\right]. \end{cases}$$

The last problem we consider is for $r \in (\frac{m-1}{m}, 1]$:

$$\begin{cases} u'(t) = A(t)u(t) + g_n(t, u_t), & \text{for a.e. } t \in \left[t_1 - m\left(r - \frac{m-1}{m}\right)t_1, t_m\right], \\ u(t) = \bar{u}(t), & \text{for } t \in \left[-\tau, t_1 - m\left(r - \frac{m-1}{m}\right)t_1\right], \\ u(t_k^+) = u(t_k) + I_k(u_{t_k}), & k \in \{2, 3, ..., m - 1\}, \\ u(t_1) = \bar{u}(t_1) + I_1(\bar{u}_{t_1}). \end{cases}$$

$$(7.34)$$

Let $\tilde{u}_{n,r}^1$ denote the unique solution of this problem. Then the function $u_{n,r}^1$ defined as:

$$u_{n,r}^1 = \begin{cases} \bar{u}(t), & t \in \left[0, t_1 - m\left(r - \frac{m-1}{m}\right)t_1\right], \\ \tilde{u}_{n,r}^1(t), & t \in \left(t_1 - m\left(r - \frac{m-1}{m}\right)t_1, t_m\right], \end{cases}$$

also belongs to $\overline{\Theta_m^n}$.

Finally we consider the following function $h_n : [0, 1] \times \overline{\Theta_m^n} \to \overline{\Theta_m^n}$:

$$h_n(r, \bar{u}) := \begin{cases} \bar{u}, & r = 0, \\ u_{n,r}^m, & r \in \left(0, \frac{1}{m}\right], \\ u_{n,r}^{m-1}, & r \in \left(\frac{1}{m}, \frac{2}{m}\right], \\ \vdots & \\ u_{n,r}^1, & r \in \left(\frac{m-1}{m}, 1\right]. \end{cases}$$

$$(7.35)$$

Here the functions $u_{n,r}^m, u_{n,r}^{m-1}, \cdots, u_{n,r}^1$ are determined by the choice of $\bar{u} \in \overline{\Theta_m^n}$. One can show that the function h_n is continuous applying a standard method, which use a continuous dependence on initial conditions, and remembering that the maps I_k are continuous, when checking a continuity in $r \in \{\frac{i}{m} : i = 1, ..., m-1\}$. The function h_n, as continuous on $[0, 1] \times \overline{\Theta_m^n}$, is a homotopy. By definition we have $h_n(0, \bar{u}) = \bar{u}$ and $h_n(1, \bar{u}) = u_{n,1}^1$, so $\overline{\Theta_m^n}$ is a contractible set for every $n \in \mathbb{N}$. Therefore, from Theorem 2.2 the set $\overline{\Theta_m^n}$ is an R_δ-set.

7.3.2 Noncompact Intervals Case

We consider the compactness of the solution set of problem (7.2) under the following assumptions:

$(A)^\infty$ $\{A(t)\}_{t \in [0,\infty)}$ is a family of linear not necessarily bounded operators $(A(t) : D(A) \subset X \rightarrow X, t \in [0, \infty), D(A)$ a dense subset of X not depending on t) generating an evolution operator $U : \Delta_\infty \rightarrow \mathcal{L}(X)$, where $\Delta_\infty = \{(t, s) \in [0, \infty) \times [0, \infty) : 0 \le s \le t\}$

and $F : [0, \infty) \times C([-\tau, 0], X) \rightarrow P_{cp,cv}(X)$ satifies

$(F_1)^\infty$ $F(\cdot, c)$ has a strongly measurable selection for every $c \in C([-\tau, 0], X)$;
$(F_2)^\infty$ $F(t, \cdot)$ is u.s.c. for a.e. $t \in [0, \infty)$;
$(F_3)^\infty$ F for a.e. $t \in [0, \infty)$ has at most a linear growth, i.e., there exists a function $\alpha \in L_{loc}^1([0, \infty))$ such that

$$|F(t, c)| \le \alpha(t)(1 + \|c\|_C) \text{ for a.e. } t \in [0, \infty);$$

$(F_4)^\infty$ for every $\varepsilon > 0$ and every bounded set $D \subset C([-\tau, 0], X)$ there exist $\delta > 0$ and a function $\mu \in L_{loc}^1([0, \infty), \mathbb{R}^+)$ such that

$$\beta(F(t, O_\delta(D))) \le \mu(t) \sup_{-\tau \le \theta \le 0} \beta(O_\varepsilon(D(\theta))) \text{ for a.e. } t \ge 0.$$

Also, it is clear that $(F_4)^\infty$ can infer the condition:
$(F_4)'^\infty$ there exists a function $\mu \in L^1([0, \infty), \mathbb{R}^+)$ such that

$$\beta(F(t, D)) \le \mu(t) \sup_{-\tau \le \theta \le 0} \beta(D(\theta)) \text{ for a.e. } t \ge 0,$$

and for every bounded $D \subset C([-\tau, 0], \mathbb{R}^+), D(\theta) = \{c(\theta) : c \in D\}$.

Theorem 7.7 *Let the hypothesis $(A)^\infty$ hold and $F : [0, \infty) \times C([-\tau, 0], X) \rightarrow P_{cp,cv}(X)$ satisfy conditions $(F_1)^\infty$-$(F_3)^\infty$, $(F_4)'^\infty$ and maps $I_k : C([-\tau, 0], X) \rightarrow X$, $k \in \mathbb{N}$, be continuous. Then the solution set for problem (7.2) is a nonempty and compact subset of $PC([0, \infty), X)[x]$.*

Proof Besides (7.2), for every $m \in \mathbb{N}^+$, we consider problem (7.14) on a compact interval $[0, t_m]$.

Let $C_m = PC([0, t_m], X)[x]$. Consider the sequence of multivalued maps $\phi_m :$ $C_m \to P(C_m)$ as follows:

$$\phi_m(u)(t) := \left\{ U(t,0)x(0) + \sum_{0 < t_k < t} U(t, t_k) I_k(u_{t_k}) + \int_0^t U(t,s) f(s) ds, \ t \in [0, t_m] \right.$$

$$\left. : f \in L^1([0, t_m], X), \ f(s) \in F(s, u_s) \text{ for a.e. } s \in [0, t_m] \right\}$$

for $t \in [0, t_m]$ and $\phi_m(u)(t) = x(t)$ for $t \in [-\tau, 0]$. Now we consider the projections $p_m^{m+1} : C_{m+1} \to C_m$, which are defined as follows $p_m^{m+1}(u) = u|_{[-\tau, t_m]}$.

We have the equalities

$$\phi_m p_m^{m+1}(u)(t) = \left\{ U(t,0)x(0) + \sum_{0 < t_k < t} U(t, t_k) I_k(u_{t_k}) + \int_0^t U(t,s) f(s) ds, \ t \in [0, t_m] \right.$$

$$\left. : f \in L^1([0, t_m], X), \ f(s) \in F(s, u_s) \text{ for a.e. } s \in [0, t_m] \right\},$$

$$p_m^{m+1} \phi_{m+1}(u)(t) = \left\{ U(t,0)x(0) + \sum_{0 < t_k < t} U(t, t_k) I_k(u_{t_k}) + \int_0^t U(t,s) f(s) ds, \ t \in [0, t_m] \right.$$

$$\left. : f \in L^1([0, t_{m+1}], X), \ f(s) \in F(s, u_s) \text{ for a.e. } s \in [0, t_{m+1}] \right\},$$

and from the observation that

$$\{ f \in L^1([0, t_m], X) : f(s) \in F(s, u_s) \text{ for a.e. } s \in [0, t_m] \}$$
$$= \{ f|_{[-\tau, t_m]}, \ f \in L^1([0, t_{m+1}], X) : f(s) \in F(s, u_s) \text{ for a.e. } s \in [0, t_{m+1}] \}$$

we obtain that $\phi_m p_m^{m+1} = p_m^{m+1} \phi_{m+1}$, so $\{id, \phi_m\}$ is the map of the inverse system $\{C_m, p_m^n\}$. The map $\{id, \phi_m\}$ induces the limit map $\phi : C \to P(C)$, where $C = PC([0, \infty), X)[x]$

$$\phi(u)(t) = \left\{ U(t,0)x(0) + \sum_{0 < t_k < t} U(t, t_k) I_k(u_{t_k}) + \int_0^t U(t,s) f(s) ds, \ t \in [0, \infty) \right.$$

$$\left. : f \in L^1_{loc}([0, \infty), X), \ f(s) \in F(s, u_s) \text{ for a.e. } s \in [0, \infty) \right\}$$

for $t \in [0, t_m]$ and $\phi(u)(t) = x(t)$ for $t \in [-\tau, 0]$. Note that $\Theta := \text{Fix}(\phi) = \lim_{\leftarrow} \Theta_m$ is the solution set of problem (7.2). It is known from Theorem 7.5 that for every $m \geq 1$ the solution set Θ_m to (7.14) is a nonempty and compact subset of $PC([0, t_m], X)[x]$. Using Theorem 1.18 we see that the set Θ is nonempty and compact.

Theorem 7.6 enables us to examine a structure of the solution set on the half-line.

Theorem 7.8 *Let X be a Banach space and the hypothesis $(A)^\infty$ hold. Suppose that the multivalued map $F : [0, \infty) \times \mathcal{C}([-\tau, 0], X) \to P_{cp,cv}(X)$ satisfies conditions $(F_1)^\infty$-$(F_4)^\infty$. Moreover, assume that the maps $I_k : \mathcal{C}([-\tau, 0], X) \to X$, $k \in \mathbb{N}$, are continuous and satisfy (7.25). Then the solution set for problem (7.2) is an R_δ-set in $PC([0, \infty), X)[x]$.*

Proof We have proved in Theorem 7.6, that solution sets on compact intervals are R_δ sets (that is, for problem (7.14)). Next we consider an inverse system like in the proof of Theorem 7.7. Using Theorem 1.19 we obtain that the solution set of problem (7.2) is an R_δ-set.

Chapter 8
Stochastic Evolution Inclusions

Abstract In this chapter, we investigate the topological structure of solution sets for stochastic evolution inclusions in Hilbert spaces in cases that semigroup is compact and noncompact, respectively. It is shown that the solution set is nonempty, compact and R_δ-set which means that the solution set may not be a singleton but, from the point of view of algebraic topology, it is equivalent to a point, in the sense that it has the same homology group as one-point space. As applications of the obtained results, an example is given.

8.1 Introduction

There has been a great deal of interest in optimal control systems described by stochastic and evolution equations. These optimal control problems lead to stochastic and evolution inclusions. In recent years, many authors have investigated the existence of solutions, continuation of solutions, dependence on initial conditions and parameters, and other qualitative behaviors of solutions to stochastic differential inclusions. For more details we refer reader to, e.g., Gawarecki and Mandrekar [111], Kisielewicz [136], Prato [171] and references cited therein.

In this chapter we are interested in developing the topological structure of the solution set for the following stochastic evolution inclusions

$$\begin{cases} du(t) \in [Au(t) + f(t, u(t))]dt + \Sigma(t, u(t))dW(t), & t \in [0, b], \\ u(0) = u_0, \end{cases} \tag{8.1}$$

where A is the infinitesimal generator of a strongly continuous semigroup $\{T(t)\}_{t\geq0}$ in a Hilbert space H with inner product $(\cdot, \cdot)$ and the norm $|\cdot|$, the state $u(\cdot)$ takes values in H, $f : J \times H \to H$, $\Sigma : J \times H \to P(\mathcal{L}_2^0)$ is a nonempty, bounded, closed, and convex multimap, $\{W(t)\}_{t\geq0}$ is a given K-valued Brownian motion or Wiener process with a finite trace nuclear covariance operator $Q \geq 0$; here K is a Hilbert space with inner product $(\cdot, \cdot)_K$ and the norm $|\cdot|_K$, $\mathcal{L}_2^0$ is given as in Sect. 1.8.1.

Throughout this chapter, let H be real separable Hilbert spaces. Denote by $\mathcal{L}(H)$ the space of all linear bounded operators on Hilbert space H with the norm $\|\cdot\|_{\mathcal{L}(H)}$.

© Springer Nature Singapore Pte Ltd. 2017 231
Y. Zhou et al., *Topological Structure of the Solution Set for Evolution Inclusions*,
Developments in Mathematics 51, https://doi.org/10.1007/978-981-10-6656-6_8

Denote by $L^2(\Omega, H)$ the Banach space of all $\mathscr{F}_b$-measurable square integrable random variables with the norm $\|u(\cdot)\| = (\mathbb{E}|u(\cdot, \omega)|^2)^{\frac{1}{2}}$. Let $C([0, b], L^2(\Omega, H))$ be Banach space of continuous functions from $[0, b]$ into $L^2(\Omega, H)$ satisfying $\sup_{t \in [0,b]} \|u(t)\|^2 < \infty$. Let C be a closed subspace of $C([0, b], L^2(\Omega, H))$ consisting of measurable and $\mathscr{F}_t$-adapted H-valued processes $u \in C([0, b], L^2(\Omega, H))$ endowed with the norm

$$\|u\|_C = \left(\sup_{t \in [0,b]} \|u(t)\|^2 \right)^{\frac{1}{2}}.$$

The chapter is divided into three sections. Section 8.2 gives the concept of mild solutions for stochastic inclusion (8.1). In Sect. 8.3, we use the weak topology approach to obtain the existence of solutions. Section 8.4.1 is devoted to proving that the solution set of stochastic inclusion (8.1) is a nonempty compact R_δ-set in the case that the semigroup is compact. Section 8.4.2 provides the existence of mild solutions for stochastic inclusion (8.1) in the case that the semigroup is noncompact, then proceed to study the R_δ-structure of the solution set of (8.1). Finally, we present an example to illustrate the obtained theory.

The results in this chapter are taken from Zhou and Peng [208].

8.2 Statement of the Problem

We study stochastic evolution inclusion (8.1) under the following assumptions:

(H_A) the operator A generates a strongly continuous semigroup $\{T(t)\}_{t \geq 0}$ in H, and there exists a constant $M_1 \geq 1$ such that $\sup_{t \in J} \|T(t)\|_{\mathcal{L}(H)} \leq M_1$;

(H_1) the function $f(t, \cdot) : H \to H$ is weakly sequentially continuous for each $t \in [0, b]$, and maps bounded sets into bounded sets;

$(H_1)'$ the function $f : [0, b] \times H \to H$ is locally Lipschitz and there exist two positive constants c_0, c_1 such that

$$|f(t, u)|^2 \leq c_0 |u|^2 + c_1.$$

The multimap $\Sigma : [0, b] \times H \to P(\mathcal{L}_2^0)$ has closed bounded and convex values, and satisfies the following conditions:

(H_2) $\Sigma(\cdot, u)$ has a measurable selection for every $u \in H$, i.e., there exists a measurable function $\sigma : [0, b] \to \mathcal{L}_2^0$ such that $\sigma(t) \in F(t, u)$ for a.e. $t \in [0, b]$;

(H_3) $\Sigma(t, \cdot)$ is weakly sequentially closed for a.e. $t \in [0, b]$, i.e., it has a weakly sequentially closed graph;

$(H_3)'$ $\Sigma(t, \cdot)$ is weakly u.s.c.;

(H_4) for every $r > 0$, there exists a function $\mu_r \in L^1([0, b], \mathbb{R}^+)$ such that for each $u \in H$, $|u|^2 \leq r$,

$$\|\Sigma(t, u)\|_{\mathcal{L}_2^0}^2 \leq \mu_r(t) \quad \text{for a.e. } t \in [0, b],$$

where $\|\Sigma(t, u)\|_{\mathcal{L}_2^0} = \sup\{\|\sigma(t)\|_{\mathcal{L}_2^0} : \sigma \in \Sigma(t, u)\}$;
$(H_4)'$ there exists a function $\alpha \in L^1([0, b], \mathbb{R}^+)$ such that

$$\|\Sigma(t, u)\|_{\mathcal{L}_2^0}^2 \leq \alpha(t)(1 + |u|^2) \quad \text{for a.e. } t \in [0, b], \ u \in H.$$

Given $u \in C$, let us denote

$$Sel_\Sigma(u) := \{\sigma \in L^2([0, b], \mathcal{L}_2^0) : \sigma(t) \in \Sigma(t, u(t)) \text{ for a.e. } t \in [0, b]\}.$$

The set $Sel_\Sigma(u)$ is always nonempty as Lemmas 8.1 and 8.2 below show.

Lemma 8.1 *Assume that the multimap Σ satisfies conditions (H_2)–(H_4). Then the $Sel_\Sigma(u)$ is nonempty for any $u \in C$.*

Proof Let $u \in C$, by the uniform continuity of u, there exists a sequence $\{u_n\}$ of step functions $u_n : [0, b] \to L^2(\Omega, H)$ such that

$$\sup_{t \in [0,b]} \|u_n(t) - u(t)\|^2 \to 0 \ \text{as } n \to \infty. \tag{8.2}$$

Hence, by (H_2), there exists a sequence of functions $\{\sigma_n\}$ such that $\sigma_n(t) \in \Sigma(t, u_n(t))$ for a.e. $t \in [0, b]$ and $\sigma_n : [0, b] \to \mathcal{L}_2^0$ is measurable for any $n \in \mathbb{N}$. From (8.2) there exists a bounded set $E \subset L^2(\Omega, H)$ such that $u_n(t), u(t) \in E$ for any $t \in [0, b]$ and $n \in \mathbb{N}$, and by (H_4) there exists $\mu_r \in L^1([0, b], \mathbb{R}^+)$ such that

$$\|\sigma_n(t)\|_{\mathcal{L}_2^0}^2 \leq \|\Sigma(t, u)\|_{\mathcal{L}_2^0}^2 \leq \mu_r(t), \quad \forall n \in \mathbb{N} \text{ and a.e. } t \in [0, b].$$

Hence $\{\sigma_n\} \subset L^2([0, b], \mathcal{L}_2^0)$ is bounded and uniformly integrable and $\{\sigma_n(t)\}$ is bounded in $\mathcal{L}_2^0$ for a.e. $t \in [0, b]$. According to the reflexivity of the space $\mathcal{L}_2^0$ and Lemma 1.23, we have the existence of a subsequence, denoted as the sequence, such that

$$\sigma_n \rightharpoonup \sigma \in L^2([0, b], \mathcal{L}_2^0).$$

By Mazur's theorem, we obtain a sequence

$$\tilde{\sigma}_n = \sum_{i=0}^{k_n} \lambda_{n,i} \sigma_{n+i}, \quad \lambda_{n,i} \geq 0, \quad \sum_{i=0}^{k_n} \lambda_{n,i} = 1$$

such that $\tilde{\sigma}_n \to \sigma$ in $L^2([0, b], \mathcal{L}_2^0)$ and, up to subsequence, $\tilde{\sigma}_n(t) \to \sigma(t)$ for all $t \in [0, b]$. By (H_3), the multimap $\Sigma(t, \cdot)$ is locally weakly compact for a.e. $t \in [0, b]$, i.e., for a.e. $t \in [0, b]$ and every $u \in H$, there is a neighbourhood V of u such that the restriction of $\Sigma(t, \cdot)$ to V is weakly compact. Hence by (H_3) and the locally weak compactness, we easily get that $\Sigma(t, \cdot) : H_w \to P((\mathcal{L}_2^0)_w)$ is u.s.c. for a.e. $t \in [0, b]$. Thus, $\Sigma(t, \cdot) : H \to P((\mathcal{L}_2^0)_w)$ is u.s.c. for a.e. $t \in [0, b]$.

To conclude we only need to prove that $\sigma(t) \in \Sigma(t, u(t))$ for a.e. $t \in [0, b]$. Indeed, let Lebesgue measure of N_0 be zero. Then $\Sigma(t, \cdot) : H \to P((L_2^0)_w)$ is u.s.c., $\sigma_n(t) \in \Sigma(t, u_n(t))$ and $\tilde{\sigma}_n(t) \to \sigma(t)$ for all $t \in [0, b] \setminus N_0$ and $n \in \mathbb{N}$. Fix $t_0 \notin N_0$ and assume, by contradiction, that $\sigma(t_0) \notin \Sigma(t_0, u(t_0))$. Since $\Sigma(t_0, u(t_0))$ is closed and convex, from Hahn-Banach theorem there is a weakly open convex set $V \supset \Sigma(t_0, u(t_0))$ satisfying $\sigma(t_0) \notin \overline{V}$. Since $\Sigma(t_0, \cdot) : H \to P((L_2^0)_w)$ is u.s.c., we can find a neighbourhood U of $u(t_0)$ such that $\Sigma(t_0, u) \subset V$ for all $u \in U$. The convergence $u_n(t_0) \to u(t_0)$ as $n \to \infty$ implies the existence of $n_0 \in \mathbb{N}$ such that $u_n(t_0) \in U$ for all $n > n_0$. Therefore $\sigma_n(t_0) \in \Sigma(t_0, u_n(t_0)) \subset V$ for all $n > n_0$. Since V is convex, we also have that $\tilde{\sigma}_n(t_0) \in V$ for all $n > n_0$ and, by the convergence, we get the contradiction $\sigma(t_0) \in \overline{V}$. Thus we conclude $\sigma(t) \in \Sigma(t, u(t))$ for a.e. $t \in [0, b]$.

Lemma 8.2 *Assume that the multimap Σ satisfies conditions (H_2), $(H_3)'$ and $(H_4)'$. Then $Sel_\Sigma(u)$ is weakly u.s.c. with nonempty, convex and weakly compact values.*

Proof Let $u \in \mathcal{C}$, by the uniform continuity of x, there exists a sequence $\{u_n\}$ of step functions, $u_n : [0, b] \to L^2(\Omega, H)$ such that

$$\sup_{t \in [0,b]} \|u_n(t) - u(t)\|^2 \to 0, \quad \text{as } n \to \infty.$$

Hence, by (H_2), there exists a sequence of functions $\{\sigma_n\}$ such that $\sigma_n(t) \in \Sigma(t, u_n(t))$ for a.e. $t \in [0, b]$ and $\sigma_n : [0, b] \to \mathcal{L}_2^0$ is measurable for any $n \in \mathbb{N}$.

Moreover, in view of $(H_4)'$ we have that $\{\sigma_n\} \subset L^2([0, b], \mathcal{L}_2^0)$ is bounded and uniformly integrable and $\{\sigma_n(t)\}$ is bounded in $\mathcal{L}_2^0$ for a.e. $t \in [0, b]$. With the same reason as Lemma 8.1, we obtain a sequence $\tilde{\sigma}_n \in \overline{\text{co}}\{\sigma_k, k \geq n\}$ for $n \geq 1$ such that $\tilde{\sigma}_n \to \sigma$ in $L^2([0, b], \mathcal{L}_2^0)$ and, up to subsequence, $\tilde{\sigma}_n(t) \to \sigma(t)$ for a.e. $t \in [0, b]$ and $\sigma_n(t) \in \Sigma(t, u_n(t))$ for all $n \geq 1$.

Denote by N_0 the set of all $t \in [0, b]$ such that $\tilde{\sigma}_n(t) \to \sigma(t)$ in $\mathcal{L}_2^0$ and $\sigma_n(t) \in \Sigma(t, u_n(t))$ for all $n \geq 1$. Let $x^* \in (\mathcal{L}_2^0)^*$, $\varepsilon > 0$. From $(H_3)'$, it follows immediately that $\langle x^*, \Sigma(t, \cdot) \rangle : H \to P(\mathbb{R})$ is u.s.c. with compact convex values, so ε-δ u.s.c. with compact convex values, here, $\langle x^*, \Sigma(t, \cdot) \rangle$ denote duality product. Accordingly, we have

$$\langle x^*, \tilde{\sigma}_n(t) \rangle \in \overline{\text{co}}\{\langle x^*, \sigma_k(t) \rangle : k \geq n\} \subset \langle x^*, \Sigma(t, u_n(t)) \rangle \subset \langle x^*, \Sigma(t, u(t)) \rangle + (-\varepsilon, \varepsilon).$$

Therefore, we obtain that $\langle x^*, \sigma(t) \rangle \in \langle x^*, \Sigma(t, u) \rangle$ for each $x^* \in (\mathcal{L}_2^0)^*$ and $t \in N_0$. Since Σ has convex and closed values, we conclude that $\sigma(t) \in \Sigma(t, u(t))$ for each $t \in N_0$, which implies $\sigma \in Sel_\Sigma(u)$. This proves the desired result.

Finally, the similar argument (with $\{u_n\} \subset \mathcal{C}$ instead of the step functions) together with Lemma 1.7 shows that $Sel_\Sigma(u)$ is weakly u.s.c. with convex and weakly compact values, completing the proof.

Definition 8.1 A stochastic process $u \in \mathcal{C}$ is said to be a mild solution of inclusion (8.1) if $u(0) = u_0$ and there exists $\sigma(t) \in Sel_\Sigma(u)(t)$ satisfying the following integral equation

$$u(t) = T(t)u_0 + \int_0^t T(t-s)f(s, u(s))ds + \int_0^t T(t-s)\sigma(s)dW(s).$$

Remark 8.1 For any $u \in C$, now define a solution multioperator $\mathscr{F} : C \to P(C)$ as follows:

$$\mathscr{F} = S \circ Sel_{\Sigma},$$

where

$$S(\sigma) = T(t)u_0 + \int_0^t T(t-s)f(s, u(s))ds + \int_0^t T(t-s)\sigma(s)dW(s).$$

It is easy to verify that the fixed points of the multioperator $\mathscr{F}$ are mild solutions of inclusion (8.1).

8.3 Existence via Weak Topology

We study the existence of solutions for stochastic evolution inclusion (8.1) under the conditions (H_A) and (H_1)–(H_4).

Fix $n \in \mathbb{N}$, consider $\mathscr{Q}_n = \{u \in C : \|u\|_C^2 \le n\}$ and denote by $\mathscr{F}_n = \mathscr{F}|_{\mathscr{Q}_n} : \mathscr{Q}_n \to P(C)$ the restriction of the multioperator $\mathscr{F}$ on the set $\mathscr{Q}_n$. We describe some properties of $\mathscr{F}_n$.

Lemma 8.3 *The multioperator $\mathscr{F}_n$ has a weakly sequentially closed graph.*

Proof Let $\{u_m\} \subset \mathscr{Q}_n$ and $\{v_m\} \subset C$ be such that $v_m \in \mathscr{F}_n(u_m)$ for all m and $u_m \rightharpoonup u$, $v_m \rightharpoonup v$ in C, we will prove that $v \in \mathscr{F}_n(u)$.

Since $u_m \in \mathscr{Q}_n$ for all m and $u_m(t) \rightharpoonup u(t)$ for every $t \in [0, b]$, it follows that $\|u(t)\| \le \liminf_{m \to \infty} \|u_m(t)\| \le n^{\frac{1}{2}}$ for all t (see [50, Proposition III.5]). The fact that $v_m \in \mathscr{F}(u_m)$ means that there exists a sequence $\{\sigma_m\}$, $\sigma_m \in Sel_{\Sigma}(u_m)$, such that for every $t \in [0, b]$,

$$v_m(t) = T(t)u_0 + \int_0^t T(t-s)f(s, u_m(s))ds + \int_0^t T(t-s)\sigma_m(s)dW(s).$$

We observe that, according to (H_4), $\|\sigma_m(t)\|_{\mathcal{L}_2^0}^2 \le \mu_n(t)$ for a.e. t and every m, i.e., $\{\sigma_m\}$ is bounded and uniformly integrable and $\{\sigma_m(t)\}$ is bounded in $\mathcal{L}_2^0$ for a.e. $t \in [0, b]$. Hence, by the reflexivity of the space $\mathcal{L}_2^0$ and Lemma 1.23, we have the existence of a subsequence, denoted as the sequence, and a function σ such that $\sigma_m \rightharpoonup \sigma$ in $L^2([0, b], \mathcal{L}_2^0)$.

Moreover, we have

$$\int_0^t T(t-s)\sigma_m(s)dW(s) \rightharpoonup \int_0^t T(t-s)\sigma(s)dW(s).$$

Indeed, let $x' : L^2(\Omega, H) \to \mathbb{R}$ be a linear continuous operator. We first prove that the operator

$$h \mapsto \int_0^t T(t-s)h(s)dW(s)$$

is linear and continuous operator from $L^2([0, b], \mathcal{L}_2^0)$ to $L^2(\Omega, H)$.

For any $h_m, h \in L^2([0, b], \mathcal{L}_2^0)$ and $h_m \to h$ $(m \to \infty)$, by (H_4), we get for each $t \in [0, b]$,

$$\mathbb{E}\left| \int_0^t T(t-s)[h_m(s) - h(s)]dW(s) \right|^2$$

$$\leq M_1^2 tr(Q) \int_0^t \mathbb{E}\|h_m(s) - h(s)\|_{\mathcal{L}_2^0}^2 ds \to 0, \text{ as } m \to \infty.$$

Hence, the operator

$$h \mapsto \int_0^t T(t-s)h(s)dW(s)$$

is continuous. Thus we have that the operator

$$h \mapsto x' \circ \int_0^t T(t-s)h(s)dW(s)$$

is a linear and continuous operator from $L^2([0, b], \mathcal{L}_2^0)$ to $\mathbb{R}$ for all $t \in [0, b]$. Then, from the definition of the weak convergence, we have for every $t \in [0, b]$,

$$x' \circ \int_0^t T(t-s)\sigma_m(s)dW(s) \to x' \circ \int_0^t T(t-s)\sigma(s)dW(s).$$

On the other hand, $f(s, u_m(s)) \rightharpoonup f(s, u(s))$ due to hypotheses (H_1). By the linearity and continuity of the integral and the evolution operator $T(t)$, thus we have that the operator

$$g \mapsto \left\langle x', \int_0^t T(t-s)g(s)ds \right\rangle$$

is a linear and continuous operator from $L^2([0, b], L^2(\Omega, H))$ to $\mathbb{R}$ for all $t \in [0, b]$. Then, from the definition of the weak convergence, we have for every $t \in [0, b]$,

$$\left\langle x', \int_0^t T(t-s)f(s, u_m(s))ds \right\rangle \to \left\langle x', \int_0^t T(t-s)f(s, u(s))ds \right\rangle.$$

Thus,

$$\int_0^t T(t-s)f(s, u_m(s))ds \rightharpoonup \int_0^t T(t-s)f(s, u(s))ds.$$

From the above arguments, we have

$$v_m(t) \rightharpoonup T(t)u_0 + \int_0^t T(t-s)f(s,u(s))ds$$

$$+ \int_0^t T(t-s)\sigma(s)dW(s) = v^*(t), \quad \forall t \in [0,b],$$

which implies, for the uniqueness of the weak limit in $L^2(\Omega, H)$, that $v^*(t) = v(t)$ for all $t \in [0, b]$.

By using the similar method in Lemma 8.1, we can prove that $\sigma(t) \in \Sigma(t, u(t))$ for a.e. $t \in [0, b]$. The proof is completed.

Lemma 8.4 *The multioperator $\mathscr{F}_n$ is weakly compact.*

Proof We first prove that $\mathscr{F}_n(\mathcal{Q}_n)$ is relatively weakly sequentially compact.

Let $\{u_m\} \subset \mathcal{Q}_n$ and $\{v_m\} \subset \mathcal{C}$ satisfying $v_m \in \mathscr{F}_n(u_m)$ for all $m \in \mathbb{N}$. By the definition of the multioperator $\mathscr{F}_n$, there exists a sequence $\{\sigma_m\}$, $\sigma_m \in Sel_\Sigma(u_m)$, such that for all $t \in [0, b]$,

$$v_m(t) = T(t)u_0 + \int_0^t T(t-s)f(s,u_m(s))ds + \int_0^t T(t-s)\sigma_m(s)dW(s).$$

Further, as the reason for Lemma 8.3, we have that there exists a subsequence, denoted as the sequence, and a function σ such that $\sigma_m \rightharpoonup \sigma$ in $L^2([0,b], \mathcal{L}_2^0)$. Since the operator f maps bounded sets into bounded sets and $\mathcal{Q}_n$ is bounded, we obtain that $f(s, u_m(s)) \rightharpoonup \overline{f}(s) \in H$ up to subsequence. Therefore,

$$v_m(t) \rightharpoonup \bar{v}(t) = T(t)u_0 + \int_0^t T(t-s)\overline{f}(s)ds$$

$$+ \int_0^t T(t-s)\sigma(s)dW(s), \quad \forall t \in [0,b].$$

Furthermore, by (H_A), (H_1) and (H_4), we have

$$\mathbb{E}|v_m(t)|^2 \le 3\mathbb{E}|T(t)u_0|^2 + 3\mathbb{E}\left|\int_0^t T(t-s)f(s,u_m(s))ds\right|^2$$

$$+ 3\mathbb{E}\left|\int_0^t T(t-s)\sigma_m(s)dW(s)\right|^2$$

$$\le 3M_1^2\mathbb{E}|u_0|^2 + 3M_1^2\int_0^t \mathbb{E}|f(s,u_m(s))|^2ds + 3M_1^2tr(Q)\int_0^t \mathbb{E}\|\sigma_m(s)\|_{\mathcal{L}_2^0}^2 ds$$

$$\le M_1^2\mathbb{E}|u_0|^2 + 3M_1^2bl + 3M_1^2tr(Q)\|\mu_n\|_{L([0,b],\mathbb{R})}$$

for all $m \in \mathbb{N}$ and $t \in [0, b]$, where $l = \max_m \mathbb{E}|f(s, u_m(s))|^2$. Thus

$$\|v_m(t)\|^2 \leq M_1^2 \mathbb{E}|u_0|^2 + 3M_1^2 bl + 3M_1^2 tr(Q)\|\mu_n\|_{L([0,b],\mathbb{R})} \leq N \quad \text{for some } N.$$

Recalling the weak convergence of $C([0, b], L^2(\Omega, H))$, it is easy to prove that $v_m \rightharpoonup \bar{v}$ in $\mathcal{C}$. Thus $\mathscr{F}_n(\mathcal{Q}_n)$ is relatively weakly sequentially compact, hence relatively weakly compact by Theorem 1.26.

Lemma 8.5 *The multioperator $\mathscr{F}_n$ has convex and weakly compact values.*

Proof Fix $u \in \mathcal{Q}_n$, since Σ is convex valued, from the linearity of the integral and the operator $T(t)$, it follows that the set $\mathscr{F}_n(u)$ is convex. The weak compactness of $\mathscr{F}_n(u)$ follows by Lemmas 8.3 and 8.4.

Now we state the main results of this section.

Theorem 8.1 *Assume that (H_A), (H_1)–(H_4) hold. Moreover,*

$$\liminf_{n \to \infty} \frac{1}{n} \int_0^t \mu_n(s)ds = 0. \tag{8.3}$$

Then inclusion (8.1) has at least one mild solution.

Proof We show that there exists $n \in \mathbb{N}$ such that the operator $\mathscr{F}_n$ maps the ball $\mathcal{Q}_n$ into itself.

Assume to the contrary, that there exist sequences $\{u_n\}$, $\{z_n\}$ such that $u_n \in \mathcal{Q}_n$, $z_n \in \mathscr{F}_n(u_n)$ and $z_n \notin \mathcal{Q}_n$, $\forall\, n \in \mathbb{N}$. Then there exists a sequence $\{\sigma_n\} \subset L^2([0, b], \mathcal{L}_2^0)$, $\sigma_n(s) \in \Sigma(s, u_n(s))$, $\forall\, n \in \mathbb{N}$ and for a.e. $s \in [0, b]$ such that

$$z_n(t) = T(t)u_0 + \int_0^t T(t-s)f(s, u_n(s))ds + \int_0^t T(t-s)\sigma_n(s)dW(s), \quad \forall\, t \in [0, b].$$

As the reason for Lemma 8.4, we have

$$1 < \frac{\|z_n\|_{\mathcal{C}}^2}{n} \leq \frac{1}{n}\left(3M_1^2\|u_0\|^2 + 3M_1^2 bl\right) + \frac{3M_1^2 tr(Q)}{n}\int_0^t \mu_n(s)ds, \quad n \in \mathbb{N},$$

which contradicts (8.3).

Now, choose $n \in \mathbb{N}$ such that $\mathscr{F}_n(\mathcal{Q}_n) \subseteq \mathcal{Q}_n$. By Lemma 8.4, the set $V_n = \overline{\mathscr{F}_n(\mathcal{Q}_n)}^w$ is weakly compact. Let now $\tilde{V}_n = \overline{co}(V_n)$, where $\overline{co}(V_n)$ denotes the closed convex hull of V_n. By Theorem 1.27, $\tilde{V}_n$ is a weakly compact set. Moreover from the fact that $\mathscr{F}_n(\mathcal{Q}_n) \subset \mathcal{Q}_n$ and that $\mathcal{Q}_n$ is a convex closed set we have that $\tilde{V}_n \subset \mathcal{Q}_n$ and hence

$$\mathscr{F}_n(\tilde{V}_n) = \mathscr{F}_n(\overline{co}(\mathscr{F}_n(\mathcal{Q}_n))) \subseteq \mathscr{F}_n(\mathcal{Q}_n) \subseteq \overline{\mathscr{F}_n(\mathcal{Q}_n)}^w = V_n \subset \tilde{V}_n.$$

In view of Lemma 8.3, $\mathscr{F}_n$ has a weakly sequentially closed graph. Thus by Theorem 1.29, inclusion (8.1) has a solution. The proof is completed.

Remark 8.2 Suppose, for example, that there exist $\alpha(t) \in L^1([0, b], \mathbb{R}^+)$ and a nondecreasing function $\rho : [0, +\infty) \to [0, +\infty)$ such that $\mathbb{E}\|\Sigma(t, u)\|^2_{\mathcal{L}^0_2} \leq \alpha(t)\rho$ $(\|u\|^2_{\mathcal{C}})$ for a.e. $t \in [0, b]$ and every $u \in \mathcal{C}$. Then condition (8.3) is equivalent to

$$\liminf_{n \to \infty} \frac{\rho(n)}{n} = 0.$$

Theorem 8.2 *Assume that (H_A), (H_1)–(H_3) and $(H_4)'$ hold. If*

$$3M_1^2 tr(Q)\|\alpha\|_{L([0,b],\mathbb{R})} < 1, \tag{8.4}$$

then inclusion (8.1) has at least one mild solution.

Proof As the reason for Theorem 8.1, assume that there exist $\{u_n\}$, $\{z_n\}$ such that $u_n \in \mathcal{Q}_n$, $z_n \in \mathscr{F}_n(u_n)$ and $z_n \notin \mathcal{Q}_n$, $\forall n \in \mathbb{N}$, we get

$$\mathbb{E}|z_n(t)|^2 \leq 3M_1^2\|u_0\|^2 + 3M_1^2 \int_0^t l\,ds + 3M_1^2 tr(Q) \int_0^t \alpha_n(t)(1 + \mathbb{E}|u_n(s)|^2)ds$$

$$\leq 3M_1^2\|u_0\|^2 + 3M_1^2 bl + 3M_1^2 tr(Q)(1 + n)\|\alpha\|_{L([0,b],\mathbb{R})}, \quad n \in \mathbb{N},$$

so

$$n < \|z_n\|^2_{\mathcal{C}} \leq 3M_1^2\|u_0\|^2 + 3M_1^2 bl + 3M_1^2 tr(Q)(1 + n)\|\alpha\|_{L([0,b],\mathbb{R})},$$

which contradicts (8.4).

The conclusion then follows by Theorem 1.29, like Theorem 8.1.

Furthermore we also consider superlinear growth condition, as next theorem shows.

Theorem 8.3 *Assume that (H_A) and (H_1)–(H_3) hold. Suppose in addition that $(H_4)''$ there exist $\alpha \in L^1([0, b], \mathbb{R}^+)$ and a nondecreasing function $\rho : [0, +\infty) \to [0, +\infty)$ such that*

$$\mathbb{E}\|\Sigma(t, u)\|^2_{\mathcal{L}^0_2} \leq \alpha(t)\rho(\|u\|^2_{\mathcal{C}}) \text{ for a.e. } t \in [0, b], \ \forall\, u \in \mathcal{C}.$$

Furthermore, there exists a constant $R > 0$ such that

$$3M_1^2\big(\|u_0\|^2 + bl + tr(Q)\rho(R)\|\alpha\|_{L([0,b],\mathbb{R})}\big) < R.$$

Then inclusion (8.1) has at least one mild solution.

Proof It is sufficient to prove that the operator $\mathscr{F}$ maps the ball $\mathscr{Q}_R$ into itself. In fact, given any $u \in \mathscr{Q}_R$ and $z \in \mathscr{F}(u)$, it holds

$$\mathbb{E}|z(t)|^2 \leq 3M_1^2|u_0\|^2 + 3M_1^2 \int_0^t l\,ds + 3M_1^2 tr(Q) \int_0^b \alpha(s)\rho(\|u\|_C^2)ds$$
$$\leq 3M_1^2\|u_0\|^2 + 3M_1^2 bl + 3M_1^2 tr(Q)\rho(R)\|\alpha\|_{L([0,b],\mathbb{R})},$$

which implies

$$\|y\|_C^2 \leq 3M_1^2\|u_0\|^2 + 3M_1^2 bl + 3M_1^2 tr(Q)\rho(R)\|\alpha\|_{L([0,b],\mathbb{R})} \leq R.$$

The conclusion then follows by Theorem 1.29, like Theorem 8.1.

8.4   Topological Structure of Solution Set

In this section, we study the topological structure of solution sets in cases that $T(t)$ is compact and noncompact, respectively. Similar to the proof of [71, Lemma 3.3], we get the following lemma.

Lemma 8.6 *Assume that (H_2), $(H_3)'$ and $(H_4)'$ hold. Then there exists a sequence $\{\Sigma_n\}$ with $\Sigma_n : [0, b] \times H \rightarrow P_{cl,cv}(\mathcal{L}_2^0)$ such that*

(i) *$\Sigma(t, u) \subset \Sigma_{n+1}(t, u) \subset \Sigma_n(t, u) \subset \overline{co}(\Sigma(t, B_{3^{1-n}}(u)))$, $n \geq 1$, for each $t \in [0, b]$ and $u \in H$;*

(ii) *$\|\Sigma_n(t, u)\|_{\mathcal{L}_2^0}^2 \leq \alpha(t)(3 + 2|u|^2)$, $n \geq 1$, for a.e. $t \in [0, b]$ and each $u \in H$;*

(iii) *there exists $E \subset [0, b]$ with $mes(E) = 0$ such that for each $x^* \in \mathcal{L}_2^0$, $\varepsilon > 0$ and $(t, u) \in [0, b] \setminus E \times H$, there exists $N > 0$ such that for all $n \geq N$,*

$$\langle x^*, \Sigma_n(t, u) \rangle \subset \langle x^*, \Sigma(t, u) \rangle + (-\varepsilon, \varepsilon);$$

(iv) *$\Sigma_n(t, \cdot) : H \rightarrow P(\mathcal{L}_2^0)$ is continuous for a.e. $t \in [0, b]$ with respect to Hausdorff metric for each $n \geq 1$;*

(v) *for each $n \geq 1$, there exists a selection $\tilde{\sigma}_n : [0, b] \times H \rightarrow \mathcal{L}_2^0$ of Σ_n such that $\tilde{\sigma}_n(\cdot, u)$ is measurable for each $u \in H$ and for any compact subset $\mathscr{D} \subset H$ there exist constants $C_V > 0$ and $\delta > 0$ for which the estimate*

$$\|\tilde{\sigma}_n(t, u_1) - \tilde{\sigma}_n(t, u_2)\|_{\mathcal{L}_2^0}^2 \leq C_V \alpha(t)|u_1 - u_2|^2$$

holds for a.e. $t \in [0, b]$ and each $u_1, u_2 \in V$ with $V := \mathscr{D} + B_\delta(0)$;

(vi) *Σ_n verifies the condition $(H_3)'$ with Σ_n instead of Σ for each $n \geq 1$.*

8.4.1 Compact Operator Case

In this subsection, we study the topological structure of solution sets in the case that $T(t)$ is compact. The following compactness characterizations of solution sets to inclusion (8.1) will be useful.

Lemma 8.7 *Suppose that $T(t)$ is compact for $t > 0$ and $(H_1)'$ holds. Let $D \subset L^2(\Omega, H)$ be relatively compact and $U \subset L^2([0, b], \mathcal{L}_2^0)$-integrable bounded, that is, $\|\sigma(t, u)\|_{\mathcal{L}_2^0}^2 \leq \gamma(t)$ for all $\sigma \in U$ and a.e. $t \in [0, b]$, where $\gamma \in L^1([0, b], \mathbb{R}^+)$. Then the set of mild solutions*

$$\{u(\cdot, u_0, \sigma) : u_0 \in D, \ \sigma \in U\}$$

is relatively compact in $\mathcal{C}$.

Proof Write

$$\Delta(D \times U) = \{u(\cdot, u_0, \sigma) : u_0 \in D, \ \sigma \in U\}.$$

Let $t \in [0, b]$ be arbitrary and $\varepsilon > 0$ be small enough. Define an operator $\Psi_\varepsilon : \Delta(D \times U)(t) \to L^2(\Omega, H)$ by

$$\Psi_\varepsilon u(t) = T(t)u_0 + \int_0^{t-\varepsilon} T(t-s)f(s, u(s))ds + \int_0^{t-\varepsilon} T(t-s)\sigma(s)dW(s).$$

Using the compactness of $T(t)$ for $t > 0$, we deduce that the set $\{\Psi_\varepsilon u(t) : u(t) \in \Delta(D \times U)(t)\}$ is relatively compact in $L^2(\Omega, H)$ for every ε, $0 < \varepsilon < t$. Moreover, for every $u \in \Delta(D \times U)$, we have

$$\mathbb{E}|\Psi_\varepsilon u(t) - u(t)|^2 \leq 2\mathbb{E}\left|\int_{t-\varepsilon}^t T(t-s)f(s, u(s))ds\right|^2 + 2\mathbb{E}\left|\int_{t-\varepsilon}^t T(t-s)\sigma(s)dW(s)\right|^2$$

$$\leq 2M_1^2 \int_{t-\varepsilon}^t \mathbb{E}|f(s, u(s))|^2 ds + 2M_1^2 tr(Q) \int_{t-\varepsilon}^t \mathbb{E}\|\sigma(s)\|_{\mathcal{L}_2^0}^2 ds$$

$$\leq 2M_1^2 \int_{t-\varepsilon}^t (c_0 \mathbb{E}|u(s)|^2 + c_1)ds + 2M_1^2 tr(Q) \int_{t-\varepsilon}^t \gamma(t)ds$$

$$\to 0 \text{ as } \varepsilon \to 0.$$

Then, we obtain $\|\Psi_\varepsilon u(t) - u(t)\| \to 0$, which proves that the identity operator $I : \Delta(D \times U)(t) \to \Delta(D \times U)(t)$ is a compact operator, which yields that the set $\Delta(D \times U)(t)$ is relatively compact in $L^2(\Omega, H)$ for each $t \in (0, b]$.

We proceed to verify that the set $\Delta(D \times U)$ is equicontinuous on $(0, b]$. Taking $0 < t_1 < t_2 \leq b$. For each $u \in \Delta(D \times U)$, we obtain

$$\mathbb{E}|u(t_2) - u(t_1)|^2 \leq 3\left[\mathbb{E}\left|(T(t_2) - T(t_1))u_0\right|^2 + \sum_{i=1}^2 \mathbb{E}|I_i(t_2) - I_i(t_1)|^2\right],$$

where

$$I_1(t) = \int_0^t T(t-s)f(s, u(s))ds,$$

$$I_2(t) = \int_0^t T(t-s)\sigma(s)dW(s).$$

From the strong continuity of $T(t)$, it is clear that the first term goes to zero as $t_2 - t_1 \to 0$.

Next, it follows from the assumption $(H_1)'$ that

$$\mathbb{E}|I_1(t_2) - I_1(t_1)|^2 \le 2\mathbb{E}\left|\int_{t_1}^{t_2} T(t_2-s)f(s, u(s))ds\right|^2$$

$$+ 2\mathbb{E}\left|\int_0^{t_1} \left(T(t_2-s) - T(t_1-s)\right)f(s, u(s))ds\right|^2$$

$$\le 2M_1^2 \int_{t_1}^{t_2} \mathbb{E}|f(s, u(s))|^2 ds$$

$$+ 2\sup_{s\in[0,t_1]} \|T(t_2-s) - T(t_1-s)\|_{\mathcal{L}(H)}^2 \int_0^{t_1} \mathbb{E}|f(s, u(s))|^2 ds.$$

Further, we obtain

$$\mathbb{E}|I_2(t_2) - I_2(t_1)|^2$$

$$\le 2\mathbb{E}\left|\int_{t_1}^{t_2} T(t_2-s)\sigma(s)dW(s)\right|^2 + 2\mathbb{E}\left|\int_0^{t_1} \left(T(t_2-s) - T(t_1-s)\right)\sigma(s)dW(s)\right|^2$$

$$\le 2M_1^2 tr(Q) \int_{t_1}^{t_2} \mathbb{E}\|\sigma(s)\|_{\mathcal{L}_2^0}^2 ds$$

$$+ 2tr(Q) \sup_{s\in[0,t_1]} \|T(t_2-s) - T(t_1-s)\|_{\mathcal{L}(H)}^2 \int_0^{t_1} \mathbb{E}\|\sigma(s)\|_{\mathcal{L}_2^0}^2 ds$$

$$\le 2M_1^2 tr(Q) \int_{t_1}^{t_2} \gamma(s)ds + 2tr(Q) \sup_{s\in[0,t_1]} \|T(t_2-s) - T(t_1-s)\|_{\mathcal{L}(H)}^2 \int_0^{t_1} \gamma(s)ds.$$

Therefore, for $t_2 - t_1$ sufficiently small, the right-hand sides of above two inequalities tend to zero, since $T(t)$ is strongly continuous, and the compactness of $T(t)$ $(t > 0)$, implies the continuity in the uniform operator topology.

Moreover, we see by the relative compactness of D that these limits remain true uniformly for $u \in \Delta(D \times U)$. That is to say

$$\|u(t_1) - u(t_2)\|^2 \to 0 \text{ as } t_2 - t_1 \to 0$$

uniformly for $u \in \Delta(D \times U)$ and hence we get the desired result.

Thus, an application of Arzela-Ascoli's theorem justifies that $\Delta(D \times U)$ is relatively compact in $\mathcal{C}$. The proof is completed.

Let us also present the following approximation result, the proof of which is similar to the proof of [197, Lemma 2.4].

Lemma 8.8 *Suppose $\{T(t)\}_{t>0}$ is compact and $f(t, u)$ is continuous. If the two sequences $\{\sigma_m\} \subset L^2([0, b], \mathcal{L}_2^0)$ and $\{u_m\} \subset \mathcal{C}$, where u_m is a mild solution of the stochastic problem*

$$\begin{cases} du_m(t) = [Au_m(t) + f(t, u_m(t))]dt + \sigma_m(t)dW(t), & t \in (0, b], \\ u_m(0) = u_0, \end{cases}$$

$\lim_{m \to \infty} \sigma_m = \sigma$ *weakly in* $L^2([0, b], \mathcal{L}_2^0)$ *and* $\lim_{m \to \infty} u_m = u$ *in* $\mathcal{C}$, *then* u *is a mild solution of the limit problem*

$$\begin{cases} du(t) = [Au(t) + f(t, u(t)]dt + \sigma(t)dW(t), & t \in (0, b], \\ u(0) = u_0. \end{cases}$$

Theorem 8.4 *Let conditions (H_A), $(H_1)'$, (H_2), $(H_3)'$ and $(H_4)'$ be satisfied. Suppose in addition that $T(t)$ is compact for $t > 0$. Then the solution set of stochastic evolution inclusion (8.1) for fixed $u_0 \in L^2(\Omega, H)$ is nonempty in $\mathcal{C}$.*

Proof Set

$$\mathcal{M}_0 = \{u \in \mathcal{C} : \|u(t)\|^2 \leq \psi(t), \ t \in [0, b]\},$$

where $\psi(t)$ is the solution of the following equation

$$\begin{cases} \psi'(t) = 3M_1^2(c_0 + tr(Q)\alpha(t))\psi(t) + 3M_1^2(c_1 + tr(Q)\alpha(t)) & \text{a.e. on } [0, b], \\ \psi(0) = 3M_1^2\|u_0\|^2. \end{cases}$$

It is clear that $\mathcal{M}_0$ is closed and convex in $\mathcal{C}$. We first show that $\mathscr{F}(\mathcal{M}_0) \subset \mathcal{M}_0$. Indeed, taking $u \in \mathcal{M}_0$ and $v \in \mathscr{F}(u)$, we have

$$\mathbb{E}|v(t)|^2 \leq 3\mathbb{E}|T(t)u_0|^2 + 3\mathbb{E}\left|\int_0^t T(t-s)f(s, u(s))ds\right|^2$$

$$+ 3\mathbb{E}\left|\int_0^t T(t-s)\sigma(s)dW(s)\right|^2$$

$$\leq 3M_1^2\mathbb{E}|u_0|^2 + 3M_1^2\int_0^t \mathbb{E}|f(s, u(s))|^2 ds + 3M_1^2 tr(Q)\int_0^t \mathbb{E}\|\sigma(s)\|_{\mathcal{L}_2^0}^2 ds$$

$$\leq 3M_1^2\|u_0\|^2 + 3M_1^2\int_0^t (c_0\psi(s) + c_1)ds + 3M_1^2 tr(Q)\int_0^t \alpha(s)(1 + \psi(s))ds$$

$$\leq \psi(t),$$

which implies that $\|v(t)\|^2 \le \psi(t)$. Thus $v \in \mathcal{M}_0$. Set $\widetilde{\mathcal{M}} = \overline{\text{co}}\mathcal{F}(\mathcal{M}_0)$, it is clear that $\widetilde{\mathcal{M}}$ is a closed, bounded and convex set. Also, $\widetilde{\mathcal{M}}$ is invariant under $\mathcal{F}$, i.e., $\mathcal{F}(\widetilde{\mathcal{M}}) \subset \widetilde{\mathcal{M}}$. Moreover, using Lemma 8.7 enables us to find that $\widetilde{\mathcal{M}}$ is relative compact in $\mathcal{C}$.

We consider $\mathcal{F} : \widetilde{\mathcal{M}} \to P(\widetilde{\mathcal{M}})$. In order to apply the fixed point theorem given by Theorem 1.17, it remains to show that $\mathcal{F}$ is u.s.c. with contractible values.

By Lemma 1.9, it suffices to show that $\mathcal{F}$ has a closed graph. Let $u_n \subset \widetilde{\mathcal{M}}$ with $u_n \to u$ and $v_n \in \mathcal{F}(u_n)$ with $v_n \to v$. We shall prove that $v \in \mathcal{F}(u)$. By the definition of $\mathcal{F}$, there exist $\sigma_n \in Sel_{\Sigma}(u_n)$ such that

$$v_n(t) = T(t)u_0 + \int_0^t T(t-s)f(s, u_n(s))ds + \int_0^t T(t-s)\sigma_n(s)dW(s), \quad \forall t \in [0, b].$$

We need to prove that there exists $\sigma \in Sel_{\Sigma}(u)$ such that for a.e. $t \in [0, b]$,

$$v(t) = T(t)u_0 + \int_0^t T(t-s)f(s, u(s))ds + \int_0^t T(t-s)\sigma(s)dW(s).$$

First, we should prove

$$\int_0^t T(t-s)f(s, u_n(s))ds \to \int_0^t T(t-s)f(s, u(s))ds.$$

Since $u_n \subset \widetilde{\mathcal{M}}$ and $u_n \to u$, using $(H_1)'$, we have $f(t, u_n(t)) \to f(t, u(t))$ as $n \to \infty$. On the other hand, we get for each $t \in [0, b]$,

$$\int_0^t \mathbb{E}|f(s, u_n(s)) - f(s, u(s))|^2 ds \le 4\left(c_0 \int_0^t \psi(s)ds + c_1 t\right) < \infty.$$

By Lebesgue's dominated convergence theorem, we get

$$\mathbb{E}\left|\int_0^t T(t-s)\big(f(s, u_n(s)) - f(s, u(s))\big)ds\right|^2$$
$$\le M_1^2 \mathbb{E}\left(\int_0^t |f(s, u_n(s)) - f(s, u(s))|ds\right)^2$$
$$\le M_1^2 \int_0^t \mathbb{E}|f(s, u_n(s)) - f(s, u(s))|^2 ds$$
$$\to 0 \text{ as } n \to \infty.$$

Therefore,

$$\int_0^t T(t-s)f(s, u_n(s))ds \to \int_0^t T(t-s)f(s, u(s))ds.$$

Noticing that Sel_Σ is weakly u.s.c. with weakly compact and convex values due to Lemma 8.2, an application of Lemma 1.7 yields that there exist $\sigma \in Sel_\Sigma(u)$ and a subsequence of σ_n, still denoted by σ_n, such that $\sigma_n \rightharpoonup \sigma$ in $L^2([0, b], \mathcal{L}_2^0)$. From this and Lemma 8.8, we see that $v(t) = S(\sigma)$ and $v \in \mathcal{F}(u)$. It follows that $\mathcal{F}$ is closed.

After that, we can show that $\mathcal{F}$ has contractible values. Given $u \in \widetilde{\mathcal{M}}$, fix $\sigma^* \in Sel_\Sigma(u)$ and put $v^* = S(\sigma^*)$. Define a function $h : [0, 1] \times \mathcal{F}(u) \to \mathcal{F}(u)$ as

$$h(\lambda, v)(t) = \begin{cases} v(t), & t \in [0, \lambda b], \\ u(t, \lambda b, v(\lambda b)), & t \in (\lambda b, b], \end{cases}$$

for each $(\lambda, v) \in [0, 1] \times \mathcal{F}(u)$, where

$$u(t, \lambda b, v(\lambda b)) = T(t - \lambda b)v(\lambda b) + \int_{\lambda b}^t T(t - s) f(s, u(s)) ds$$
$$+ \int_{\lambda b}^t T(t - s)\sigma^*(s) dW(s).$$

It is easy to see that h is well defined. Also, it is clear that

$$h(0, v) = v^*, \quad h(1, v) = v \text{ on } \mathcal{F}(u).$$

Moreover, it follows readily that h is continuous. Thus, we have proved that $\mathcal{F}(u)$ is contractible.

Remark 8.3 Now, we let $\Theta(u_0)$ denote the solution set of inclusion (8.1), then $\Theta(u_0)$ is a compact subset of $\mathcal{C}$. Indeed, $\Theta(u_0) = \text{Fix}(\mathcal{F})$, so $\Theta(u_0) \subset \mathcal{F}(\Theta(u_0))$. Assume that $\{v_n\} \subset \Theta(u_0)$, then one can choose $\sigma_n \in Sel_\Sigma(v_n)$ such that $v_n = S(\sigma_n)$. By Lemma 8.7, we obtain that $\{v_n\}$ is relatively compact.

Theorem 8.5 *Under the conditions in Theorem 8.4, $\Theta(u_0)$ is a compact R_δ-set.*

Proof To this aim, let us consider the following stochastic evolution inclusion

$$\begin{cases} du(t) \in [Au(t) + f(t, u(t))]dt + \Sigma_n(t)dW(t), & t \in [0, b], \\ u(0) = u_0, \end{cases} \tag{8.5}$$

where multivalued functions $\Sigma_n : [0, b] \times H \to P_{cl,cv}(\mathcal{L}_2^0)$ are established in Lemma 8.6. Let $\Theta_n(u_0)$ denote the set of all mild solutions of inclusion (8.5).

From Lemma 8.6 (ii) and (vi), it follows that $\{\Sigma_n\}$ verifies the conditions (H_2), $(H_3)'$ and $(H_4)'$ for each $n \geq 1$. Then from Lemma 8.2 one finds that Sel_{Σ_n} is nonempty weakly u.s.c. with convex and weakly compact values. Moreover, one can see from Theorem 8.4 and Remark 8.3 that the solution set of inclusion (8.5) is nonempty and compact in $\mathcal{C}$ for each $n \geq 1$.

We show that $\Theta_n(u_0)$ is contractible for all $n \geq 1$. To do this, let $u \in \Theta_n(u_0)$ and for any $\lambda \in [0, 1)$, we deal with the existence and uniqueness of solutions to the integral equation

$$v_n(t, \lambda b, u_n(\lambda b)) = T(t - \lambda b)u_n(\lambda b) + \int_{\lambda b}^{t} T(t - s)f(s, u(s))ds$$
$$+ \int_{\lambda b}^{t} T(t - s)\tilde{\sigma}_n(s)dW(s), \tag{8.6}$$

where $\tilde{\sigma}_n$ is the selection of Σ_n. Since the function $\tilde{\sigma}_n$ is locally Lipschitz in Lemma 8.6 (v), together with $(H_1)'$, we know that equation (8.6) has a unique solution, denoted by $v_n^{\lambda}(t)$. Moreover, $v_n^{\lambda}(t)$ depends continuously on (λ, u_n). The functions

$$z_n^{\lambda}(t) = \begin{cases} u_n(t), & t \in [0, \lambda b], \\ v_n^{\lambda}(t), & t \in [\lambda b, b] \end{cases}$$

belong to $\Theta_n(u_0)$. Define a function $h : [0, 1] \times \Theta_n(u_0) \to \Theta_n(u_0)$ as

$$h(\lambda, u_n) = \begin{cases} z_n^{\lambda}(t), & \lambda \in [0, 1), \\ u_n, & \lambda = 1. \end{cases}$$

It is easy to see that h is well defined and continuous. Also, it is clear that $h(0, \cdot) = v_n^0(t)$ and $h(1, \cdot)$ is the identity, hence $\Theta_n(u_0)$ is contractible for each $n \geq 1$.

Finally, by Lemma 8.6 (i), it is easy to verify that $\Theta(u_0) \subset \cdots \subset \Theta_n(u_0) \cdots \subset \Theta_2(u_0) \subset \Theta_1(u_0)$, this implies that $\Theta(u_0) \subset \bigcap_{n \geq 1} \Theta_n(u_0)$. To prove the reverse inclusion, we take $u \in \bigcap_{n \geq 1} \Theta_n(u_0)$. Therefore, there exists a sequence $\{\sigma_n\} \subset L^2([0, b], \mathcal{L}_2^0)$ such that $\sigma_n \in Sel_{\Sigma_n}(u)$, $u = S(\sigma_n)$ and for $n \geq 1$,

$$\|\sigma_n(t)\|_{\mathcal{L}_2^0}^2 \leq \alpha(t)(3 + 2|u|^2) \text{ for a.e. } t \in [0, b]$$

in view of Lemma 8.6 (ii). According to the reflexivity of the space $\mathcal{L}_2^0$ and Lemma 1.23, we have the existence of a subsequence, denoted as the sequence, such that $\sigma_n \rightharpoonup \sigma \in L^2([0, b], \mathcal{L}_2^0)$. By using the similar method in Lemma 8.2, together with Lemma 8.6 (iii), we know $\sigma \in Sel_{\Sigma}(u)$. Moreover, we deduce, thanks to Lemma 8.8, that $u = S(\sigma)$. This proves that $u \in \bigcap_{n \geq 1} \Theta(u_0)$. We conclude that $\Theta(u_0) = \bigcap_{n \geq 1} \Theta_n(u_0)$.

Consequently, we conclude that $\Theta(u_0)$ is an R_{δ}-set, completing this proof.

Example 8.1 An example is presented for the existence results to the following stochastic evolution inclusion:

$$\begin{cases} \dfrac{\partial z(t,\xi)}{\partial t} \in \dfrac{\partial^2 z(t,\xi)}{\partial \xi^2} + \hat{f}(t,z(t,\xi)) + G(t,z(t,\xi))\dfrac{dW(t)}{dt}, & t \in [0,1],\ \xi \in [0,\pi], \\[2mm] z(t,0) = z(t,\pi) = 0, & t \in [0,1], \\[2mm] z(0,\xi) = u_0(\xi), & \xi \in [0,\pi], \end{cases}$$
$$(8.7)$$

where $W(t)$ is a real standard one dimensional Brownian motion in H defined on the filtered probability space $(\Omega, \mathscr{F}, \mathbb{P})$, $\hat{f} : [0,1] \times H \to H$ satisfies $(H_1)'$.

Setting $H = L^2([0,1], \mathbb{R}^+)$ and $K = \mathbb{R}$. Consider the operator $A : D(A) \subset H \to H$ defined by

$$D(A) = \{z \in H : z, z' \text{ are absolutely continuous, } z'' \in H,\ z(0) = z(\pi) = 0\},$$
$$Az = z''.$$

Then

$$Az = \sum_{n=1}^{\infty} n^2(z, z_n)z_n,$$

where $z_n(t) = \sqrt{\frac{2}{\pi}}\sin(nt)$, $n = 1, 2, \ldots$ is the orthogonal basis of eigenvectors of A. It is well known that A generates a compact, analytic semigroup $\{T(t)\}_{t \geq 0}$ in H (see [167]).

Now, we assume that:

$$g_i : [0,1] \times H \to \mathcal{L}_2^0,\quad i = 1, 2$$

are such that
(F_1) g_1 is l.s.c. and g_2 is u.s.c.;
(F_2) $g_1(t,z) \leq g_2(t,z)$ for each $(t,z) \in [0,1] \times H$;
(F_3) there exist $\alpha_1, \alpha_2 \in L^\infty([0,1], \mathbb{R}^+)$ such that

$$\|g_i(t,z)\|_{\mathcal{L}_2^0}^2 \leq \alpha_1(t)|z|^2 + \alpha_2(t),\quad i = 1, 2,$$

for each $(t,z) \in [0,1] \times H$.

Let $\Sigma(t,z) = [g_1(t,z), g_2(t,z)]$. From our assumptions on (F_1)–(F_3), it follows readily that the multivalued function $\Sigma(\cdot, \cdot) : [0,1] \times \overline{\Omega} \to P(\mathcal{L}_2^0)$ satisfies (H_2), $(H_3)'$ and $(H_4)'$.

Then inclusion (8.7) can be reformulated as

$$\begin{cases} du(t) \in [Au(t) + f(t,u(t))]dt + \Sigma(t,z(t,u(t)))dW(t), & t \in [0,1], \\ u(0) = u_0, \end{cases}$$

where $u(t)(\xi) = z(t,\xi)$, $\Sigma(t,u(t))(\xi) = G(t,z(t,\xi))$.

Thus, all the assumptions in Theorem 8.4 are satisfied, our result can be used to inclusion (8.7).

8.4.2 Noncompact Operator Case

We study the stochastic evolution inclusion (8.1) under the following assumptions:
(H_5) there exist functions $k_1, k_2 \in L^2([0, b], \mathbb{R}^+)$ such that

$$\beta(\Sigma(t, D)) \le k_1(t)\beta(D) \quad \text{and} \quad \beta(f(t, D)) \le k_2(t)\beta(D)$$

for every bounded set $D \subset L^2(\Omega, H)$.

Define the generalized Cauchy operators $\Pi_1 : L^2([0, b], L^2(\Omega, H)) \to \mathcal{C}$ and $\Pi_2 : L^2([0, b], \mathcal{L}_2^0) \to \mathcal{C}$ as follows:

$$\Pi_1(f)(t) = \int_0^t T(t - s)f(s)ds,$$

$$\Pi_2(\sigma)(t) = \int_0^t T(t - s)\sigma(s)dW(s).$$

Lemma 8.9 *The operator Π_2 satisfies the properties* (i) *and* (ii):
(i) *there exists a constant $c_2 > 0$ such that*

$$\mathbb{E}|\Pi_2(\sigma_1)(t) - \Pi_2(\sigma_2)(t)|^2 \le c_2 \int_0^t \mathbb{E}\|\sigma_1(s) - \sigma_2(s)\|_{\mathcal{L}_2^0}^2 ds, \quad t \in [0, b],$$

for every $\sigma_1, \sigma_2 \in L^2([0, b], \mathcal{L}_2^0)$;
(ii) *for each compact set $K \subset \mathcal{L}_2^0$ and sequence $\{\sigma_n\} \subset L^2([0, b], \mathcal{L}_2^0)$ such that $\{\sigma_n(t)\} \subset K$ for a.e. $t \in [0, b]$, the weak convergence $\sigma_n \rightharpoonup \sigma$ implies the convergence $\Pi_2(\sigma_n) \to \Pi_2(\sigma)$.*

Proof (i) By calculation, we have

$$\mathbb{E}|\Pi_2(\sigma_1)(t) - \Pi_2(\sigma_2)(t)|^2 = \mathbb{E}\left| \int_0^t T(t - s)(\sigma_1(s) - \sigma_2(s))dW(s) \right|^2$$

$$\le M_1^2 tr(Q) \int_0^t \mathbb{E}\|\sigma_1(s) - \sigma_2(s)\|_{\mathcal{L}_2^0}^2 ds$$

$$= c_2 \int_0^t \mathbb{E}\|\sigma_1(s) - \sigma_2(s)\|_{\mathcal{L}_2^0}^2 ds,$$

where $c_2 = M_1^2 tr(Q)$.

(ii) First we prove for every $t \in [0, b]$, the sequence $\{\Pi_2(\sigma_n)(t)\} \subset L^2(\Omega, H)$ is relatively compact. Indeed, notice that $K \subset \mathcal{L}_2^0$ is compact and $\{\sigma_n(t)\} \subset K$. Therefore $\{\sigma_n(t)\}$ has a convergent subsequence $\{\sigma_{n_k}(t)\}$, denoted by $\sigma_{n_k}(t) \to \sigma_0(t) \in K$.

Since

$$\int_0^t \mathbb{E}\|\sigma_{n_k}(s) - \sigma_0(s)\|_{\mathcal{L}_2^0}^2 ds \leq 2\left(\int_0^t \mathbb{E}\|\sigma_{n_k}(s)\|_{\mathcal{L}_2^0}^2 ds + \int_0^t \mathbb{E}\|\sigma_0(s)\|_{\mathcal{L}_2^0}^2 ds \right) < \infty,$$

and by Lebesgue's dominated convergence theorem, we have

$$\mathbb{E}|\Pi_2(\sigma_{n_k})(t) - \Pi_2(\sigma_0)(t)|^2 \leq M_1^2 tr(Q) \int_0^t \mathbb{E}\|\sigma_{n_k}(s) - \sigma_0(s)\|_{\mathcal{L}_2^0}^2 ds \to 0 \ \text{as } k \to \infty.$$

Hence, the sequence $\{\Pi_2(\sigma_n)(t)\}$ has a convergent subsequence, which implies that $\{\Pi_2(\sigma_n)(t)\} \subset L^2(\Omega, H)$ is relatively compact for every $t \in [0, b]$.

On the other hand, we have

$$\mathbb{E}|\Pi_2(\sigma_n)(t_2) - \Pi_2(\sigma_n)(t_1)|^2$$

$$\leq 2\mathbb{E}\left| \int_{t_1}^{t_2} T(t_2 - s)\sigma_n(s)dW(s) \right|^2 + 2\mathbb{E}\left| \int_0^{t_1} [T(t_2 - s) - T(t_1 - s)]\sigma_n(s)dW(s) \right|^2$$

$$\leq 2M_1^2 tr(Q) \int_{t_1}^{t_2} \mathbb{E}\|\sigma_n(s)\|_{\mathcal{L}_2^0}^2 ds$$

$$+ 2tr(Q) \sup_{s \in [0, t_1]} \|T(t_2 - s) - T(t_1 - s)\|_{\mathcal{L}(H)}^2 \int_0^{t_1} \mathbb{E}\|\sigma_n(s)\|_{\mathcal{L}_2^0}^2 ds.$$

According to $\{\sigma_n(t)\} \subset K$ for a.e. $t \in [0, b]$, and $\{T(t)\}$ is strong continuous, the right-hand side of this inequality tends to zero as $t_2 \to t_1$ uniformly with respect to n. So $\{\Pi_2(\sigma_n)\}$ is an equicontinuous set. Thus from Arzela-Ascoli's theorem, we obtain that the sequence $\{\Pi_2(\sigma_n)\} \subset C$ is relatively compact.

Property (i) ensures that $\Pi_2 : L^2([0, b], \mathcal{L}_2^0) \to C$ is a bounded linear operator. Then it is continuous with respect to the topology of weak sequential convergence, that is the weak convergence $\sigma_n \rightharpoonup \sigma$ ensuring $\Pi_2(\sigma_n) \rightharpoonup \Pi_2(\sigma)$. Taking into account that $\{\Pi_2(\sigma_n)\}$ is relatively compact, we arrive at the conclusion that $\Pi_2(\sigma_n) \to \Pi_2(\sigma)$ in C.

Lemma 8.10 *Let the sequence $\{\sigma_n\} \subset L^2([0, b], \mathcal{L}_2^0)$ be integrably bounded:*

$$\|\sigma_n(t))\|_{\mathcal{L}_2^0} \leq \zeta(t) \ \text{for a.e. } t \in [0, b],$$

where $\zeta \in L^2([0, b], \mathbb{R}^+)$. Assume that

$$\beta(\{\sigma_n(t)\}) \leq \eta(t)$$

for a.e. $t \in [0, b]$, where $\eta \in L^2([0, b], \mathbb{R}^+)$. Then we have

$$\beta(\{\Pi_2(\sigma_n)(t)\}_{n=1}^{\infty}) \leq 2\left(c_2 \int_0^t |\eta(s)|^2 ds\right)^{\frac{1}{2}}$$

for all $t \in [0, b]$.

Proof For any $\varepsilon > 0$, choose $\delta \in (0, \varepsilon)$ such that for every set $m \subset [0, b]$, $\mathrm{mes}(m) < \delta$, it is

$$\int_m |\zeta(s)|^2 ds < \varepsilon.$$

Take m_δ and G_δ from Lemma 1.13. Then any sequence $\{g_n\}_{n=1}^{\infty}$, $g_n \in G_\delta$, is weakly compact by

$$\|g_n(t)\|_{\mathcal{L}_2^0} \leq [\zeta(t) + 2\eta(t)] + \delta \quad \text{for a.e. } t \in [0, b].$$

Therefore from Lemma 8.9 (ii), it follows that the sequence $\{\Pi_2(\sigma_n)(t)\}_{n=1}^{\infty}$ is relatively compact in $\mathcal{C}$. This means that $\Pi_2(G_\delta)$ is relatively compact in $\mathcal{C}$.

By Lemma 8.9 (i), we have σ_n and corresponding function g_n satisfying

$$\mathbb{E}|\Pi_2(\sigma_n)(t) - \Pi_2(g_n)(t)|^2$$

$$\leq M_1^2 tr(Q) \int_0^t \mathbb{E}\|\sigma_n(s) - g_n(s)\|_{\mathcal{L}_2^0}^2 ds$$

$$\leq M_1^2 tr(Q)\left(\int_{[0,t]\backslash m_\delta} \mathbb{E}\|\sigma_n(s) - g_n(s)\|_{\mathcal{L}_2^0}^2 ds + \int_{[0,t]\cap m_\delta} \mathbb{E}\|\sigma_n(s)\|_{\mathcal{L}_2^0}^2 ds\right)$$

$$\leq M_1^2 tr(Q)\left(\int_{[0,t]\backslash m_\delta} (2\eta(s) + \delta)^2 ds + \varepsilon\right)$$

$$\leq c_2\left(\int_0^t (2\eta(s) + \varepsilon)^2 ds + \varepsilon\right).$$

Therefore the relatively compact set $\Pi_2(G_\delta)$ forms a $\left[c_2\left(\int_0^t (2\eta(s) + \varepsilon)^2 ds + \varepsilon\right)\right]^{\frac{1}{2}}$ -net for the set $\{\Pi_2(\sigma_n)(t)\}_{n=1}^{\infty}$, proving the theorem due to the arbitrary choice of ε.

With the similar proof of Lemma 8.9, we get the following result.

Lemma 8.11 *Let $\{\sigma_n\}$ be a semicompact sequence in $L^2([0, b], \mathcal{L}_2^0)$. Then $\{\sigma_n\}$ is weakly compact in $L^2([0, b], \mathcal{L}_2^0)$, and $\{\Pi_2(\sigma_n)\}$ is relatively compact in $\mathcal{C}$. Moreover, if $\sigma_n \rightharpoonup \sigma$, then $\Pi_2(\sigma_n) \to \Pi_2(\sigma)$.*

Lemma 8.12 *Let the sequence $\{f_n\} \subset L^2([0, b], L^2(\Omega, H))$ be L^2-integrably bounded, i.e.,*

$$\|f_n(t)\| \leq \zeta_1(t)$$

for all n = 1, 2, ..., and a.e. t ∈ [0, b], where $\zeta_1 \in L^2([0, b], \mathbb{R}^+)$. Assume that

$$\beta(f_n(t)) \leq \eta_1(t)$$

for a.e. t ∈ [0, b], where $\eta_1 \in L^2([0, b], \mathbb{R}^+)$. Then we have

$$\beta(\{\Pi_1(f_n)(t)\}) \leq 2\left(c_3 \int_0^t |\eta_1(s)|^2 ds\right)^{\frac{1}{2}}$$

for any t ∈ [0, b], where $c_3 = M_1^2$.

Theorem 8.6 *Let conditions (H_A), $(H_1)'$, (H_2), $(H_3)'$, $(H_4)'$ and (H_5) be satisfied. Then the solution set of inclusion (8.1) is a nonempty compact set for each initial value $u_0 \in L^2(\Omega, H)$.*

Proof For the same $\mathcal{M}_0$, as the reason for Theorem 8.4, we see that $\mathcal{M}_0$ is a closed and convex subset of $\mathcal{C}$.

  Claim 1 The multioperator $\mathscr{F} = S \circ Sel_\Sigma$ has a closed graph with compact values. Let $u_n \subset \mathcal{M}_0$ with $u_n \to u$ and $v_n \in \mathscr{F}(u_n)$ with $v_n \to v$. We shall prove that $v \in \mathscr{F}(u)$. By the definition of $\mathscr{F}$, there exist $\sigma_n \in Sel_\Sigma(u_n)$ such that

$$v_n(t) = T(t)u_0 + \Pi_2(\sigma_n)(t) + \int_0^t T(t-s)f(s, u_n(s))ds.$$

We need to prove that there exists $\sigma \in Sel_\Sigma(u)$ such that for a.e. t ∈ [0, b],

$$v(t) = T(t)u_0 + \Pi_2(\sigma)(t) + \int_0^t T(t-s)f(s, u(s))ds.$$

  In view of $(H_4)'$ we have that $\{\sigma_n\}$ is bounded in $L^2([0, b], \mathcal{L}_2^0)$, one obtains $\sigma_n \rightharpoonup \sigma$ in $L^2([0, b], \mathcal{L}_2^0)$ (see Lemma 1.23). Since Sel_Σ is weakly u.s.c. with weakly compact and convex values (see Lemma 8.2), in view of Lemma 1.7, we have $\sigma \in Sel_\Sigma(u)$.

  We see that $\{\sigma_n\}$ is integrably bounded by $(H_4)'$, and the following inequality holds by (H_5),

$$\beta(\{\sigma_n(t)\}) \leq k_1(t)\beta(\{u_n(t)\}).$$

For the sequence $\{u_n\}$ converging in $\mathcal{C}$, thus $\beta(\{\sigma_n(t)\}) = 0$ for a.e. t ∈ [0, b], then $\{\sigma_n\}$ is a semicompact sequence. By Lemma 8.11, we may assume, without loss of generality, that there exists $\sigma \in Sel_\Sigma(u)$ such that

$$\sigma_n \rightharpoonup \sigma \quad \text{and} \quad \Pi_2(\sigma_n) \to \Pi_2(\sigma).$$

By the same methods as in Theorem 8.4, we have

$$\int_0^t T(t-s) f(s, u_n(s)) ds \rightarrow \int_0^t T(t-s) f(s, u(s)) ds,$$

which implies

$$v_n(t) = T(t) u_0 + \Pi_2(\sigma_n)(t) + \int_0^t T(t-s) f(s, u_n(s)) ds$$

$$\rightarrow T(t) u_0 + \Pi_2(\sigma)(t) + \int_0^t T(t-s) f(s, u(s)) ds = v(t).$$

It remains to show that, for $u \in \mathcal{M}_0$ and $\{\sigma_n\}$ chosen in $Sel_\Sigma(u)$, the sequence $\{\Pi_2(\sigma_n)\}$ is relatively compact in $\mathcal{C}$. Hypotheses $(H_4)'$ and (H_5) imply that $\{\sigma_n\}$ is semicompact. Using Lemma 8.11, we obtain that $\{\Pi_2(\sigma_n)\}$ is relatively compact in $\mathcal{C}$. Thus $\mathcal{F}(u)$ is relatively compact in $\mathcal{C}$, together with the closeness of $\mathcal{F}$, then $\mathcal{F}(u)$ has compact values.

Claim 2 The multioperator $\mathcal{F}$ is u.s.c. In view of Lemma 1.9, it suffices to check that $\mathcal{F}$ is a quasicompact multimap. Let U be a compact set. We prove that $\mathcal{F}(U)$ is a relatively compact subset of $\mathcal{C}$.

Assume that $\{v_n\} \subset \mathcal{F}(U)$. Then

$$v_n(t) = T(t) u_0 + \Pi_2(\sigma_n)(t) + \int_0^t T(t-s) f(s, u_n(s)) ds,$$

where $\sigma_n \in Sel_\Sigma(u_n)$ for a certain sequence $\{u_n\} \subset U$. Hypotheses $(H_4)'$ and (H_5) yield the fact that $\{\sigma_n\}$ is semicompact and then it is a weakly compact sequence in $L^2([0, b], \mathcal{L}_2^0)$. Similar arguments as the previous proof of closeness imply that $\{\Pi_2(\sigma_n)\}$ is relatively compact in $\mathcal{C}$. Thus, $\{v_n\}$ converges in $\mathcal{C}$, so the multioperator $\mathcal{F}$ is u.s.c.

Claim 3 The multioperator $\mathcal{F}$ is a condensing multioperator. We first need an MNC constructed suitably for our problem. For a bounded subset $\Lambda \subset \mathcal{M}_0$, let $\mathrm{mod}_\mathcal{C}(\Lambda)$ be the modulus of equicontinuity of the set of functions Λ given by

$$\mathrm{mod}_\mathcal{C}(\Lambda) = \lim_{\delta \to 0} \sup_{u \in \Lambda} \max_{|t_2 - t_1| < \delta} \| u(t_2) - u(t_1) \|.$$

Given Hausdorff MNC β, let χ be the real MNC defined on bounded set $D \subset \mathcal{C}$ by

$$\chi(D) = \sup_{t \in [0,b]} e^{-Lt} \beta(D(t)).$$

Here, the constant L is chosen such that

$$l_1 := 2 \sup_{t \in [0,b]} \left(c_2 \int_0^t e^{-2L(t-s)} k_1^2(s) ds \right)^{\frac{1}{2}} + 2 \sup_{t \in [0,b]} \left(c_3 \int_0^t e^{-2L(t-s)} k_2^2(s) ds \right)^{\frac{1}{2}} < 1,$$

where $k_1(t)$, $k_2(t)$ are the functions from condition (H_5).

Consider the function $\nu(\Lambda) = \max_{D \in \Delta'(\Lambda)} \left(\chi(D), \mathrm{mod}_C(D) \right)$ in the space of C, where $\Delta'(\Lambda)$ is the collection of all countable subsets of Λ.

To show that $\mathscr{F}$ is ν-condensing, let $\Lambda \subset \mathcal{M}_0$ be a bounded set in $\mathcal{M}_0$ such that

$$\nu(\Lambda) \leq \nu(\mathscr{F}(\Lambda)). \tag{8.8}$$

We will show that Λ is relatively compact. Let $\nu(\mathscr{F}(\Lambda))$ be achieved on a sequence $\{v_n\} \subset \mathscr{F}(\Lambda)$, i.e.,

$$\nu(\{v_n\}) = \left(\chi(\{v_n\}), \mathrm{mod}_C(\{v_n\}) \right).$$

Then

$$v_n(t) = T(t)u_0 + \Pi_2(\sigma_n) + \int_0^t T(t-s)f(s, u_n(s))ds,$$

where $\{u_n\} \subset \Lambda$ and $\sigma_n \in Sel_\Sigma(u_n)$. Now inequality (8.8) implies

$$\chi(\{v_n\}) \geq \chi(\{u_n\}). \tag{8.9}$$

It follows from (H_5) that

$$\beta(\{\sigma_n(t)\}) \leq k_1(t)\beta(u_n(t)) \text{ and } \beta(\{f(t, u_n(t))\}) \leq k_2(t)\beta(u_n(t))$$

for $t \in [0, b]$. Then

$$\beta(\{\sigma_n(t)\}) \leq k_1(t)e^{Lt}\left(\sup_{s\in[0,t]} e^{-Ls}\beta(u_n(s)) \right) \leq k_1(t)e^{Lt}\chi(\{u_n\}),$$

$$\beta(\{f(t, u_n(t))\}) \leq k_2(t)e^{Lt}\left(\sup_{s\in[0,t]} e^{-Ls}\beta(u_n(s)) \right) \leq k_2(t)e^{Lt}\chi(\{u_n\}).$$

Now the applications of Lemma 8.10 for Π_2 and Lemma 8.12 for Π_1 yield

$$e^{-Lt}\beta(\{\Pi_2(\sigma_n)(t)\}) \leq 2e^{-Lt}\left(c_2 \int_0^t e^{2Ls}k_1^2(s)ds \right)^{\frac{1}{2}} \chi(\{u_n\})$$

$$\leq 2\left(c_2 \int_0^t e^{-2L(t-s)}k_1^2(s)ds \right)^{\frac{1}{2}} \chi(\{u_n\}),$$

$$e^{-Lt}\beta(\{\Pi_1(f)(t)\}) \leq 2e^{-Lt}\left(c_3 \int_0^t e^{2Ls}k_2^2(s)ds \right)^{\frac{1}{2}} \chi(\{u_n\})$$

$$\leq 2\left(c_3 \int_0^t e^{-2L(t-s)}k_2^2(s)ds \right)^{\frac{1}{2}} \chi(\{u_n\})$$

for any $t \in [0, b]$. Putting this relation together with (8.9), we obtain

$$
\begin{aligned}
\chi(\{u_n\}) \le \chi(\{v_n\}) = \; & \sup_{t \in [0,b]} e^{-Lt} \beta(v_n(t)) \\
\le \; & \sup_{t \in [0,b]} e^{-Lt} \beta(\{\Pi_2(\sigma_n)(t)\}) + \sup_{t \in [0,b]} e^{-Lt} \beta(\{\Pi_1(f)(t)\}) \\
\le \; & l_1 \chi(\{u_n\}).
\end{aligned}
$$

Therefore $\chi(\{u_n\}) = 0$. This implies $\beta(u_n(t)) = 0$. Thus for every $t \in [0, b]$, $\{\Pi_1(f)(t)\}$ is relatively compact in $L^2(\Omega, H)$, this together with the equicontinuity of $\{\Pi_1(f)(t)\}$ (due to Lemma 8.7), implies that $\{\Pi_1(f)\}$ is relatively compact in $\mathcal{C}$.

Using $(H_4)'$ and (H_5) again, one gets that $\{\sigma_n\}$ is a semicompact sequence. Then, Lemma 8.11 ensures that $\{\Pi_2(\sigma_n)\}$ is relatively compact in $\mathcal{C}$. This yields that $\{v_n\}$ is relatively compact in $\mathcal{C}$. Hence $\mathrm{mod}_C(\{v_n\}) = 0$. Finally, $\nu(\{v_n\}) = 0$, and so the map $\mathscr{F}$ is ν-condensing.

From Theorem 1.12, we deduce that the fixed point set $\mathrm{Fix}(\mathscr{F})$ is a nonempty compact set.

Theorem 8.7 *Under the conditions in Theorem 8.6, $\Theta(u_0)$ is a compact R_δ-set.*

Proof We also consider inclusion (8.5). Clearly, its solution set $\Theta_n(u_0)$ is nonempty in $\mathcal{C}$.

We show that each sequence $\{v_n\}$ such that $v_n \in \Theta_n(u_0)$ for all $n \ge 1$ has a convergent subsequence $v_{n_k} \to v \in \Theta(u_0)$.

At first we notice that

$$
v_n(t) = T(t)u_0 + \int_0^t T(t-s) f(s, v_n(s)) ds + \int_0^t T(t-s) \tilde{\sigma}_n(s) dW(s)
$$

for $t \in [0, b]$, where $\tilde{\sigma}_n \in L^2([0, b], \mathcal{L}_2^0)$ is such that $\tilde{\sigma}_n(s) \in \Sigma_n(s, v_n(s))$ for almost every $s \in [0, b]$.

For any $r \ge 1$ we obtain

$$
\begin{aligned}
\beta(\{\tilde{\sigma}_n(s)\}_{n \ge 1}) = \; & \beta(\{\tilde{\sigma}_n(s)\}_{n \ge r}) \\
\le \; & \beta[\Sigma(s, B(\{v_n(s)\}_{n \ge r}, 3d_r))] \\
\le \; & k_1(s) \beta[B(\{v_n(s)\}_{n \ge r}, 3d_r)] \\
\le \; & k_1(s)[\beta(\{v_n(s)\}_{n \ge r}) + 3d_r] \\
= \; & k_1(s)(\bar{\rho}(s) + 3d_r),
\end{aligned}
$$

where $\bar{\rho}(s) = \beta(\{v_n(s)\}_{n \ge 1})$. Now

$$
\begin{aligned}
\beta(\{v_n(t)\}_{n \ge 1}) \le \; & \beta\left(\int_0^t T(t-s) f(s, v_n(s)) ds \right) \\
& + \beta\left(\int_0^t T(t-s) \tilde{\sigma}_n(s) dW(s) \right)
\end{aligned}
$$

$$\leq 2\left(c_3 \int_0^t k_2^2(s)\bar{\rho}^2(s)ds\right)^{\frac{1}{2}} + 2\left(c_2 \int_0^t k_1^2(s)(\bar{\rho}(s) + 3d_r)^2 ds\right)^{\frac{1}{2}}$$

for every $r \geq 1$. Since $d_r \to 0$ as $r \to \infty$, we obtain

$$\bar{\rho}(s) \leq 2\left(c_3 \int_0^t k_2^2(s)\bar{\rho}^2(s)ds\right)^{\frac{1}{2}} + 2\left(c_2 \int_0^t k_1^2(s)\bar{\rho}^2(s)ds\right)^{\frac{1}{2}}.$$

Thus,

$$\bar{\rho}^2(s) \leq 2^3\left(c_3 \int_0^t k_2^2(s)\bar{\rho}^2(s)ds + c_2 \int_0^t k_1^p(s)\bar{\rho}^2(s)ds\right)$$

$$\leq 2^3 \int_0^t \left(c_3 k_2^2(s) + c_2 k_1^2(s)\right)\bar{\rho}^2(s)ds.$$

By Gronwall's inequality we get $\bar{\rho}(s) = 0$ and, as a consequence, $\beta(\{v_n(t)\}_{n\geq 1}) = 0$. This also implies that $\beta(\{\tilde{\sigma}_n(s)\}_{n\geq 1}) = 0$.

Since the equicontinuity of $\{v_n\}$ is proved in Lemma 8.7. This implies the existence of a subsequence $\{v_{n_k}\}$ which is convergent on $[0, b]$. Denote the limit by v.

Since $\beta(\{\tilde{\sigma}_n(s)\}_{n\geq 1}) = 0$, we can assume, up to subsequence, that $\tilde{\sigma}_n(s) \to \tilde{\sigma}_0(s) \in \mathcal{L}_2^0$ for $s \in [0, t]$. Together with the above discussion, therefore,

$$v(t) = T(t)u_0 + \int_0^t T(t-s)f(s, v(s))ds + \int_0^t T(t-s)\tilde{\sigma}_0(s)dW(s).$$

Since $\{\Sigma_n\}$ satisfies the conditions (H_2), $(H_3)'$ and $(H_4)'$ for each $n \geq 1$, from (ii) and (vi) in Lemma 8.6. By using the similar method in Lemma 8.2, we can prove that $\tilde{\sigma}_0(t) \in \Sigma(t, v(t))$ for a.e. $t \in [0, b]$.

It follows that $\sup\{\text{dist}(v, \Theta(u_0)) : v \in \Theta_n(u_0)\} \to 0$ (an easy proof by contradiction). Therefore $\sup\{\text{dist}(v, \Theta(u_0)) : v \in \overline{\Theta_n(u_0)}\} \to 0$ as well. Since $\Theta(u_0)$ is compact and $\Theta_{n+1}(u_0) \subset \Theta_n(u_0)$, $\beta(\Theta_n(u_0)) = \beta(\overline{\Theta_n(u_0)}) \to 0$ as $n \to \infty$ and

$$\Theta(u_0) = \bigcap_{n=1}^{\infty} \overline{\Theta_n(u_0)}.$$

By the same methods as in Theorem 8.4, we can know that $\Theta_n(u_0)$ is contractible for all $n \geq 1$. Consequently, we conclude that $\Theta(u_0)$ is an R_δ-set. The proof is completed.

References

1. N. Abada, M. Benchohra, H. Hammouche, Existence and controllability results for nondensely defined impulsive semilinear functional differential inclusions. J. Differ. Equ. **246**, 3834–3863 (2009)
2. P. Acquistapace, Evolution operators and strong solution of abstract parabolic equations. Differ. Integral Equ. **1**, 433–457 (1988)
3. P. Acquistapace, B. Terreni, A unified approach to abstract linear non-autonomous parabolic equations. Rend. Semin. Mat. Univ. Padova **78**, 47–107 (1987)
4. V.N. Afanasiev, V.B. Kolmanovskii, V.R. Nosov, *Mathematical Theory of Control Systems Design*. Math. Appl., vol. 341 (Kluwer Academic Publishers, Dordrecht, 1995)
5. S. Aizicovici, M. McKibben, Existence results for a class of abstract nonlocal Cauchy problems. Nonlinear Anal. **39**, 649–668 (2000)
6. S. Aizicovici, M. McKibben, S. Reich, Anti-periodic solutions to non-monotone evolution equations with discontinuous nonlinearities. Nonlinear Anal. **43**, 233–251 (2001)
7. S. Aizicovici, S. Reich, Anti-periodic solutions to a class of non-monotone evolution equations. Discret. Contin. Dyn. Syst. **5**, 35–42 (1999)
8. S. Aizicovici, V. Staicu, Multivalued evolution equations with nonlocal initial conditions in Banach spaces. Nonlinear Differ. Equ. Appl. **14**, 361–376 (2007)
9. R.R. Akhmerov, M.I. Kamenskii, A.S. Potapov et al., *Measures of Noncompactness and Condensing Operators* (Birkhäser, Boston, 1992)
10. J. Andres, L. Górniewicz, *Topological Fixed Point Principles for Boundary Value Problems* (Kluwer, Dordrecht, 2003)
11. J. Andres, G. Gabor, L. Górniewicz, Boundary value problems on infinite intervals. Trans. Am. Math. Soc. **351**, 4861–4903 (1999)
12. J. Andres, G. Gabor, L. Górniewicz, Topological structure of solution sets to multi-valued asymptotic problems. Z. Anal. Anwendungen **19**(1), 35–60 (2000)
13. J. Andres, G. Gabor, L. Górniewicz, Acyclicity of solution sets to functional inclusions. Nonlinear Anal. **49**, 671–688 (2002)
14. J. Andres, M. Pavlačková, Topological structure of solution sets to asymptotic boundary value problems. J. Differ. Equ. **248**, 127–150 (2010)
15. C.T. Anh, N.M. Chuong, T.D. Ke, Global attractors for the m-semiflow generated by a quasilinear degenerate parabolic equation. J. Math. Anal. Appl. **363**, 444–453 (2010)
16. C.T. Anh, T.D. Ke, On quasilinear parabolic equations involving weighted p-Laplacian operators. Nonlinear Diff. Equa. Appl. **17**, 195–212 (2010)
17. W. Arendt, Vector valued Laplace transforms and Cauchy problems. Israel J. Math. **59**, 327–352 (1987)
18. L. Arnold, *Stochastic Differential Equations: Theory and Applications* (Wiley, New York, 1974)

19. N. Aronszajn, Le correspondant topologique de l'unicité dans la théorie des équations différentielles. Ann. Math. **43**, 730–738 (1942)
20. J.-P. Aubin, A. Cellina, *Differential Inclusions* (Springer, Berlin, 1984)
21. A. Augustynowicz, Some remarks on comparison functions. Ann. Polon. Math. **96**(2), 97–106 (2009)
22. R.J. Aumann, Integrals of set-valued functions. J. Math. Anal. Appl. **12**(1), 1–12 (1965)
23. E. Avgerinos, N. Papageorgiou, Nonconvex perturbations of evolution equations with m-dissipative operators in Banach spaces. Comment Math. Univ. Carolin. **30**, 657–664 (1989)
24. R. Bader, W. Kryszewski, On the solution sets of differential inclusions and the periodic problem in Banach spaces. Nonlinear Anal. **54**, 707–754 (2003)
25. A. Bakowska, G. Gabor, Topological structure of solution sets to differential problems in Fréchet spaces. Ann. Polon. Math. **95**(1), 17–36 (2009)
26. J.M. Ball, Continuity properties and global attractor of generalized semiflows and the Navier-Stokes equations. J. Nonlinear Sci. **7**, 475–502 (1997)
27. J.M. Ball, Global attractor for damped semilinear wave equations. Discret. Contin. Dyn. Syst. **10**, 31–52 (2004)
28. J. Banas, K. Goebel, *Measure of Noncompactness in Banach Spaces* (Marcel Dekker, Inc., New York, 1980)
29. J. Banasiak, L. Arlotti, *Perturbations of Positive Semigroups with Applications* (Springer, New York, 2006)
30. P. Baras, Compacité de l'opérateur definissant la solution d'une équation d'évolution non linéaire $(du/dt) + Au \ni f$. C. R. Acad. Sci. Sér. I Math. **286**, 1113–1116 (1978)
31. V. Barbu, *Nonlinear Semigroups and Differential Equation in Banach Spaces* (Editura Academiei/Noordhoff, Bucuresti, 1976)
32. V. Barbu, *Nonlinear Differential Equations of Monotone Types in Banach Spaces* (Springer, New York, 2010)
33. G. Bartuzel, A. Fryszkowski, Filippov lemma for cetain second order differential inclusions. Cent. Eur. J. Math. **10**, 1944–1952 (2012)
34. M. Benchohra, S. Abbas, *Advanced Functional Evolution Equations and Inclusions* (Springer, Berlin, 2015)
35. I. Benedetti, L. Malaguti, V. Taddei, Semilinear evolution equations in abstract spaces and applications. Rend. Ist. Matem. Un. Trieste (Special issue dedicated to the 60th birthday of Prof. F. Zanolin) **44**, 371–388 (2012)
36. I. Benedetti, V. Obukhovskii, V. Taddei, Controllability for systems governed by semilinear evolution inclusions without compactness. Nonlinear Differ. Equ. Appl. **21**, 795–812 (2014)
37. I. Benedetti, P. Rubbioni, Existence of solutions on compact and non-compact intervals for semilinear impulsive differential inclusions with delay. Topol. Methods Nonlinear Anal. **32**, 227–245 (2008)
38. I. Benedetti, V. Obukhovskii, P. Zecca, Controllability for impulsive semilinear functional differential inclusions with a non-compact evolution operator. Discussiones Math. Differ. Inclusions Control Optim. **31**(1), 39–69 (2011)
39. P. Bénilan, Solutions intégrales d'équations d'évolution dans un espace de Banach. C. R. Acad. Sci. Paris Sér. A-B **274**, A47–A50 (1972)
40. A. Bensoussan, G. Da Prato, M.C. Delfour, S.K. Mitter, *Representation and Control of Infinite Dimensional Systems* (Birkhäuser, Boston, 2007)
41. S. Bochner, A.E. Taylor, Linear functionals on certain spaces of abstractly valued functions. Ann. Math. **39**, 913–944 (1938)
42. Y.G. Borisovich, B.D. Gelman, A.D. Myshkis, V.V. Obukhovskii, *Introduction to Theory of Multivalued Maps and Differential Inclusions*, 2nd edn. (Librokom, Moscow, 2011). (in Russian)
43. K. Borsuk, Sur les groupes des classes de transformations continues. CR Acad. Sci. Paris **202**(2), 1400–1403 (1936)
44. K. Borsuk, *Theory of Retracts*. Monografie Mat., vol. 44 (PWN, Warsaw, 1967)

45. D. Bothe, Multi-valued perturbations of m-accretive differential inclusions. Israel J. Math. **108**, 109–138 (1998)
46. D. Bothe, *Nonlinear Evolutions in Banach Spaces* (Habilitationsschrift, Paderborn, 1999)
47. D. Bothe, Semilinear differential inclusions with application to hybrid system. Submitted to SIAM J. Math. Anal
48. A. Bressan, Z.P. Wang, Classical solutions to differential inclusions with totally disconnected right-hand side. J. Differ. Equ. **246**, 629–640 (2009)
49. H. Brezis, Problèmes unilatéraux. J. Math. Pures Appl. **51**, 1–168 (1972)
50. H. Brezis, *Analyse Fonctionelle* (Théorie et Applications. Masson Editeur, Paris, 1983)
51. F.E. Browder, C.P. Gupta, Topological degree and nonlinear mappings of analytic type in Banach spaces. J. Math. Anal. Appl. **26**, 390–402 (1969)
52. L. Byszewski, Theorems about the existence and uniqueness of solutions of semilinear evolution nonlocal Cauchy problems. J. Math. Anal. Appl. **162**, 494–505 (1991)
53. I. Căpraru, Approximate weak invariance and relaxation for fully nonlinear differential inclusions. Mediterr. J. Math. **10**(1), 201–212 (2013)
54. T. Caraballo, M.J. Garrido-Atienza, B. Schmalfuss, J. Valero, Non-autonomous and random attractors for delay random semilinear equations without uniqueness. Discret. Contin. Dyn. Syst. **21**, 415–443 (2008)
55. T. Caraballo, P.E. Kloeden, Non-autonomous attractors for integro-differential evolution equations, Discret. Contin. Dyn. Syst. Ser.S **2**, 17–36 (2009)
56. T. Caraballo, J.A. Langa, V.S. Melnik, J. Valero, Pullback attractors of nonautonomous and stochastic multivalued dynamical systems. Set-Valued Anal. **11**, 153–201 (2003)
57. T. Cardinali, P. Rubbioni, On the existence of mild solutions of semilinear evolution differential inclusions. J. Math. Anal. Appl. **308**, 620–635 (2005)
58. T. Cardinali, P. Rubbion, Impulsive semilinear differential inclusions: topological structure of the solution set and solutions on non-compact domains. Nonlinear Anal. **69**, 73–84 (2008)
59. O. Cârjă, T. Donchev, A.I. Lazu, Generalized solutions of semilinear evolution inclusions. SIAM J. Opti. **26**, 1365–1378 (2016)
60. O. Cârjă, T. Donchev, V. Postolache, Nonlinear evolution inclusions with one-sided Perron right-hand side. J. Dyn. Control Syst. **19**(3), 439–456 (2013)
61. O. Cârjă, T. Donchev, V. Postolache, Relaxation results for nonlinear evolution inclusions with one-sided Perron right-hand side. Set-Valued Var. Anal. **22**(4), 657–671 (2014)
62. O. Cârjă, A.I. Lazu, Lower semi-continuity of the solution set for semilinear differential inclusions. J. Math. Anal. Appl. **385**, 865–873 (2012)
63. O. Cârjă, A. Lazu, Approximate weak invariance for differential inclusions in Banach spaces. J. Dyn. Control Syst. **18**(2), 215–227 (2012)
64. O. Cârjă, M. Necula, I.I. Vrabie, *Viability, Invariance and Applications*. Math. Stud., vol. 207 (Elsevier, North-Holland, 2007)
65. O. Cârjă, M. Necula, I.I. Vrabie, Necessary and sufficient conditions for viability for nonlinear evolution inclusions. Set-Valued Anal. **16**(5–6), 701–731 (2008)
66. O. Cârjă, I.I. Vrabie, Differential Equations on Elosed Sets. In: Cañada, A., Drábek, P., Fonda, A. (eds.): *Ordinary Differential Equations*, vol. 2 (Elsevier B. V., 2005), pp. 147–238
67. A.N. Carvalho, J.A. Langa, Non-autonomous perturbation of autonomous semilinear differential equations: continuity of local stable and unstable manifolds. J. Differ. Equ. **233**(2), 622–653 (2007)
68. A.N. Carvalho, J.A. Langa, An extension of the concept of gradient semigroups which is stable under perturbation. J. Differ. Equ. **246**(7), 2646–2668 (2009)
69. C. Castaing, D.P. Monteiro-Marques, Periodic solutions of evolution problems associated with a moving convex set. C.R. Acad. Sci. Paris Sr. A **321**, 531–536 (1995)
70. S.S. Ceron, O. Lopes, α-Contractions and attractors for dissipative semilinear hyperbolic equations and systems. Ann. Mat. Pura Appl. **160**, 193–206 (1991)
71. D.H. Chen, R.N. Wang, Y. Zhou, Nonlinear evolution inclusions: topological characterizations of solution sets and applications. J. Funct. Anal. **265**, 2039–2073 (2013)

72. V.V. Chepyzhov, M.I. Vishik, Evolution equations and their trajectory attractors. J. Math. Pures Appl. **76**, 913–964 (1997)

73. V.V. Chepyzhov, M.I. Vishik, Attractors for equations of mathematics physics, in *American Mathematical Society Colloquium Publications*, vol. **49** (American Mathematical Society, Providence, 2002)

74. G. Conti, V. Obukhovskii, P. Zecca, On the topological structure of the solution set for a semilinear functional-differential inclusion in a Banach space. Banach Center Publ. **35**, 159–169 (1996)

75. M.G. Crandall, A. Pazy, Nonlinear evolution equations in Banach spaces. Israel J. Math. **11**, 57–94 (1972)

76. R.F. Curtain, H.J. Zwart, *An Introduction to Infinite Dimensional Linear Systems Theory* (Springer, New York, 1995)

77. N.V. Dac, T.D. Ke, Pullback attractor for differential evolution inclusions with infinite delays. Appl. Math. Comput. **265**, 667–680 (2015)

78. F.S. De Blasi, J. Myjak, On the solutions sets for differential inclusions. Bull. Pol. Acad. Sci. Math. **12**, 17–23 (1985)

79. F.S. De Blasi, J. Myjak, On the structure of the set of solutions of the Darboux problem for hyperbolic equations. Proc. Edinb. Math. Soc. **29**, 7–14 (1986)

80. K. Deimling, *Multivalued Differential Equtions* (de Gruytr, Berlin, 1992)

81. K. Deimling, *Nonlinear Functional Analysis* (Springer, Berlin, 1985)

82. K. Deng, Exponential decay of solutions of semilinear parabolic equations with nonlocal initial conditions. J. Math. Anal. Appl. **179**, 630–637 (1993)

83. Z. Denkowski, S. Migórski, N.S. Papageorgiou, *An Introduction to Nonlinear Analysis: Theory* (Springer Science & Business Media, Berlin, 2013)

84. J. Diestel, W.M. Ruess, W. Schachermayer, Weak compactness in $L^1(\mu; X)$. Proc. Am. Math. Soc. **118**, 447–453 (1993)

85. J. Diestel, J.J. Uhl, *Vector Measures. Mathematical Surveys*, vol. 15 (American Mathematical Society, Providence, 1977)

86. Q. Din, T. Donchev, D. Kolev, Filippov-Pliss lemma and m-dissipative differential inclusions. J. Glob. Optim. **56**(4), 1707–1717 (2013)

87. S. Djebali, L. Górniewicz, A. Ouahab, Filippov-Ważewski theorems and structure of solution sets for first order impulsive semilinear functional differential inclusions. Topol. Methods Nonlinear Anal. **32**, 261–312 (2008)

88. S. Djebali, L. Górniewicz, A. Ouahab, Topological structure of solution sets for impulsive differential inclusions in Fréchet spaces. Nonlinear Anal. **74**, 2141–2169 (2011)

89. S. Djebali, L. Górniewicz, A. Ouahab, *Existence and Structure of Solution Sets for Impulsive Differential Inclusions: A Survey*, vol. 13, Lecture Notes in Nonlinear Analysis (Juliusz Schauder Center for Nonlinear Studies, Toruń, 2012)

90. S. Djebali, L. Górniewicz, A. Ouahab, *Solution Sets for Differential Equations and Inclusions*. De Gruyter Series in Nonlinear Analysis and Applications, vol. 18 (Walter de Gruyter & Co., Berlin, 2013)

91. T. Donchev, Functional differential inclusions with monotone right-hand side. Nonlinear Anal. **16**, 533–542 (1991)

92. T. Donchev, Generic properties of multifunctions: application to differential inclusions. Nonlinear Anal. **74**(7), 2585–2590 (2011)

93. T. Donchev, E. Farkhi, On the theorem of Filippov-Pliś and some applications. Control Cybernet. **38**, 1251–1271 (2009)

94. T. Donchev, E. Farkhi, B.S. Mordukhovich, Discrete approximations, relaxation, and optimization of one-sided Lipschitzian differential inclusions in Hilbert spaces. J. Differ. Equ. **243**, 301–328 (2007)

95. T. Donchev, A.I. Lazu, A. Nosheen, One-sided Perron differential inclusions. Set-Valued Var. Anal. **21**(2), 283–296 (2013)

96. R. Dragoni, J.W. Macki, P. Nistri, P. Zecca, *Solution Sets of Differential Equations in Abstract Spaces*. Pitman Res. Notes math. Ser., vol. 342 (Longman Sci. Tech., Harlow, 1996)

97. N. Dunford, J.T. Schwartz, *Linear Operators* (Wiley, New York, 1988)
98. J. Dugundji, An extension of Tietze's theorem. Pac. J. Math. **1**, 353–367 (1951)
99. M. Efendiev, *Evolution Equations Arising in the Modelling of Life Sciences*. vol. 163 (Springer Science & Business Media, Berlin, 2013)
100. K.J. Engel, R. Nagel, *One-Parameter Semigroups for Linear Evolution Equations*. Grad. Texts in Math., vol. 194 (Springer, New York, 2000)
101. R. Engelking, *Outline of General Topology* (North-Holland, Amsterdam, 1968)
102. H. Frankowska, A priori estimates for operator differential inclusions. J. Differ. Equ. **84**, 100–128 (1990)
103. G. Gabor, On the acyclicity of fixed point sets of multivalued maps. Topol. Methods Nonlinear Anal. **14**(2), 327–343 (1999)
104. G. Gabor, Acyclicity of solution sets of inclusions in metric spaces. Topol. Methods Nonlinear Anal. **14**, 327–343 (1999)
105. G. Gabor, Some results on existence and structure of solution sets to differential inclusions on the halfline. Boll. Unione Mat. Ital. Sez. B Artic. Ric. Mat. **5**(8), 431–446 (2002)
106. G. Gabor, A. Grudzka, Structure of the solution set to impulsive functional differential inclusions on the half-line. Nonlinear Differ. Equ. Appl. **19**, 609–627 (2012)
107. G. Gabor, A. Grudzka, Erratum to: Structure of the solution set to impulsive functional differential inclusions on the half-line. Nonlinear Differ. Equ. Appl. **22**, 175–183 (2015)
108. G. Gabor, M. Quincampoix, On existence of solutions to differential equations or inclusions remaining in a prescribed closed subset of a finite-dimensional space. J. Differ. Equ. **185**, 483–512 (2002)
109. T. Caraballo, P. Marin-Rubio, J.C. Robinson, A comparision between to theories for multivalued semiflows and their asymptotic behaviour. Set-Valued Anal. **11**, 297–322 (2003)
110. J. Garcí-Falset, Existence results and asymptotic behaviour for nonlocal abstract Cauchy problems. J. Math. Anal. Appl. **338**, 639–652 (2008)
111. L. Gawarecki, V. Mandrekar, *Stochastic Differential Equations in Infinite Dimensions: with Applications to Stochastic Partial Differential Equations* (Springer, Berlin, 2011)
112. J.A. Goldstein, *Semigroups of Linear Operators and Applications*, Oxford Mathematical Monographs (Oxford University Press, New York, 1985)
113. L. Górniewicz, *Topological Fixed Point Theory of Multivalued Mappings*, 2nd edn. (Springer, Berlin, 2006)
114. L. Górniewicz, M. Lassonde, Approximation and fixed points for compositions of R_δ-maps. Topol. Appl. **55**(3), 239–250 (1994)
115. L. Górniewicz, T. Pruszko, On the set of solutions of the Darboux problem for some hyperbolic equations. Bull. Acad. Polon. Math. **28**(5–6), 279–286 (1980)
116. J.R. Graef, J. Henderson, A. Ouahab, *Impulsive Differential Inclusions: A Fixed Point Approach* (Walter de Gruyter GmbH, Berlin, 2013)
117. L. Guedda, Some remarks in the study of impulsive differential equations and inclusions with delay. Fixed Point Theory **12**(2), 349–354 (2011)
118. P. Hájek, V. Montesinos, V. Zizler, Geometry and Gateaux smoothness in separable Banach spaces. Oper. Matrices **6**(2), 201–232 (2012)
119. A. Halanay, *Differential Equations, Stability, Oscillations, Time Lags* (Academic Press, New York, 1996)
120. J.K. Hale, J. Kato, Phase spaces for retarded equations with infinite delay. Funkcial. Ekvac. **21**, 11–41 (1978)
121. J.K. Hale, S.M.V. Lunel, *Theory of Functional Differential Equations* (Springer, New York, 1993)
122. H.-P. Heinz, On the behaviour of measure of noncompactness with respect to differentiation and integration of vector-valued functions. Nonlinear Anal.: TMA **7**, 1351–1371 (1983)
123. E. Hernández, D. O'Regan, Existence results for abstract neutral functional differential equations. Proc. Am. Math. Soc. **137**, 3309–3318 (2009)
124. N. Hirano, Existence of periodic solutions for nonlinear evolution equations in Hilbert spaces. Proc. Am. Math. Soc. **120**, 185–192 (1994)

125. S. Hu, N.S. Papageorgiou, *Handbook of Multivalued Analysis, Applications*, vol. II (Kluwer, Dordrecht, 2000)

126. S.C. Hu, N.S. Papageorgiou, On the topological regularity of the solution set of differential inclusions with constraints. J. Differ. Equ. **107**, 280–289 (1994)

127. M. Hukuhara, Sur les systémes des équations differentielles ordinaires. Jpn. J. Math. **5**, 345–350 (1928)

128. D.M. Hyman, ANR divisors and absolute neighborhood contractibility. Fund. Math. **62**, 61–73 (1968)

129. D.M. Hyman, On decreasing sequence of compact absolute retracts. Fund. Math. **64**, 91–97 (1969)

130. M. Kamenskii, V. Obukhovskii, P. Zecca, Condensing Multivalued Maps and Semilinear Differential Inclusions in Banach Spaces. de Gruyter Series in Nonlinear Analysis and Applicaitons, vol. 7 (Walter de Gruyter, Berlin, New York, 2001)

131. L.V. Kantorovich, G.P. Akilov, *Functional Analysis* (Pergamon Press, Oxford, 1982)

132. T.D. Ke, D. Lan, Global attractor for a class of functional differential inclusions with Hille-Yosida operators. Nonlinear Anal. **103**, 72–86 (2014)

133. T.D. Ke, V. Obukhovskii, N.C. Wong, J.C. Yao, Approximate controllability for system governed by nonlinear Volterrra type equations. Differ. Equ. Dyn. Syst. **20**(1), 35–52 (2012)

134. H. Kellerman, M. Hieber, Integrated semigroup. J. Funct. Anal. **84**, 160–180 (1989)

135. M. Kisielewicz, *Differential Inclusions and Optimal Control*, vol. 44 (Kluwer, Academic Publishers, 1991)

136. M. Kisielewicz, *Stochastic Differential Inclusions and Applications* (Springer, New York, 2013)

137. A. Kneser, Unteruchung und asymptotische darstellung der integrale gewisser differentialgleichungen bei grossen werthen des arguments. J. Reine Angew. Math. **116**, 178–212 (1896)

138. H. Kneser, Über die Lösungen eines Systems gewöhnlicher Differentialgleichungen. das der Lipschitzschen Bedingung nicht genügt, S. B. Pruss. Akad **4**, 171–174 (1923)

139. V. Lakshmikantham, S. Leela, *Nonlinear Differential Equations in Abstract Spaces* (Pergamon Press, Oxford, 1981)

140. A. Lazu, V. Postolache, Approximate weak invariance for semilinear differential inclusions in Banach spaces. Cent. Eur. J. Math. **9**(5), 1143–1155 (2011)

141. E. Mahmudov, *Approximation and Optimization of Discrete and Evolution Inclusions* (Elsevier, New York, 2011)

142. N.I. Mahmudov, A. Denker, On controllability of linear stochastic systems. Int. J. Control **73**, 144–151 (2000)

143. P. Marin-Rubio, J. Real, On the relation between two different concepts of pullback attractors for non-autonomous dynamical systems. Nonlinear Anal. **71**, 3956–3963 (2009)

144. R. Martin, *Nonlinear Operators and Differential Equations in Banach Spaces* (Wiley, New York, 1976)

145. V.S. Melnik, J. Valero, On attractors of multivalued semi-flows and differential inclusions. Set-Valued Anal. **6**, 83–111 (1998)

146. V.S. Melnik, J. Valero, On global attractors of multivalued semiprocesses and nonautonomous evolution inclusions. Set-Valued Anal. **8**, 375–403 (2000)

147. S. Migórski, Existence and relaxation results for nonlinear evolution inclusions revisited. J. Appl. Math. Stoch. Anal. **8**(2), 143–149 (1995)

148. V.D. Milman, A. Myshkis, On the stability of motion in the presence of impulses. Sib. Math. J. **1**, 233–237 (1960)

149. B. Mordukhovich, Discrete approximations and refined Euler-Lagrange conditions for differential inclusions. SIAM J. Control Optim. **33**, 882–915 (1995)

150. B. Mordukhovich, Variational Analysis and Generalized Differentiation, II: Applications. Grundlehren Math. Wiss. 331 (Springer, Berlin, 2006)

151. B. Mordukhovich, D. Wang, Optimal control of semilinear evolution inclusions via discrete approximations. Control Cybern. **34**, 849–870 (2005)

152. B. Mordukhovich, D. Wang, Optimal control of semilinear unbounded differential inclusions. Nonlinear Anal. **63**, 847–853 (2005)
153. K. Naito, On controllability for a nonlinear Volterra equation. Nonlinear Anal. **18**(1), 99–108 (1992)
154. K. Naito, J.Y. Park, Approximate controllability for trajectories of a delay Volterra control system. J. Optim. Theory Appl. **61**(2), 271–279 (1989)
155. S.K. Ntouyas, D. O'Regan, Existence results for semilinear neutral functional differential inclusions via analytic semigroups. Acta Applicandae Mathematicae **98**, 223–253 (2007)
156. V. Obukhovskii, J.-C. Yao, On impulsive functional differential inclusions with Hille-Yosida operators in Banach spaces. Nonlinear Anal. **73**(6), 1715–1728 (2010)
157. V. Obukhovskii, P. Rubbioni, On a controllability problem for systems governed by semilinear functional differential inclusions in Banach spaces. Topol. Methods Nonlinear Anal. **15**, 141–151 (2000)
158. D. O'Regan, Fixed point theorems for weakly sequentially closed maps. Arch. Math. **36**, 61–70 (2000)
159. D. O'Regan, Topological structure of solution sets in Fréchet spaces: the projective limit approach. J. Math. Anal. Appl. **324**, 1370–1380 (2006)
160. Y.V. Rogovchenko, Impulsive evolution systems: main results and new trends. Dyn. Contin. Discret. Impuls. Syst. **3**(1), 57–88 (1997)
161. A. Paicu, Periodic solutions for a class of differential inclusions in general Banach spaces. J. Math. Anal. Appl. **337**, 1238–1248 (2008)
162. A. Paicu, I.I. Vrabie, A class of nonlinear evolution equations subjected to nonlocal initial conditions. Nonlinear Anal. **72**, 4091–4100 (2010)
163. N.S. Papageorgiou, Optimal control of nonlinear evolution inclusions. J. Optim. Theory Appl. **67**, 321–354 (1990)
164. N.S. Papageorgiou, Extremal solutions of evolution inclusions associated with time dependent convex subdifferentials. Math. Nachr. **158**, 219–232 (1992)
165. N.S. Papageorgiou, Existence of solutions of boundary value problems of semilinear evolution inclusions. Indian J. Pure Appl. Math. **23**, 477–488 (1992)
166. N.S. Papageorgiou, A continuous version of the relaxation theorem for nonlinear evolution inclusions. Kodai Math. J. **18**(1), 169–186 (1995)
167. A. Pazy, *Semigroups of Linear Operators and Applications to Partial Differential Equations*. Applied Mathematical Sciences, vol. 44 (Springer, New York, 1983)
168. B.J. Pettis, On the integration in vector spaces. Trans. Am. Math. Soc. **44**, 277–304 (1938)
169. G. Pianigiani, On the fundamental theory of multivalued differential equations. J. Differ. Equa. **25**(1), 30–38 (1977)
170. A. Pliś, Trajectories and quasitrajectories of an orientor field. Bull. Acad. Polon. Sci. Sér. Sci. Math. Astronom. Phys. **11**, 369–370 (1963)
171. G.D. Prato, J. Zabczyk, *Stochastic Equations in Infinite Dimensions* (Cambridge University Press, Cambridge, 1992)
172. W. Rudin, *Functional Analysis*, 2nd ed. (McGraw-Hill Science, 1991)
173. V. Šeda, Š. Belohorec, A remark on second order functional differential systems. Arch. Math. **36**(3–4), 169–176 (1993)
174. T.I. Seidman, Invariance of the reachable set under nonlinear perturbations. SIAM J. Control Optim. **25**, 1173–1191 (1987)
175. R.S. Schatten, *Norm Ideals of Continuous Operators* (Springer, New York, 1970)
176. G.V. Smirnov, *Introduction to the Theory of Differential Inclusions*, vol. 41 (American Mathematical Society, Providence, 2002)
177. V. Staicu, On the solution sets to nonconvex differential inclusions of evolution type. Discret. Contin. Dynam. Syst. **2**, 244–252 (1998)
178. V. Staicu, On the solution sets to differential inclusions on unbounded interval. Proc. Edinb. Math. Soc. **43**, 475–484 (2000)
179. I.M. Stamova, *Stability Analysis of Impulsive Functional Differential Equations* (De Gruyter, 2009)

180. H.B. Stewart, Generation of analytic semigroups by strongly elliptic operators. Trans. Am. Math. Soc. **199**, 141–162 (1974)

181. J. Tabor, Generalized differential inclusions in Banach spaces. Set-Valued Anal. **14**, 121–148 (2006)

182. H. Tateishi, A relaxation theorem for differential inclusions: infinite-dimensional case. Math. Jpn. **45**(3), 411–421 (1997)

183. R. Temam, *Infinite Dimensional Dynamical Systems in Mechanics and Physics*, 2nd ed. (Springer, Berlin, 1997)

184. H.R. Thieme, Semiflows generated by Lipschitz perturbations of non-densely defined operators. Differ. Integral Equ. **3**, 1001–1229 (1990)

185. A. Tolstonogov, *Differential Inclusions in a Banach Space* (Kluwer, Dordrecht, 2000)

186. J. Valero, Finite and infinite-dimensional attractor of multivalued reaction-diffusion equations. Acta Math. Hungar. **88**, 239–258 (2000)

187. J. Valero, Attractors of parabolic equations without uniqueness. J. Dyn. Differ. Equ. **13**, 711–744 (2001)

188. I.I. Vrabie, Periodic solutions for nonlinear evolution equations in a Banach space. Proc. Am. Math. Soc. **109**(3), 653–661 (1990)

189. I.I. Vrabie, *Compactness Methods for Nonlinear Evolutions*. Pitman Monographs and Surveys in Pure and Applied Mathematics, 2nd ed., vol. 75 (Longman and Wiley, 1995)

190. I.I. Vrabie, C_0-semigroups and applications. in *North-Holland Mathematics Studies*, vol. 191 (North-Holland Publishing Co., Amsterdam, 2003)

191. I.I. Vrabie, Existence for nonlinear evolution inclusions with nonlocal retarded initial conditions. Nonlinear Anal. **74**, 7047–7060 (2011)

192. I.I. Vrabie, Existence in the large for nonlinear delay evolution inclusions with nonlocal initial conditions. J. Funct. Anal. **262**, 1363–1391 (2012)

193. W. Wang, Generalized Halanay inequality for stability of nonlinear neutral functional differential equations. J. Inequal. Appl. ArtID 475019 (2010)

194. R.N. Wang, T.J. Xiao, J. Liang, A note on the fractional Cauchy problems with nonlocal initial conditions. Appl. Math. Lett. **24**, 1435–1442 (2011)

195. R.N. Wang, D.H. Chen, T.J. Xiao, Abstract fractional Cauchy problems with almost sectorial operators. J. Differ. Equ. **252**, 202–235 (2012)

196. R.N. Wang, Q.H. Ma, Y. Zhou, Topological theory of non-autonomous parabolic evolution inclusions on a noncompact interval and applications. Math. Ann. **362**, 173–203 (2015)

197. R.N. Wang, P.X. Zhu, Non-autonomous evolution inclusions with nonlocal history conditions: global integral solutions. Nonlinear Anal. **85**, 180–191 (2013)

198. R.N. Wang, P.X. Zhu, Q.H. Ma, Multi-valued nonlinear perturbations of time fractional evolution equations in Banach spaces. Nonlinear Dyn. **80**, 1745–1759 (2015)

199. J.H. Wu, *Theory and Applications of Partial Functional Differential Equations*, vol. 119 (Springer Science & Business Media, Berlin, 2012)

200. Z.Y. Yang, M. Blanke, A unified approach to controllability analysis for hybrid systems. Nonlinear Anal. Hybrid Syst. **1**, 21–222 (2007)

201. N.C. Yannelis, On the upper and lower semicontinuity of the Aumann integral. J. Math. Econ. **19**(4), 373–389 (1990)

202. A. Yagi, Abstract quasilinear evolution equations of parabolic type in Banach spaces. Boll. Unione Mat. Ital. B **5**(7), 341–368 (1991)

203. J. Yorke, *Spaces of Solutions*. Lect. Notes Op. Res. Math. Econ., vol. 12 (Springer, Berlin, 1969), pp. 383–403

204. H. You, R. Yuan, Global attractor for some partial functional differential equations with finite delay. Nonlinear Anal. **72**, 3566–3574 (2010)

205. C. Zălinescu, *Convex analysis in General Vector Spaces* (World Scientific Publishing Co., Inc, River Edge, 2002)

206. M.C. Zelati, P. Kalita, Minimality properties of set-valued processes and their pullback attractors. SIAM J. Math. Anal. **47**, 1530–1561 (2015)

207. M.Z. Zgurovsky, P. Kasyanov, O.V. Kapustyan, J. Valero, N.V. Zadoianchuk, *Evolution Inclusions and Variation Inequalities for Earth Data Processing III: Long-Time Behavior of Evolution Inclusions Solutions in Earth Data Analysis* (Springer Science & Business Media, Berlin, 2012)
208. Y. Zhou, L. Peng, B. Ahmad, Topological characterizations of solution sets for stochastic evolution inclusions, Submitted to Stochastic Anal. Appl
209. Y. Zhou, L. Peng, Topological properties of solutions set for partial functional evolution inclusions. C. R. Acad. Sci. Paris, Ser. I **355**, 45–64 (2017)
210. Q.J. Zhu, On the solution set of differential inclusions in Banach space. J. Differ. Equ. **93**, 213–237 (1991)

Index

© Springer Nature Singapore Pte Ltd. 2017
Y. Zhou et al., *Topological Structure of the Solution Set for Evolution Inclusions*,
Developments in Mathematics 51, https://doi.org/10.1007/978-981-10-6656-6